T0309298

Lectures on
Differential
Geometry

Lectures on
Differential
Geometry

Rui Loja Fernandes

University of Illinois Urbana-Champaign, USA

World Scientific

NEW JERSEY · LONDON · SINGAPORE · BEIJING · SHANGHAI · HONG KONG · TAIPEI · CHENNAI · TOKYO

Published by

World Scientific Publishing Co. Pte. Ltd.
5 Toh Tuck Link, Singapore 596224
USA office: 27 Warren Street, Suite 401-402, Hackensack, NJ 07601
UK office: 57 Shelton Street, Covent Garden, London WC2H 9HE

Library of Congress Control Number: 2024022143

British Library Cataloguing-in-Publication Data
A catalogue record for this book is available from the British Library.

About the cover: A detail from an untitled work (c. 2017) by surrealist painter Luiz Morgadinho.

LECTURES ON DIFFERENTIAL GEOMETRY

ISBN 978-981-12-5264-8 (hardcover)
ISBN 978-981-12-5336-2 (paperback)
ISBN 978-981-12-5265-5 (ebook for institutions)
ISBN 978-981-12-5266-2 (ebook for individuals)

For any available supplementary material, please visit
https://www.worldscientific.com/worldscibooks/10.1142/12733#t=suppl

Desk Editors: Nambirajan Karuppiah/Rok Ting Tan

Typeset by Stallion Press
Email: enquiries@stallionpress.com

Preface

These lecture notes have been utilized for several years in the courses I taught at both UIUC and IST-Lisbon.

Originally written in Portuguese, these notes were designed for an advanced undergraduate course at IST-Lisbon, targeting students with a basic knowledge of analysis, algebra, and topology. The initial motivation behind writing these notes was the lack of suitable Portuguese textbooks on the subject.

When I moved to UIUC and started teaching a first-year graduate course on differentiable manifolds, I still found it valuable to rely on my old lecture notes as a teaching guide. Furthermore, despite the availability of numerous excellent books on the subject, I recognized the importance of providing my students with a written version of the lectures, enabling them to fully concentrate on the class without the burden of extensive note-taking. As a result, I decided to develop an English version of the notes.

The lecture notes are organized into four parts, with each part containing sections corresponding roughly to lectures lasting approximately 1 hour and 30 minutes of classroom time. However, some lectures include more material than others, reflecting different rhythms in the class. The exercises at the end of each lecture play a vital role in the course, as they provide valuable opportunities for students to deepen their understanding through practical problem-solving. Additionally, some exercises may contain results that were mentioned in class but not proven, and these results might be used in later lectures. I always caution the students to consider that the exercises come with varying degrees of difficulty, which aligns with the nature of encountering new mathematical problems. At first, it's often unclear whether a problem has an easy solution, a hard one, or if it remains an open problem.

Throughout my classes, I make it a point to emphasize that the lecture notes do not serve as a replacement for the recommended textbooks. On the contrary, I encourage my students to explore more comprehensive works, and I often tell them: "Read the classics!" Certain parts in these notes are inspired or follow closely ideas that can be found in classic textbooks such as Bott and Tu (1982), (Dubrovin *et al.*, 1992), Helgason (2001), Kobayashi and Nomizu (1996), Lee (2013), Sharpe (1997), Spivak (1979), and Taubes (2011).

The development of these lecture notes benefited significantly from the feedback and input received from many students and colleagues. In particular, I am grateful to David Altizio, Raquel Caseiro, Ricardo Inglês, Ricardo Joel, Georgios Kydonakis, Daan Michiels, Miguel Negrão, Miguel Olmos, João Pimentel Nunes, Roger Picken, Ana Rita Pires, Matthew Romney, Olivier Massicot, Wilmer Smilde, and Joel Villatoro. Furthermore, I would like to extend special thanks to my colleagues from IST, Sílvia Anjos and José Natário, who diligently pointed out many typos and mistakes and provided several helpful suggestions for corrections. As an ongoing process, I continue to update and refine these lecture notes, and I am always appreciative of any corrections and suggestions for improvement that are sent my way.

Finally, I would like to express my gratitude for the generous support provided by the Portuguese Science Foundation and the National Science Foundation throughout the many years it took to shape these notes.

About the Author

Rui Loja Fernandes received his Ph.D. in mathematics from the University of Minnesota in 1994, under the direction of Peter J. Olver. He joined the Department of Mathematics at Instituto Superior Técnico in Lisbon, where he progressed through the tenured ranks, becoming Professor Catedrático in 2007. In 2012, he moved back to the USA and has since held the position of Lois M. Lackner Professor of Mathematics at the University of Illinois Urbana–Champaign. In 2016, he became a Fellow of the American Mathematical Society "for contributions to the study of Poisson geometry and Lie algebroids, and for service to the mathematical community." His research focuses on differential geometry, particularly on Poisson and symplectic geometry, and Lie groupoid theory. He is the author of more than 50 research papers and has supervised 9 graduate students.

Contents

PART 1
Basic Concepts

Lecture 1

Manifolds as Subsets of Euclidean Space

Recall that the *Euclidean space of dimension n* is

$$\mathbb{R}^n := \left\{ (x^1, \ldots, x^n) : x^1, \ldots, x^n \in \mathbb{R} \right\}.$$

We will also denote by $x^i : \mathbb{R}^n \to \mathbb{R}$ the ith coordinate function in $\mathbb{R}^n$. If $U \subset \mathbb{R}^n$ is an open set, a map $f : U \to \mathbb{R}^m$ is called a **smooth map** if all its partials derivatives of every order

$$\frac{\partial^r f^j}{\partial x^{i_1} \cdots \partial x^{i_r}}(x),$$

exist and are continuous functions in U. More generally, given any subset $X \subset \mathbb{R}^n$ and a map $f : X \to \mathbb{R}^m$, where X is not necessarily an open set, we say that f is a **smooth map** if for each $x \in X$, there is an open neighborhood $U \subset \mathbb{R}^n$ and a smooth map $F : U \to \mathbb{R}^m$ such that $f|_{X \cap U} = F|_{X \cap U}$.

A very basic property which we leave as an exercise is the following.

Proposition 1.1. *Let $X \subset \mathbb{R}^n$, $Y \subset \mathbb{R}^m$, and $Z \subset \mathbb{R}^p$. If $f : X \to Y$ and $g : Y \to Z$ are smooth maps, then $g \circ f : X \to Z$ is also a smooth map.*

A bijection $f : X \to Y$, where $X \subset \mathbb{R}^n$ and $Y \subset \mathbb{R}^m$, with inverse map $f^{-1} : Y \to X$, such that both f and f^{-1} are smooth, is called

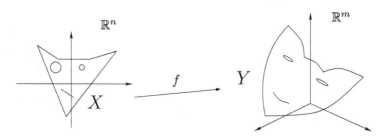

Fig. 1.1.   Diffeomorphic subsets.

a **diffeomorphism** and we say that X and Y are **diffeomorphic subsets** (see Figure 1.1).

One would like to study properties of sets which are invariant under diffeomorphisms, characterize classes of sets invariant under diffeomorphisms, etc. However, in this definition, the sets X and Y are just too general and it is hopeless to try to say anything interesting about classes of such diffeomorphic subsets. One must consider nicer subsets of Euclidean space. For example, it is desirable that the subset has at each point a tangent space and that the tangent spaces vary smoothly.

Recall that a subset $X \subset \mathbb{R}^n$ has an induced topology, called the *subset topology* or *relative topology*. For this topology, the open sets are just the sets of the form $U \cap X$, where $U \subset \mathbb{R}^n$ is an open set.

Definition 1.1. A subset $M \subset \mathbb{R}^n$ is called a **smooth manifold of dimension** d if each $p \in M$ has a neighborhood $U \cap M$ which is diffeomorphic to an open set $V \subset \mathbb{R}^d$.

The diffeomorphism $\phi : U \cap M \to V$ in this definition is called a **coordinate system** or a **chart** (see Figure 1.2). The inverse map $\phi^{-1} : V \to U \cap M$, which by assumption is smooth, is called a **parameterization**.

We have the **category of smooth manifolds** where

- the objects are smooth manifolds;
- the morphisms are smooth maps.

The reason they form a category is that the composition of smooth maps yields a smooth map, and the identity is also a smooth map.

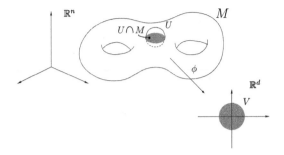

Fig. 1.2.   A coordinate system.

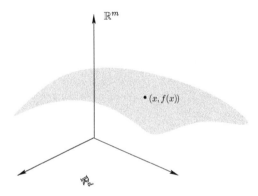

Fig. 1.3.   A graph of a map $f : \mathbb{R}^d \to \mathbb{R}^m$.

Example 1.1. An open subset $U \subset \mathbb{R}^d$ is itself a smooth manifold of dimension d: the inclusion $i : U \hookrightarrow \mathbb{R}^d$ gives a globally defined chart.

Example 1.2. If $f : \mathbb{R}^d \to \mathbb{R}^m$ is any smooth map, its graph

$$\mathrm{Graph}(f) := \{(x, f(x)) : x \in \mathbb{R}^d\} \subset \mathbb{R}^{d+m}$$

is a smooth manifold of dimension d: the map $x \mapsto (x, f(x))$ is a diffeomorphism $\mathbb{R}^d \to \mathrm{Graph}(M)$, so gives a global parametrization of $\mathrm{Graph}(f)$ (see Figure 1.3).

Example 1.3. The **unit d-sphere** is the subset of $\mathbb{R}^{d+1}$ formed by all vectors of length 1

$$\mathbb{S}^d := \{x \in \mathbb{R}^{d+1} : ||x|| = 1\}.$$

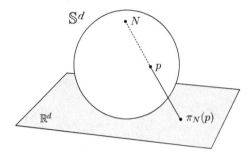

Fig. 1.4.   Stereographic projection.

Fig. 1.5.   1-dimensional manifolds.

Fig. 1.6.   A 2-torus.

This is a d-dimensional manifold which does not admit a global chart. However, we can cover the sphere by two coordinate systems: if we let $N = (0, \ldots, 0, 1)$ and $S = (0, \ldots, 0, -1)$ denote the north and south poles, then stereographic projection relative to N and S give two charts $\pi_N : \mathbb{S}^d - \{N\} \to \mathbb{R}^d$ and $\pi_S : \mathbb{S}^d - \{S\} \to \mathbb{R}^d$ (see Figure 1.4).

Example 1.4. The only connected manifolds of dimension 1 are the line $\mathbb{R}$ and the unit circle $\mathbb{S}^1$. What this statement means is that any connected manifold of dimension 1 is diffeomorphic to $\mathbb{R}$ or to $\mathbb{S}^1$ (see Figure 1.5).

Example 1.5. The manifolds of dimension 2 include the compact surfaces of genus g. For $g = 0$ this is the sphere $\mathbb{S}^2$, while for $g = 1$ this is the 2-torus. For $g > 1$ the **compact surface of genus** g is a g-holed torus as in Figures 1.6 and 1.7.

Fig. 1.7. A surface of genus g.

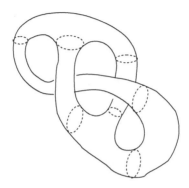

Fig. 1.8. An embedded surface of genus 2.

A compact surface of genus g can be "embedded" in $\mathbb{R}^3$ in many different ways. This statement will be made precise later, but here is one example (see Figure 1.8).

In the definitions, we have adopted so far we have chosen the smooth category, where differentiable maps have all partial derivatives of all orders. We could have chosen other classes, such as continuous maps, C^k-maps, or analytic maps.[1] This would lead us to the categories of **topological manifolds**, C^k **manifolds** or **analytic manifolds**. Note that in each such category we have an appropriate notion of equivalence: for example, two topological manifolds X and Y are equivalent if and only if there exists a homeomorphism between them, i.e., a continuous bijection $f : X \to Y$ such that the inverse is also continuous.

---

[1]We shall also use the term C^k-**map**, $k = 1, 2, \ldots, +\infty$, for a map whose partial derivatives of all orders up to k exist and are continuous. A C^0-map is simply a continuous map and a C^ω-**map** means an analytic map. A C^k-map which is invertible and whose inverse is also a C^k-map is called a C^k-**equivalence** or a C^k-**isomorphism**.

Most of the times we will be working with smooth manifolds. However, there are many situations where it is desirable to consider other categories of manifolds, so you should keep them in mind.

Example 1.6. Let $I = [-1, 1]$. The unit cube d-dimensional cube is the set:

$$I^d = \{(x^1, \ldots, x^d) \in \mathbb{R}^{d+1} : x^i \in I, \text{ for all } i = 1, \ldots, n\}.$$

The boundary of the cube

$$\partial I^d = \{(x^1, \ldots, x^d) \in I^d : x^i = -1 \text{ **or** } 1, \text{ for some } i = 1, \ldots, n\}.$$

is a topological manifold of dimension $d - 1$, which is not a smooth manifold (see Figure 1.9).

Example 1.7. If $f : \mathbb{R}^d \to \mathbb{R}^l$ is any map of class C^k, its graph:

$$\text{Graph}(f) := \{(x, f(x)) : x \in \mathbb{R}^d\} \subset \mathbb{R}^{d+l}$$

is a C^k-manifold of dimension d. Similarly, if f is any analytic map then $\text{Graph}(f)$ is an analytic manifold.

You may wonder if the dimension d that appears in the definition of a manifold is a well-defined integer, in other words, if a manifold $M \subset \mathbb{R}^n$ could be of dimension d and d' for distinct integers $d \neq d'$. The reason that this cannot happen is due to the following important result:

Theorem 1.1 (Invariance of Domain). *Let $U \subset \mathbb{R}^n$ be an open set and let $\phi : U \to \mathbb{R}^n$ be a 1:1, continuous map. Then $\phi(U)$ is open.*

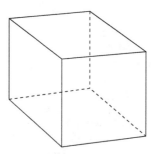

Fig. 1.9. A cube.

The reason for calling this result "invariance of domain" is that a domain is a connected open set of $\mathbb{R}^n$, so the result says that the property of being a domain remains invariant under a continuous, 1:1 map. The proof of this result requires some methods from algebraic topology and can be found in any good book on the subject (see, e.g., Hatcher, 2002, Theorem 2B.3). We leave it as an exercise to show that the invariance of domain implies that the dimension of a manifold is a well-defined integer.

Exercises

Exercise 1.1
Let $X \subset \mathbb{R}^n$, $Y \subset \mathbb{R}^m$ and $Z \subset \mathbb{R}^p$. If $f : X \to Y$ and $g : Y \to Z$ are smooth maps, show that $g \circ f : X \to Z$ is also a smooth map.

Exercise 1.2
Let $f : \mathbb{R}^d \to \mathbb{R}^m$ be a map of class $C^k, k = 0, \ldots, \omega$. Show that $\phi : \mathbb{R}^d \to \text{Graph}(f)$, $x \mapsto (x, f(x))$, is a C^k-equivalence.

Exercise 1.3
Show that the sphere $\mathbb{S}^d$ and the boundary of the cube ∂I^{d+1} are equivalent topological manifolds.

Exercise 1.4
Consider the set $\text{SL}(2, \mathbb{R})$ formed by all 2×2 matrices with real entries and determinant 1:

$$\text{SL}(2, \mathbb{R}) = \left\{ \begin{bmatrix} a & b \\ c & d \end{bmatrix} : ad - bc = 1 \right\} \subset \mathbb{R}^4.$$

Show that $\text{SL}(2, \mathbb{R})$ is a 3-dimensional smooth manifold.

Exercise 1.5
Use invariance of domain to show that the notion of dimension of a topological manifold is well defined.

Lecture 2

Abstract Manifolds

In many situations, manifolds do not arise naturally as subsets of Euclidean space. We will see several examples of this later. For that reason, the definition of manifold that we have seen in the previous lecture is often not the most useful one. We need a different definition of a manifold, where M is not assumed *a priori* to be a subset of some $\mathbb{R}^n$. We will see later that these abstract manifolds can always be embedded in some Euclidean space, and so, the abstract definition is actually equivalent to the more concrete definition of the previous lecture.

For this more abstract definition of manifold, we need to start with a set M where we have a notion of "proximity". In other words, we need M to be furnished *a priori* with a topology. At this point, it maybe useful to remind yourself of the basics of point set topology.

Definition 2.1. A topological space M is called a **topological manifold of dimension** d if for every $p \in M$, there is a neighborhood $U \subset M$ and a homeomorphism $\phi : U \to V$ onto some open subset $V \subset \mathbb{R}^d$.

Sometimes, one also calls a topological manifold a **locally Euclidean space**. In this more general context, we still use the same notation as before. We call $\phi : U \to \mathbb{R}^d$ a **coordinate system** or a **chart** (see Figure 2.1), and the functions $\phi^i = x^i \circ \phi$ are called **coordinate functions**. We shall denote a chart by (U, ϕ). Often, we write x^i instead of ϕ^i for the coordinate functions, in which case we

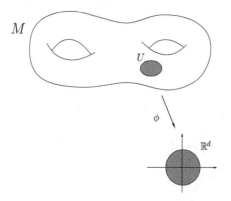

Fig. 2.1.   A topological manifold.

may denote the chart by $(U, x^1, \ldots, x^d)$. We say that a chart (U, ϕ) is *centered at a point* $p \in M$ if $\phi(p) = 0$.

Example 2.1. On $\mathbb{R}^3 \setminus \{0\}$, consider the equivalence relation $\sim$, where $v \sim w$ if and only if $v = \lambda w$ for some real number $\lambda \neq 0$. The set of equivalence classes

$$\mathbb{RP}^2 := (\mathbb{R}^3 \setminus \{0\})/\sim$$

can be identified with the set of straight lines in $\mathbb{R}^3$ that pass through the origin and it is called the **projective plane**. Denoting by $[x : y : z]$, the equivalence class of $(x, y, z) \in \mathbb{R}^3 \setminus \{0\}$, we have the quotient map

$$\pi : \mathbb{R}^3 \setminus \{0\} \to \mathbb{RP}^2, \ (x, y, z) \mapsto [x : y : z].$$

On $\mathbb{RP}^2$, we consider the quotient topology. This means that $U \subset \mathbb{RP}^2$ is open if and only if $\pi^{-1}(U) \subset \mathbb{R}^2 - \{0\}$ is open. The maps given by

$$\phi_1 : U_1 \to \mathbb{R}^2, \ [x : y : z] \mapsto \left(\frac{y}{x}, \frac{z}{x}\right), \ U_1 := \{[x : y : z] \in \mathbb{RP}^2 : x \neq 0\},$$

$$\phi_2 : U_2 \to \mathbb{R}^2, \ [x : y : z] \mapsto \left(\frac{x}{y}, \frac{z}{y}\right), \ U_2 := \{[x : y : z] \in \mathbb{RP}^2 : y \neq 0\},$$

$$\phi_3 : U_3 \to \mathbb{R}^2, \ [x : y : z] \mapsto \left(\frac{x}{z}, \frac{y}{z}\right), \ U_3 := \{[x : y : z] \in \mathbb{RP}^2 : z \neq 0\},$$

are homeomorphisms from opens in $\mathbb{RP}^2$ onto $\mathbb{R}^2$. Since $\{U_1, U_2, U_3\}$ is an open cover of $\mathbb{RP}^2$, we conclude that the projective plane is a topological manifold of dimension 2.

We make the following **tacit assumption** about the underlying topology of a manifold:

Manifolds are assumed to be Hausdorff and second countable.

This assumption has significant implications, as we shall see shortly, which are very useful in the study of manifolds (e.g., the existence of partitions of unity or Riemannian metrics). On the other hand, it means that when constructing a manifold, we have to demonstrate that the underlying topology satisfies these assumptions. This is often easy since, for example, any metric space satisfies these assumptions.

It should be noted, however, that non-Hausdorff manifolds do appear sometimes, for example, when one forms quotients of (Hausdorff) manifolds (see Lecture 10). Manifolds which are not second countable can also appear (e.g., in sheaf theory), although we will not meet them in the course of these lectures. We limit ourselves here to give two such examples.

Example 2.2. On $\mathbb{R}^2 \setminus \{0\}$, consider the connected components of the horizontal lines $y = c$. This defines a partition of $\mathbb{R}^2 \setminus \{0\}$ and so defines an equivalence relation $\sim$. The quotient space $M = \mathbb{R}^2\setminus\{0\}/\sim$ (with the quotient topology) is a topological 1-dimensional manifold: we can cover M by two open sets

$$U_+ = \{[(1,y)] : y \in \mathbb{R}\}, \ U_- = \{[(-1,y)] : y \in \mathbb{R}\},$$

for which we have homeomorphisms

$$\phi_\pm : U_\pm \to \mathbb{R}, \quad [(\pm1,y)] \mapsto y.$$

However, M is not a Hausdorff topological space since the points $[(1,0)]$ and $[(-1,0)]$ cannot be separated. One calls M the **line with two origins**.

Example 2.3. Consider on $M = \mathbb{R}^2$ the topology generated by all sets of the form $U \times \{y\}$, where $U \subset \mathbb{R}$ is open and $y \in \mathbb{R}$. This topology does not have a countable basis. However, M is a topological 1-dimensional manifold with charts $(U \times \{y\}, \phi_y)$ given by $\phi_y(x,y) = x$. In this example, M is the disjoint, uncountable, union of copies of the real line, and it is not connected. It is possible to give examples of connected, Hausdorff, manifolds M which are not second countable, such as the **long line** (see Exercise 2.13).

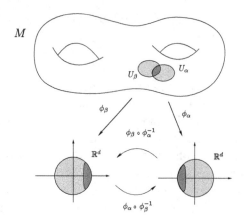

Fig. 2.2.   Transition functions between two charts.

Of course, we are interested in smooth manifolds. The definition is slightly more involved.

Definition 2.2. A **smooth structure** on a topological d-manifold M is a collection of charts $\mathcal{C} = \{(U_\alpha, \phi_\alpha) : \alpha \in A\}$ which satisfies the following properties:

(i) The collection $\mathcal{C}$ covers M: $\bigcup_{\alpha \in A} U_\alpha = M$.
(ii) For all $\alpha, \beta \in A$, the **transition function** $\phi_\alpha \circ \phi_\beta^{-1}$ is a smooth map (see Figure 2.2).
(iii) The collection $\mathcal{C}$ is maximal: if (U, ϕ) is a chart such that for all $\alpha \in A$, the maps $\phi \circ \phi_\alpha^{-1}$ and $\phi_\alpha \circ \phi^{-1}$ are smooth, then $(U, \phi) \in \mathcal{C}$.

The pair $(M, \mathcal{C})$ is called a **smooth manifold** of dimension d.

Given a topological manifold M, a collection of charts which satisfies conditions (i) and (ii) in the previous definition is called an **atlas**. Given any such atlas $\mathcal{C}_0 = \{(U_\alpha, \phi_\alpha) : \alpha \in A\}$, there exists a unique maximal atlas $\mathcal{C}$ which contains $\mathcal{C}_0$: it is enough to define $\mathcal{C}$ to be the collection of all **smooth charts** relative to $\mathcal{C}$, i.e., all charts (U, ϕ) such that $\phi \circ \phi_\alpha^{-1}$ and $\phi_\alpha \circ \phi^{-1}$ are both smooth for all $(U_\alpha, \phi_\alpha) \in \mathcal{C}_0$. For this reason, one often defines a smooth structure on a topological manifold M by specifying some atlas, and it is then implicit that the smooth structure is the one associated with the corresponding maximal atlas.

It should be clear from this definition that one can define in a similar fashion **manifolds of class** C^k for any $k = 1, \ldots, +\infty, \omega$, by requiring the transition functions to be of class C^k. In these lectures, we shall concentrate on the case $k = +\infty$.

Example 2.4. The **standard differential structure on Euclidean space** $\mathbb{R}^d$ is the maximal atlas that contains the coordinate system $(\mathbb{R}^d, i)$, where $i : \mathbb{R}^d \to \mathbb{R}^d$ is the identity map. It is a non-trivial fact that the Euclidean space $\mathbb{R}^4$ has an infinite number of smooth structures, with the same underlying topology, but which are not equivalent to this one, in a sense to be made precise later (see, e.g., Freedman and Luo, 1989). These are called **exotic smooth structures**. It is also known that $\mathbb{R}^d$, for $d \neq 4$, has no exotic smooth structures.

Example 2.5. If $M \subset \mathbb{R}^n$ is a d-dimensional manifold in the sense of Definition 1.1, then M carries a natural smooth structure: the coordinate systems in Definition 1.1 form a maximal atlas (exercise) for the topology on M induced from the usual topology on $\mathbb{R}^n$. We shall see later in Lecture 8 that the Whitney Embedding Theorem shows that, conversely, any smooth manifold M arises in this way. Henceforth, we shall refer to a manifold $M \subset \mathbb{R}^n$ in the sense of Definition 1.1 as an **embedded manifold** in $\mathbb{R}^n$.

Example 2.6. If M is a d-dimensional smooth manifold with smooth structure $\mathcal{C}$ and $U \subset M$ is an **open subset**, then U with the relative topology is also a smooth d-dimensional manifold with smooth structure given by

$$\mathcal{C}_U = \{(U_\alpha \cap U, \phi_\alpha|_{U_\alpha \cap U}) : (U, \phi_\alpha) \in \mathcal{C}\}.$$

Example 2.7. If M and N are smooth manifolds, then the **Cartesian product** $M \times N$, with the product topology, is a smooth manifold: in $M \times N$, we consider the maximal atlas that contains all coordinate systems of the form $(U_\alpha \times V_\beta, \phi_\alpha \times \psi_\beta)$, where (U_α, ϕ_α) and (V_β, ψ_β) are smooth coordinate systems of M and N, respectively. It should be clear that $\dim M \times N = \dim M + \dim N$. More generally, if $M_1, \ldots, M_k$ are smooth manifolds, then $M_1 \times \cdots \times M_k$ is a smooth manifold of dimension $\dim M_1 + \cdots + \dim M_k$. For example, the d-**torus** $\mathbb{T}^d = \mathbb{S}^1 \times \cdots \times \mathbb{S}^1$ and the **cylinders** $\mathbb{R}^n \times \mathbb{S}^m$ are smooth manifolds of dimensions d and $n + m$, respectively.

Example 2.8. Generalizing the projective plane, one defines the **real projective space** as the set

$$\mathbb{RP}^d := \left\{ L \subset \mathbb{R}^{d+1} : L \text{ is a straight line through the origin} \right\},$$

which we can think of as the quotient space of $\mathbb{R}^{d+1} \setminus \{0\} /\!\sim$ by the equivalence relation

$$(x^0, \ldots, x^d) \sim (y^0, \ldots, y^d) \quad \text{if and only if } (x^0, \ldots, x^d) = \lambda(y^0, \ldots, y^d)$$

for some $\lambda \in \mathbb{R} \setminus \{0\}$. On $\mathbb{RP}^d$, we take the quotient topology, so it becomes a topological manifold of dimension d: if we denote by $[x^0 : \cdots : x^d]$ the equivalence class of $(x^0, \ldots, x^d) \in \mathbb{R}^{d+1} \setminus \{0\}$, then for each $\alpha = 0, \ldots, n$, we have the coordinate system (U_α, ϕ_α), where

$$U_\alpha = \left\{ [x^0 : \cdots : x^d] : x^\alpha \neq 0 \right\},$$

$$\phi_\alpha : U_\alpha \to \mathbb{R}^d, \quad [x^0 : \cdots : x^d] \mapsto \left(\frac{x^0}{x^\alpha}, \ldots, \frac{\widehat{x^\alpha}}{x^\alpha}, \ldots, \frac{x^d}{x^\alpha} \right)$$

(the symbol $\widehat{a}$ means that we omit the term a). We leave it as an exercise to check that the transition functions between these coordinate functions are smooth, so they form an atlas on $\mathbb{RP}^d$. Note that $\mathbb{RP}^d$ does not arise naturally as a subset of some Euclidean space.

We have established what our objects are. Now, we turn to the morphisms.

Definition 2.3. Let M and N be smooth manifolds.

(i) A function $f : M \to \mathbb{R}$ is called a **smooth function** if $f \circ \phi^{-1}$ is smooth for all smooth coordinate systems (U, ϕ) of M.
(ii) A map $\Psi : M \to N$ is called a **smooth map** if $\tau \circ \Psi \circ \phi^{-1}$ is smooth for all smooth coordinate systems (U, ϕ) of M and (V, τ) of N.

A smooth map $\Psi : M \to N$ which is invertible and whose inverse is smooth is called a **diffeomorphism**. In this case, we say that M and N are **diffeomorphic manifolds**.

In order to check that a map $\Psi : M \to N$ is smooth, it is enough to verify that, for each $p \in M$, there exists a smooth chart (U, ϕ) of M with $p \in U$ and a smooth chart (V, τ) of N with $\Psi(p) \in V$, such that $\tau \circ \Psi \circ \phi^{-1}$ is a smooth map. Also, a smooth function $f : M \to \mathbb{R}$ is just a smooth map where $\mathbb{R}$ has its standard smooth structure.

Clearly, the composition of two smooth maps, whenever defined, is a smooth map. The identity map is also a smooth map. So, we have the **category of smooth manifolds**, whose objects are the smooth manifolds and whose morphisms are the smooth maps.

Just as we did for maps between subsets of Euclidean space, when $X \subset M$ and $Y \subset N$ are arbitrary subsets of some smooth manifolds, we will say that $\Psi : X \to Y$ is a **smooth map** if for each $p \in X$, there is an open neighborhood $U \subset M$ and a smooth map $F : U \to N$ such that $F|_{U \cap X} = \Psi|_{U \cap X}$. The set of smooth maps from X to Y will be denoted as $C^\infty(X, Y)$. When $Y = \mathbb{R}$, we use $C^\infty(X)$ instead of $C^\infty(X, \mathbb{R})$.

Example 2.9. Let $M \subset \mathbb{R}^n$ be an embedded manifold (recall our convention from Example 2.5), any smooth function $F : U \to \mathbb{R}$ defined on an open $\mathbb{R}^n \supset U \supset M$ induces, by restriction, a smooth function $f : M \to \mathbb{R}$. Conversely, every smooth function $f : M \to \mathbb{R}$ is the restriction of some smooth function $F : U \to \mathbb{R}$ defined on some open set $\mathbb{R}^n \supset U \supset M$. To see this, we will need the partitions of unity to be introduced in Lecture 4.

You should also check that if $M \subset \mathbb{R}^n$ and $N \subset \mathbb{R}^m$ are embedded manifolds, then $\Psi : M \to N$ is a smooth map if and only if for every $p \in M$, there exists an open neighborhood $U \subset \mathbb{R}^n$ of p and a smooth map $F : U \to \mathbb{R}^m$ such that $\Psi|_{U \cap M} = F|_{U \cap M}$. This shows that the notion of smooth map in Definition 2.3 extends the notion we have introduced in the previous lecture.

Example 2.10. The map $\pi : \mathbb{S}^d \to \mathbb{RP}^d$ defined by

$$\pi(x^0, \ldots, x^d) = [x^0 : \cdots : x^d]$$

is a smooth map. Moreover, any smooth function $F : \mathbb{S}^d \to \mathbb{R}$ which is invariant under inversion, i.e., such that $F(-x) = F(x)$, induces a unique smooth function $f : \mathbb{RP}^d \to \mathbb{R}$ that makes the following diagram commute:

$$\begin{array}{ccc} \mathbb{S}^d & \xrightarrow{F} & \mathbb{R} \\ {\scriptstyle \pi}\downarrow & \nearrow \; {\scriptstyle f} & \\ \mathbb{RP}^d & & \end{array}$$

Conversely, every smooth function $f \in C^\infty(\mathbb{RP}^d)$ arises in this way.

If we are given two smooth structures $\mathcal{C}_1$ and $\mathcal{C}_2$ on the same manifold M, we say that they are **equivalent smooth structures** if there is a diffeomorphism $\Psi : (M, \mathcal{C}_1) \to (M, \mathcal{C}_2)$.

Example 2.11. On the line $\mathbb{R}$, the identity map $\mathbb{R} \to \mathbb{R}$, $x \mapsto x$ gives a chart which defines a smooth structure $\mathcal{C}_1$. We can also consider the chart $\mathbb{R} \to \mathbb{R}$, $x \mapsto x^3$, and this defines a distinct smooth structure $\mathcal{C}_2$ on $\mathbb{R}$ (why?). However, these two smooth structures are equivalent since the map $x \mapsto x^3$ gives a diffeomorphism from $(M, \mathcal{C}_2)$ to $(M, \mathcal{C}_1)$.

It is known that every topological manifold of dimension less than or equal to 3 has a unique smooth structure. For dimensions greater than 3, the situation is much more complicated, and not much is known. However, as we have mentioned before, the smooth structures on $\mathbb{R}^d$, compatible with the usual topology, are all equivalent if $d \neq 4$, and there are uncountably many inequivalent **exotic smooth structures** on $\mathbb{R}^4$. On the other hand, for the sphere $\mathbb{S}^d$, there are no exotic smooth structures for $d \leq 6$, but Milnor found that $\mathbb{S}^7$ has 27 inequivalent smooth structures. It is known, e.g., that $\mathbb{S}^{31}$ has more than 16 million inequivalent smooth structures! One can read more about Milnor's work in Milnor (2015).

Exercises

Exercise 2.1

Let M be a topological manifold. Show that M is locally compact, i.e., every point of M has a compact neighborhood.

Exercise 2.2

The Urysohn's metrization theorem states that a Hausdorff, regular, topological space with a countable basis is metrizable (see, e.g., Kelley, 1975). Use this to show that every topological manifold M is metrizable.

Exercise 2.3

Let M be a *connected* topological manifold. Show that M is path connected. If, additionally, M is a smooth manifold, show that for any $p, q \in M$, there exists a smooth path $c : [0,1] \to M$ with $c(0) = p$ and $c(1) = q$.

Exercise 2.4

Let $\phi : \mathbb{R}^m \to \mathbb{R}^n$ be a diffeomorphism. Use the chain rule to deduce that one must have $m = n$. Use this result to conclude that if M and N are diffeomorphic smooth manifolds, then $\dim M = \dim N$ without appealing to invariance of domain.

Exercise 2.5

Compute the transition functions for the atlas of real projective space $\mathbb{RP}^d$ and show that they are smooth. Show also that

(a) $\mathbb{RP}^1$ is diffeomorphic to $\mathbb{S}^1$;
(b) $\mathbb{RP}^d \setminus \mathbb{RP}^{d-1}$ is diffeomorphic to the open ball $B^n = \{x \in \mathbb{R}^d : \|x\| < 1\}$, where we identify $\mathbb{RP}^{d-1}$ with the subset $\{[x^0 : \cdots : x^d] : x^d = 0\} \subset \mathbb{RP}^d$.

Exercise 2.6

The **complex projective d-dimensional space** is the set

$$\mathbb{CP}^d = \left\{ L \subset \mathbb{C}^{d+1} : L \text{ is a complex line through the origin} \right\}.$$

Construct a smooth structure of a $2d$-dimensional manifold on $\mathbb{CP}^d$ similar to the construction of a smooth structure on real projective space $\mathbb{RP}^d$.

Exercise 2.7

Show that if $M \subset \mathbb{R}^n$ is an embedded manifold in the sense of Definition 1.1, then M carries a natural smooth structure.

Exercise 2.8

Let $M \subset \mathbb{R}^n$ be a subset with the following property: for each $p \in M$, there exists an open set $U \subset \mathbb{R}^n$ containing p and diffeomorphism $\Phi : U \to V$ onto an open set $V \subset \mathbb{R}^n$, such that

$$\Phi(U \cap M) = \left\{ q \in V : q^{d+1} = \cdots = q^n = 0 \right\}.$$

Show that M is a smooth manifold of dimension d (in fact, M is an embedded manifold in $\mathbb{R}^n$; see the previous exercise).

Exercise 2.9

Let M be a set and assume that one has a collection $\mathcal{C} = \{(U_\alpha, \phi_\alpha) : \alpha \in A\}$, where $U_\alpha \subset M$ and $\phi_\alpha : U_\alpha \to \mathbb{R}^d$, satisfy the following properties:

(a) For each $\alpha \in A$, $\phi_\alpha(U_\alpha) \subset \mathbb{R}^n$ is open and $\phi_\alpha : U_\alpha \to \phi_\alpha(U_\alpha)$ is a bijection.
(b) For each $\alpha, \beta \in A$, the sets $\phi_\alpha(U_\alpha \cap U_\beta) \subset \mathbb{R}^n$ are open.
(c) For each $\alpha, \beta \in A$, with $U_\alpha \cap U_\beta \neq \emptyset$, the map $\phi_\beta \circ \phi_\alpha^{-1} : \phi_\alpha (U_\alpha \cap U_\beta) \to \phi_\beta(U_\alpha \cap U_\beta)$ is smooth.
(d) There is a countable set of U_α that covers M.
(e) For any $p, q \in M$, with $p \neq q$, either there exists a U_α such that $p, q \in U_\alpha$, or there exists U_α and U_β, with $p \in U_\alpha$, $q \in U_\beta$ and $U_\alpha \cap U_\beta = \emptyset$.

Show that there exists a unique smooth structure on M such that the collection $\mathcal{C}$ is an atlas.

Exercise 2.10

Let $M = \mathbb{C} \cup \{\infty\}$. Let $U := M \setminus \{\infty\} = \mathbb{C}$ and $\phi_U : U \to \mathbb{C}$ be the identity map and let $V := M \setminus \{0\}$ and $\phi_V : V \to \mathbb{C}$ be the map $\phi_V(z) = 1/z$, with the convention that $\phi(\infty) = 0$. Use the previous exercise to show that M has a unique smooth structure with atlas $\mathcal{C} := \{(U, \phi_U), (V, \phi_V)\}$. Show that M is diffeomorphic to $\mathbb{S}^2$.

Hint: Be careful with item (e)!

Exercise 2.11

Let M and N be smooth manifolds and let $\Psi : M \to N$ be a map. Show that the following statements are equivalent:

(i) $\Psi : M \to N$ is smooth.

(ii) For every $p \in M$, there are smooth coordinate systems (U, ϕ) of M and (V, τ) of N, with $p \in U$ and $\Psi(p) \in V$, such that $\tau \circ \Psi \circ \phi^{-1}$ is smooth.

(iii) There exist atlases $\{(U_\alpha, \phi_\alpha) : \alpha \in A\}$ and $\{(U_\beta, \psi_\beta) : \beta \in B\}$ of M and N, such that for each $\alpha \in A$ and $\beta \in B$, $\psi_\beta \circ \Psi \circ \phi_\alpha^{-1}$ is smooth.

Exercise 2.12

Let M and N be smooth manifolds and let $\Phi : M \to N$ be a map. Show the following:

(i) If Φ is smooth, then for every open set $U \subset M$, the restriction $\Phi|_U : U \to N$ is a smooth map.

(ii) If every $p \in M$ has an open neighborhood U such that the restriction $\Phi|_U : U \to N$ is a smooth map, then $\Phi : M \to N$ is smooth.

Exercise 2.13

Let ω_1 denote the first uncountable ordinal (see, e.g., Levy, 2002) and let $M = (\omega_1 \times [0, 1)) \backslash \{(0, 0)\}$ with the topology induced by the lexicographic order. Show that M has a smooth structure compatible with this topology, so that it is a connected, Hausdorff, manifold which is not second countable. This manifold is known as the **long line**. Intuitively, it is obtained by gluing an uncountable number of copies of $(0, 1)$.

Lecture 3

Manifolds with Boundary

There are many spaces, such as the closed unit disk, a solid donut, or the Möbius strip, which just fail to be a manifold because they have a "boundary". One can remedy this situation by trying to enlarge the notion of manifold so that it includes this possibility. The clue to be able to include boundary points is to understand what is the local model around points in the "boundary" and this turns out to be the **closed half-space** $\mathbb{H}^d$:

$$\mathbb{H}^d := \{(x^1, \ldots, x^d) \in \mathbb{R}^d : x^d \geq 0\}.$$

We will denote the **open half-space** by

$$\operatorname{Int} \mathbb{H}^d =: \{(x^1, \ldots, x^d) \in \mathbb{R}^d : x^d > 0\}.$$

and the **boundary of the closed half-space** by

$$\partial \mathbb{H}^d =: \{(x^1, \ldots, x^d) \in \mathbb{R}^d : x^d = 0\}.$$

When $n = 0$, we have $\mathbb{H}^0 = \mathbb{R}^0 = \{0\}$, so $\operatorname{Int} H^0 = \mathbb{R}^0$ and $\partial \mathbb{H}^0 = \emptyset$.

Definition 3.1. A topological manifold with boundary of dimension d is a topological space M such that every $p \in M$ has a neighborhood U which is homeomorphic to some open set $V \subset \mathbb{H}^d$ (see Figure 3.1).

Just as we do for manifolds without boundary, we shall assume that all manifolds with boundary are Hausdorff and have a countable basis of open sets.

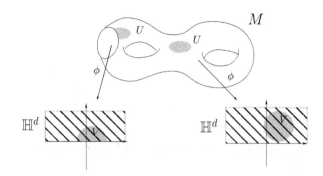

Fig. 3.1.   Charts in a manifold with boundary.

We shall use the same notations as before, so we call a homeo-morphism $\phi : U \to V$ as in the definition of a system of coordinates or a coordinate chart. Note that there are two types of open sets in $\mathbb{H}^d$ according to whether they intersect $\partial \mathbb{H}^d$ or not. These give rise to two types of coordinate systems $\phi : U \to V$, according to whether V intersects $\partial \mathbb{H}^d$ or not. In the first case, when $V \cap \partial \mathbb{H}^d = \emptyset$, we just have a coordinate system of the same sort as for manifolds without boundary, and we call it an **interior chart**. In the second case, when $V \cap \partial \mathbb{H}^d \neq \emptyset$, we call it a **boundary chart**.

Using Invariance of Domain (Theorem 1.1), one shows that

Lemma 3.1. *Let M be a topological manifold with boundary of dimension d. If for some chart (U, ϕ) we have $\phi(p) \in \partial \mathbb{H}^d$, then this is also true for every other chart.*

Proof. Exercise.   □

This justifies the following definition:

Definition 3.2. Let M be a topological manifold with boundary of dimension d. A point $p \in M$ is called a **boundary point** if there exists some chart (U, ϕ) with $p \in U$, such that $\phi(p) \in \partial \mathbb{H}^d$. Otherwise, p is called an **interior point**.

The set of boundary points of M will be denoted by ∂M and is called the **boundary of M** and the set of interior points of M will be denoted by Int M and is called the **interior of M**. If on both sets we consider the topology induced from M, we have:

Proposition 3.1. *Let M be a topological manifold with boundary of dimension $d > 0$. Then $\mathrm{Int}\, M$ and ∂M are topological manifolds without boundary of dimension d and $d - 1$, respectively. If N is another manifold with boundary and $\Psi : M \to N$ is a homeomorphism then Ψ restricts to homeomorphisms $\Psi|_{\partial M} : \partial M \to \partial N$ and $\Psi|_{\mathrm{Int}\, M} : \mathrm{Int}\, M \to \mathrm{Int}\, N$.*

Proof. Let $p \in \mathrm{Int}\, M$ and let $\phi : U \to V$ be a chart with $p \in U$ and $V \subset \mathbb{H}$. Then if we set $V_0 := V \setminus \partial \mathbb{H}$ and $U_0 := \phi^{-1}(V_0)$, we have that U_0 is an open neighborhood of M, V_0 is open in $\mathbb{R}^d$, and $\phi|_{U_0} : U_0 \to V_0$ is a homeomorphism. This shows that $\mathrm{Int}\, M$ is a topological manifold without boundary of dimension d.

On the other hand, let $p \in \partial M$ and let $\phi : U \to V$ be a chart with $p \in U$ and $\phi(p) \in \partial \mathbb{H}$. Then if we set $V_0 := V \cap \partial \mathbb{H}$ and $U_0 := \phi^{-1}(V_0)$, we have that $U_0 = U \cap \partial M$ is an open neighborhood of ∂M, V_0 is open in $\partial \mathbb{H} \simeq \mathbb{R}^{d-1}$, and $\phi|_{U_0} : U_0 \to V_0$ is a homeomorphism. This shows that ∂M is a topological manifold without boundary of dimension $d - 1$. $\qquad \square$

It is important not to confuse the notions of interior and boundary point for manifolds with boundary with the usual notions of interior and boundary point of a subset of a topological space. If M happens to be a manifold with boundary embedded in some $\mathbb{R}^n$ then the two notions may or may not coincide.

Example 3.1. $M = \mathbb{H}^d$ is itself a topological manifold with boundary of dimension d, where $\mathrm{Int}\, M = \mathrm{Int}\, \mathbb{H}^d$ and $\partial M = \partial \mathbb{H}^d$, so our notations are consistent. If we think of $\mathbb{H}^d \subset \mathbb{R}^d$, then these notions coincide with the usual notions of boundary and interior of $\mathbb{H}^d$ as a topological subspace of $\mathbb{R}^d$.

Example 3.2. The closed unit disk:

$$D^k = \overline{B^d} := \{x \in \mathbb{R}^d : \|x\| \leq 1\},$$

is a topological manifold with boundary of dimension d with interior the open unit ball B^d and boundary the unit sphere $\mathbb{S}^{d-1}$. If we think of $D^d \subset \mathbb{R}^d$, then these notions coincide with the usual notions of boundary and interior of D^d as a topological subspace of $\mathbb{R}^d$.

Example 3.3. The cube I^d is a topological manifold with boundary of dimension d. I^d and D^d are homeomorphic topological manifolds with boundary.

Example 3.4. The Möbius strip $M \subset \mathbb{R}^3$ is a topological manifold with boundary $\partial M = \mathbb{S}^1$. Note that, as a topological subspace of $\mathbb{R}^3$, all points of M are boundary points!

Now that we have the notion of chart for a topological manifold with boundary, we can define a **smooth structure** on a topological d-manifold with boundary M by exactly the same procedure as we did for manifolds without boundary: it is a collection of charts $\mathcal{C} = \{(U_\alpha, \phi_\alpha) : \alpha \in A\}$ which satisfies the following properties:

(i)   The collection $\mathcal{C}$ is an open cover of M: $\bigcup_{\alpha \in A} U_\alpha = M$;
(ii) For all $\alpha, \beta \in A$, the **transition function** $\phi_\alpha \circ \phi_\beta^{-1}$ is a smooth map;
(iii) The collection $\mathcal{C}$ is maximal: if (U, ϕ) any coordinate system such that $\phi \circ \phi_\alpha^{-1}$ and $\phi_\alpha \circ \phi^{-1}$ are smooth maps for all $\alpha \in A$, then $(U, \phi) \in \mathcal{C}$.

The pair $(M, \mathcal{C})$ is called a **smooth d-manifold with boundary**.

Again, given an atlas $\mathcal{C}_0 = \{(U_\alpha, \phi_\alpha) : \alpha \in A\}$ (i.e., a collection satisfying (i) and (ii)), there exists a unique maximal atlas $\mathcal{C}$ which contains $\mathcal{C}_0$: it is enough to define $\mathcal{C}$ to be the collection of all **smooth charts** relative to $\mathcal{C}_0$, i.e., all coordinate systems (U, ϕ) such that $\phi \circ \phi_\alpha^{-1}$ and $\phi_\alpha \circ \phi^{-1}$ are both smooth for all $(U_\alpha, \phi_\alpha) \in \mathcal{C}_0$.

The notion of smooth map $\Psi : M \to N$ between two manifolds with boundary is also defined in exactly the same way as in the case of manifolds without boundary.

Proposition 3.2. *Let M be a smooth manifold with boundary of dimension $d > 0$. Then $\text{Int } M$ and ∂M are smooth manifolds without boundary of dimension d and $d - 1$, respectively. If N is another smooth manifold with boundary and $\Psi : M \to N$ is a diffeomorphism then Ψ restricts to diffeomorphisms $\Psi|_{\partial M} : \partial M \to \partial N$ and $\Psi|_{\text{Int } M} : \text{Int } M \to \text{Int } N$.*

Proof. Exercise.   □

You should check that the half space $\mathbb{H}^d$, the closed disk D^d and the Möbius strip, are all *smooth* manifolds with boundary, while the cube I^d is not.

Although often one can work with manifolds with boundary much the same way as one can work with manifolds without boundary, some care must be taken. For example, the Cartesian product of two half-spaces is not a manifold with boundary (it is rather a *manifold with corners*, a notion we will not discuss). So the cartesian product of manifolds with boundary may not be a manifold with boundary. However, we do have the following result:

Proposition 3.3. *If M is a smooth manifold without boundary and N is a smooth manifold with boundary, then $M \times N$ is a smooth manifold with $\partial (M \times N) = M \times \partial N$ and $\mathrm{Int}(M \times N) = M \times \mathrm{Int}\, N$.*

Proof. Exercise. □

Example 3.5. If M is a manifold without boundary and $I = [0,1]$ then $M \times I$ is a manifold with boundary for which:

$$\mathrm{Int}(M \times I) = M \times]0,1[, \quad \partial(M \times I) = M \times \{0\} \cup M \times \{1\}.$$

It is very cumbersome to write always "manifold without boundary", so we agree to refer to these simply as "manifolds", and add the qualitative "with boundary", whenever that is the case. You should be aware that in the literature it is also common to use **nonbounded manifold** for a manifold in our sense, and to call a **closed manifold** a compact manifold without boundary and **open manifold** a manifold without boundary and with no compact connected component.

Exercises

Exercise 3.1
Use Invariance of Domain to show that if for a chart (U, ϕ) of a topological manifold with boundary one has $\phi(p) \in \partial\, \mathbb{H}^d$, then this also holds for every other chart.

Exercise 3.2

Let $M \subset \mathbb{R}^d$ have the induced topology. Show that if M is a closed subset and a d-dimensional manifold with boundary then the topological boundary of M coincides with ∂M. Give a counterexample to this statement when M is not a closed subset.

Exercise 3.3

Give the details of the proofs of Propositions 3.2 and 3.3.

Exercise 3.4

Let $M = D^2 \times \mathbb{S}^1$ be the solid torus (a 3-manifold with boundary). What is the boundary of the solid torus? How does this generalize to dimension > 3?

Lecture 4

Partitions of Unity

From now on, we will be dealing almost exclusively with smooth manifolds. Hence, we often write simply "manifold", we will use the term "chart" (or "coordinate system") to mean "smooth chart" (or "smooth coordinate system"). When M is a manifold and $f \in C^\infty(M)$, we define the **support of** f to be the closed set:

$$\operatorname{supp} f := \overline{\{p \in M : f(p) \neq 0\}}.$$

Also, given a collection $\mathcal{C} = \{U_\alpha : \alpha \in A\}$ of subsets of M we say that

- $\mathcal{C}$ is **locally finite** if, for all $p \in M$, there exists a neighborhood $p \in O \subset M$ such that $O \cap U_\alpha \neq \emptyset$ for only a finite number of $\alpha \in A$.
- $\mathcal{C}$ is a **cover** of M if $\bigcup_{\alpha \in A} U_\alpha = M$.
- $\mathcal{C}_0 = \{U_\beta : \beta \in B\}$ is a **subcover** if $\mathcal{C}_0 \subset \mathcal{C}$ and $\mathcal{C}_0$ still covers M.
- $\mathcal{C}' = \{V_i : i \in I\}$ is a **refinement of a cover** $\mathcal{C}$ if it is itself a cover and for each $i \in I$ there exists $\alpha_i = \alpha(i) \in A$ such that $V_i \subset U_{\alpha_i}$.

Definition 4.1. A **partition of unity** in a manifold M is a collection $\{\phi_i : i \in I\} \subset C^\infty(M)$ such that

(i) the collection of supports $\{\operatorname{supp} \phi_i : i \in I\}$ is locally finite;
(ii) $\phi_i(p) \geq 0$ and $\sum_{i \in I} \phi_i(p) = 1$ for every $p \in M$.

A partition of unity $\{\phi_i : i \in I\}$ is called *subordinated* to a cover $\{U_\alpha : \alpha \in A\}$ of M if for each $i \in I$ there exists $\alpha_i \in A$ such that $\operatorname{supp} \phi_i \subset U_{\alpha_i}$.

Note that the sum in (ii) is actually finite: by (i) for each $p \in M$ there is only a finite number of functions ϕ_i with $\phi_i(p) \neq 0$.

The existence of partitions of unity is not obvious, but we will see in this lecture that there are many partitions of unity on a manifold.

Theorem 4.1 (Existence of Partitions of Unity). *Let M be a manifold and let $\{U_\alpha : \alpha \in A\}$ be an open cover of M. Then there exists a countable partition of unity $\{\phi_i : i = 1, 2, \dots\}$, subordinated to the cover $\{U_\alpha : \alpha \in A\}$ and with $\operatorname{supp} \phi_i$ compact for all i.*

If we do not care about compact supports for any open cover, we can get partitions of unity with the same set of indices:

Corollary 4.1. *Let M be a manifold and let $\{U_\alpha : \alpha \in A\}$ be an open cover of M. Then there exists a partition of unity $\{\phi_\alpha : \alpha \in A\}$ such that $\operatorname{supp} \phi_\alpha \subset U_\alpha$ for each $\alpha \in A$.*

Proof. By Theorem 4.1 there exists a countable partition of unity

$$\{\psi_i : i = 1, 2, \dots\}$$

subordinated to the cover $\{U_\alpha : \alpha \in A\}$. For each i, we can choose a $\alpha = \alpha(i)$ such that $\operatorname{supp} \psi_i \subset U_{\alpha(i)}$. Then the functions

$$\phi_\alpha = \begin{cases} \sum_{\alpha(i)=\alpha} \psi_i, & \text{if } \{i : \alpha(i) = \alpha\} \neq \emptyset, \\ 0, & \text{otherwise}, \end{cases}$$

form a partition of unity with $\operatorname{supp} \phi_\alpha \subset U_\alpha$ for all $\alpha \in A$. $\qquad\square$

Example 4.1. For the sphere $\mathbb{S}^d$, consider the cover with the two opens sets $U_N := \mathbb{S}^d \setminus N$ and $U_S := \mathbb{S}^d \setminus S$. Then the corollary says that there exists a partition of unity subordinated to this cover with the same indices, i.e., a pair of non-negative smooth functions $\phi_N, \phi_S \in C^\infty(\mathbb{S}^d)$ with $\operatorname{supp} \phi_N \subset U_N$ and $\operatorname{supp} \phi_S \subset U_S$, such that $\phi_N(p) + \phi_S(p) = 1$ for all $p \in \mathbb{S}^d$.

Corollary 4.2. *Let $A \subset O \subset M$, where O is an open subset and A is a closed subset of a manifold M. There exists a smooth function $\phi \in C^\infty(M)$ such that*

(i) $0 \leq \phi(p) \leq 1$ *for each* $p \in M$;
(ii) $\phi(p) = 1$ *if* $p \in A$;
(iii) $\operatorname{supp} \phi \subset O$.

Proof. The open sets $\{O, M \setminus A\}$ give an open cover of M. Therefore, by the previous corollary, there is a partition of unity $\{\phi, \psi\}$ with $\operatorname{sup}\phi \subset O$ and $\operatorname{sup}\psi \subset M \setminus A$. The function ϕ satisfies (i)–(iii). $\qquad\square$

Roughly speaking, partitions of unity are used to "glue" local properties (i.e., properties that hold on domains of local coordinates), giving rise to global properties of a manifold, as shown in the proof of the following result.

Corollary 4.3 (Extension Lemma for smooth maps). *Let M be a manifold, $A \subset M$ a closed subset and $\Psi : A \to \mathbb{R}^n$ a smooth map. For any open set $A \subset U \subset M$ there exists a smooth map $\widetilde{\Psi} : M \to \mathbb{R}^n$ such that $\widetilde{\Psi}|_A = \Psi$ and $\operatorname{supp}\widetilde{\Psi} \subset U$.*

Proof. For each $p \in A$ we can find an open neighborhood $U_p \subset M$, such that we can extend $\Psi|_{U_p \cap A}$ to a smooth function $\widetilde{\Psi}_p : U_p \to \mathbb{R}^n$. By replacing U_p by $U_p \cap U$ we can assume that $U_p \subset U$. The sets $\{U_p, M \setminus A; p \in A\}$ form an open cover of M so we can find a partition of unit $\{\phi_p : p \in A\} \cup \{\phi_0\}$, subordinated to this cover with $\operatorname{supp}\phi_p \subset U_p$. Now define $\widetilde{\Psi} : M \to \mathbb{R}^n$ by setting

$$\widetilde{\Psi} := \sum_{p \in A} \phi_p \widetilde{\Psi}_p.$$

Clearly, $\widetilde{\Psi}$ has the required properties. $\qquad\square$

We now turn to the proof of Theorem 4.1. There are two main ingredients in the proof. The first one is that topological manifolds are **paracompact**, i.e., every open cover has an open locally finite refinement. This is in fact a consequence of our assumption that manifolds are Hausdorff and second countable, and we will use the following more precise versions:

(a) Every open cover of a topological manifold M has a countable subcover.

(b) Every open cover of a topological manifold M has a countable, locally finite refinement consisting of open sets with compact closures.

The proofs are left to the exercises. The second ingredient is the existence of "very flexible" smooth functions, sometimes called **bump functions**:

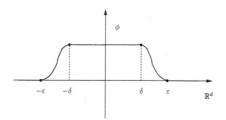

Fig. 4.1.   A bump function.

(c) for any $\varepsilon > \delta > 0$, there exists a function $\phi \in C^\infty(\mathbb{R}^d)$ such that $\phi(x) = 1$, if $x \in \overline{B_\delta(0)}$, and $\phi(x) = 0$, if $x \in B_\varepsilon(0)^c$ (see Figure 4.1).

This can be proved by observing that

- The function $f : \mathbb{R} \to \mathbb{R}$ defined by

$$f(x) = \begin{cases} \exp(-\frac{1}{x^2}), & x > 0, \\ 0, & x \leq 0. \end{cases}$$

 is a smooth function.
- If $\delta > 0$, the function $g : \mathbb{R} \to \mathbb{R}$ defined by

$$g(x) = f(x)f(\delta - x),$$

 is smooth, $g(x) > 0$ if $x \in]0, \delta[$ and $g(x) = 0$ otherwise.
- The function $h : \mathbb{R} \to \mathbb{R}$ defined by

$$h(x) := \frac{\int_0^x g(t)\,dt}{\int_0^\delta g(t)\,dt},$$

 is smooth, non-decreasing, $h(x) = 0$ if $x \leq 0$ and $h(x) = 1$ if $x \geq \delta$.

Using these functions, you should now be able to show that (c) holds.

Proof of Theorem 4.1. By (b) above, we can assume that the open cover $\{U_\alpha : \alpha \in A\}$ is countable, locally finite, and the sets $\overline{U}_\alpha$ are compact. If $p \in U_\alpha$, we can choose a smooth chart (V_p, τ), centered in p, with $V_p \subset U_\alpha$, and such that $\overline{B_\varepsilon(0)} \subset \tau(V_p)$ for some

$\varepsilon > 0$. Now if ϕ is the function defined in (c) above, we set

$$\psi_p := \begin{cases} \phi \circ \tau, & \text{in } V_p, \\ 0, & \text{in } M \setminus V_p. \end{cases}$$

Then $\psi_p \in C^\infty(M)$ is a non-negative function, taking the value 1 in an open set $W_p \subset V_p$ which contains p. Since $\{W_p : p \in M\}$ is an open cover of M, by (a) above, there exists a countable subcover $\{W_{p_1}, W_{p_2}, \dots\}$ of M. Then the open cover $\{V_{p_1}, V_{p_2}, \dots\}$ is locally finite and subordinated to the cover $\{U_\alpha : \alpha \in A\}$. Moreover, the closures $\overline{V}_{p_i}$ are compact.

The sum $\sum_i \psi_{p_i}$ may not be equal to 1. To fix this we observe that

$$\psi = \sum_{i=1}^{+\infty} \psi_{p_i},$$

is well defined, of class C^∞ and $\psi(p) > 0$ for every $p \in M$. If we define

$$\phi_i = \frac{\psi_{p_i}}{\psi},$$

then the functions $\{\phi_1, \phi_2, \dots\}$ give a partition of unity, subordinated to the cover $\{U_\alpha : \alpha \in A\}$, with supp ϕ_i compact for each $i = 1, 2, \dots$.

This completes the proof of Theorem 4.1. □

Exercises

Exercise 4.1
Show that $f : \mathbb{R} \to \mathbb{R}$, defined by $f(x) = \exp(-1/x^2)$ is a smooth function.

Exercise 4.2
Given any $\varepsilon > \delta > 0$, show that there exists a function $\phi \in C^\infty(\mathbb{R}^d)$ such that $0 \leq \phi(x) \leq 1$, $\phi(x) = 1$ if $|x| \leq \delta$ and $\phi(x) = 0$ if $|x| > \varepsilon$.

Exercise 4.3
Show that for a second countable topological space X, every open cover of X has a countable subcover.

Hint: If $\{U_\alpha : \alpha \in A\}$ is an open cover of X and $\mathcal{B} = \{V_j \in J\}$ is a countable basis of the topology of X, show that the collection $\mathcal{B}'$

formed by $V_j \in \mathcal{B}$ such that $V_j \subset U_\alpha$ for some α, is also a basis. Now, for each $V_j \in \mathcal{B}'$ choose some U_{α_j} containing V_j, and show that $\{U_{\alpha_j}\}$ is a countable subcover.

Exercise 4.4

Show that a topological manifold is paracompact, in fact, show that every open cover of a topological manifold M has a countable, locally finite refinement consisting of open sets with compact closures.

Hint: Show first that M can be covered by open sets $O_1, O_2, \ldots$, with compact closures and $\overline{O}_i \subset O_{i+1}$. Then given an arbitrary open cover $\{U_\alpha : \alpha \in A\}$ of M, choose for each $i \geq 3$ a finite subcover of the cover $\{U_\alpha \cap (O_{i+1} \setminus \overline{O}_{i-2}) : \alpha \in A\}$ of the compact set $\overline{O}_i \setminus O_{i-1}$, and a finite subcover of the cover $\{U_\alpha \cap O_3 : \alpha \in A\}$ of the compact set $\overline{O}_2$. The collection of such open sets will do it.

Exercise 4.5

Show that if $M \subset \mathbb{R}^n$ is an embedded manifold then a function $f : M \to \mathbb{R}$ is smooth if and only if there exists an open set $M \subset U \subset \mathbb{R}^n$ and a smooth function $F : U \to \mathbb{R}$ such that $F|_M = f$.

Exercise 4.6

Show that the conclusion of the Extension Lemma for Smooth Maps may fail if $A \subset M$ is not assumed to be closed.

Exercise 4.7

Show that Theorem 4.1 still holds for manifolds with boundary.

Lecture 5

The Tangent Space

The tangent space to $\mathbb{R}^d$ at $p \in \mathbb{R}^d$ is by definition the set

$$T_p\mathbb{R}^d := \left\{ (p, \vec{v}) : \vec{v} \in \mathbb{R}^d \right\}.$$

Note that this tangent space is a vector space over $\mathbb{R}$ where addition is defined by

$$(p, \vec{v}_1) + (p, \vec{v}_2) := (p, \vec{v}_1 + \vec{v}_2),$$

while scalar multiplication is given by

$$a(p, \vec{v}) := (p, a\vec{v}).$$

Of course there is a natural isomorphism $T_p\mathbb{R}^d \simeq \mathbb{R}^d$, but in many situations, it is better to think of $T_p\mathbb{R}^d$ as the set of vectors with origin at p (see Figure 5.1).

This distinction becomes even more clear in the case of an embedded manifold $M \subset \mathbb{R}^n$ (see Figure 5.2). For such an embedded manifold we can define the tangent space to M at $p \in M$ to be the subspace $T_pM \subset T_p\mathbb{R}^n$ consisting of those tangent vectors $(p, \vec{v}) \in \mathbb{R}^n$ for which there exists a smooth curve $c : (-\varepsilon, \varepsilon) \to \mathbb{R}^n$, with $c(t) \in M$, $c(0) = p$ and $c'(0) = \vec{v}$.

A tangent vector $(p, \vec{v}) \in T_pM$ acts on smooth functions defined in a neighborhood of p as follows. If $f : U \to \mathbb{R}$ is a smooth function

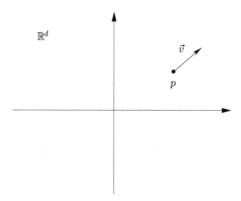

Fig. 5.1. Tangent space to $\mathbb{R}^d$.

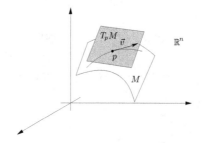

Fig. 5.2. Tangent space to an embedded manifold $M \subset \mathbb{R}^n$.

defined on a open set U containing p then we can choose a smooth curve $c : (-\varepsilon, \varepsilon) \to U$, with $c(0) = p$ and $c'(0) = \vec{v}$, and set

$$(p, \vec{v})(f) := \frac{d}{dt}(f \circ c)(0).$$

This operation does not depend on the choice of smooth curve c (exercise). In fact, this is just the usual notion of **directional derivative** of f at p in the direction $\vec{v}$.

We will now define the tangent space to an abstract manifold M at $p \in M$. There are several different approaches to define the tangent space at $p \in M$, which correspond to different points of view, all of them very useful. We shall give here three distinct descriptions and we leave it to the exercises to show that they are actually equivalent.

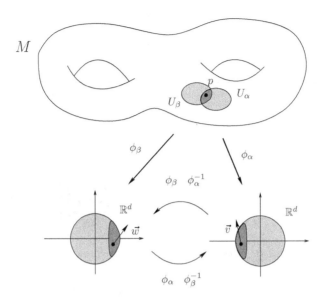

Fig. 5.3. First description of tangent space to M.

Description 1.

Let M be a smooth d-dimensional manifold with an atlas $\mathcal{C} = \{(U_\alpha, \phi_\alpha) : \alpha \in A\}$. To each point $p \in M$, we would like to associate a copy of $\mathbb{R}^d$, so that each element $\vec{v} \in \mathbb{R}^d$ should represent a tangent vector. Of course, if $p \in U_\alpha$, the system of coordinates ϕ_α gives an identification of an open neighborhood of p with $\mathbb{R}^d$. Distinct smooth charts will give different identifications, but they are all related by transition functions, as in Figure 5.3.

This suggests one should consider triples $(p, \alpha, \vec{v}) \in M \times A \times \mathbb{R}^d$, with $p \in U_\alpha$, and that two such triples should be declared to be equivalent if

$$[p, \alpha, \vec{v}] = [q, \beta, \vec{w}] \quad \text{iff} \quad p = q \quad \text{and} \quad (\phi_\alpha \circ \phi_\beta^{-1})'(\phi_\beta(p)) \cdot \vec{w} = \vec{v}.$$

Hence, we define a **tangent vector** to M at a point $p \in M$ to be an equivalence class $[p, \alpha, \vec{v}]$, and the **tangent space** at p to be the set of all such equivalence classes:

$$T_pM := \left\{ [p, \alpha, \vec{v}] : \alpha \in A, \vec{v} \in \mathbb{R}^d \right\}.$$

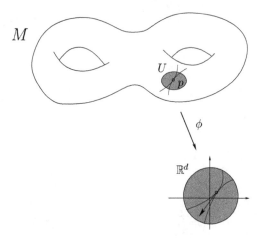

Fig. 5.4.   Second description of tangent space to M.

We leave it as an exercise to check that the operations:

$$[p, \alpha, \vec{v}_1] + [p, \alpha, \vec{v}_2] := [p, \alpha, \vec{v}_1 + \vec{v}_2], \quad a[p, \alpha, \vec{v}] := [p, \alpha, a\vec{v}],$$

are well defined and give $T_p M$ the structure of vector space over $\mathbb{R}$. Note that we still have an isomorphism $T_p M \simeq \mathbb{R}^d$, but this isomorphism now depends on the choice of a chart.

Description 2.

Again, fix $p \in M$. For this second description, we will consider all smooth curves $c : (-\varepsilon, \varepsilon) \to M$, with $c(0) = p$. Two such smooth curves c_1 and c_2 will be declared equivalent if there exists some smooth chart (U, ϕ) with $p \in U$, such that (see Figure 5.4)

$$\frac{d}{dt}(\phi \circ c_1)(0) = \frac{d}{dt}(\phi \circ c_2)(0).$$

It should be clear that if this condition holds for some smooth chart around p, then it also holds for every other smooth chart around p belonging to the smooth structure.

We call a **tangent vector** at $p \in M$ an equivalence class of smooth curves $[c]$, and the set of all such classes is called the **tangent**

space T_pM at the point p. Again, you should check that this tangent space has the structure of vector space over $\mathbb{R}$ and that T_pM is isomorphic to $\mathbb{R}^d$, through an isomorphism that depends on a choice of smooth chart.

Description 3.

The two previous descriptions use smooth charts. Our third description has the advantage of not using charts, and it will be our official definition of the tangent space.

Again we fix $p \in M$ and we look at the set of all smooth functions defined in some open neighborhood of p. Given two smooth functions $f : U \to \mathbb{R}$ and $g : V \to \mathbb{R}$, where U and V are open sets that contain p, we say that f and g define the same **germ at** p if there is an open set $W \subset U \cap V$ containing p and such that

$$f|_W = g|_W.$$

We denote by $\mathcal{G}_p$ the set of all germs of smooth functions at p. This set has the structure of an $\mathbb{R}$-algebra, where addition, product, and multiplication by scalars are defined in the obvious way:

$$[f] + [g] := [f + g], \quad [f][g] := [fg], \quad a[f] := [af].$$

Note also that it makes sense to talk of the value of a germ $[f] \in \mathcal{G}_p$ at the point p, which is $f(p)$. On the other hand, the value of $[f] \in \mathcal{G}_p$ at any other point $q \neq p$ is not defined.

Definition 5.1. A **tangent vector** at a point $p \in M$ is a linear derivation of $\mathcal{G}_p$, i.e., a linear map $\mathbf{v} : \mathcal{G}_p \to \mathbb{R}$ satisfying

$$\mathbf{v}([f][g]) = \mathbf{v}([f])g(p) + f(p)\mathbf{v}([g]).$$

The **tangent space** at a point $p \in M$ is the set of all such tangent vectors and is denoted by T_pM.

Since linear derivations can be added and multiplied by real numbers, it is clear that the tangent space T_pM has the structure of a real vector space.

Example 5.1. Let $(U, \phi) = (U, x^1, \ldots, x^d)$ be a coordinate system in M with $p \in U$. We define the tangent vectors $\frac{\partial}{\partial x^i}\big|_p \in T_p M$, $i = 1, \ldots, d$, to be the derivations

$$\frac{\partial}{\partial x^i}\bigg|_p ([f]) = \frac{\partial (f \circ \phi^{-1})}{\partial x^i}\bigg|_{\phi(p)}.$$

Note that the tangent vector $\frac{\partial}{\partial x^i}\big|_p$ corresponds to the direction one obtains by freezing all coordinates but the ith coordinate.

In order to check that $T_p M$ is a vector space with dimension equal to $\dim M$, consider the set of all germs that vanish at p

$$\mathcal{M}_p := \{[f] \in \mathcal{G}_p : f(p) = 0\},$$

It is immediate to check that $\mathcal{M}_p \subset \mathcal{G}_p$ is a maximal ideal in $\mathcal{G}_p$. The kth power of this ideal

$$\mathcal{M}_p^k = \underbrace{\mathcal{M}_p \cdots \mathcal{M}_p}_{k}.$$

consists of germs that vanish to order k at p: if $[f] \in \mathcal{M}_p^k$ and (U, ϕ) is a coordinate system centered at p, then the smooth function $f \circ \phi^{-1}$ has vanishing partial derivatives at $\phi(p)$ up to order $k - 1$. These powers form a tower of ideals

$$\mathcal{G}_p \supset \mathcal{M}_p \supset \mathcal{M}_p^2 \supset \cdots \supset \mathcal{M}_p^k \supset \cdots$$

Theorem 5.1. *The tangent space $T_p M$ is naturally isomorphic to $(\mathcal{M}_p/\mathcal{M}_p^2)^*$ and has dimension equal to $\dim M$.*

Proof. First, we check that if $[c] \in \mathcal{G}_p$ is the germ of the constant function $f(x) = c$ then $\mathbf{v}([c]) = 0$ for any tangent vector $\mathbf{v} \in T_p M$. In fact, we have that

$$\mathbf{v}([c]) = c\mathbf{v}([1]),$$

and that

$$\mathbf{v}([1]) = \mathbf{v}([1][1]) = 1\mathbf{v}([1]) + 1\mathbf{v}([1]) = 2\mathbf{v}([1]),$$

hence $\mathbf{v}([1]) = 0$.

Now if $[f] \in \mathcal{G}_p$ and $c = f(p)$, we remark that

$$\mathbf{v}([f]) = \mathbf{v}([f] - [c]),$$

so the derivation $\mathbf{v}$ is completely determined by its effect on $\mathcal{M}_p$. On the other hand, any derivation vanishes on $\mathcal{M}_p^2$, because if $f(p) = g(p) = 0$, then

$$\mathbf{v}([f][g]) = \mathbf{v}([f])g(p) + f(p)\mathbf{v}([g]) = 0.$$

We conclude that every tangent vector $\mathbf{v} \in T_pM$ determines a unique linear transformation $\mathcal{M}_p \to \mathbb{R}$, which vanishes on $\mathcal{M}_p^2$. Conversely, if $L \in (\mathcal{M}_p/\mathcal{M}_p^2)^*$ is a linear transformation, we can define a linear transformation $\mathbf{v} : \mathcal{G}_p \to \mathbb{R}$ by setting

$$\mathbf{v}([f]) := L([f] - [f(p)]).$$

This is actually a derivation (exercise), so we conclude that $T_pM \simeq (\mathcal{M}_p/\mathcal{M}_p^2)^*$.

In order to verify the dimension of T_pM, we choose some system of coordinates $(U, x^1, \ldots, x^d)$ centered at p, and we show that the tangent vector

$$\left.\frac{\partial}{\partial x^i}\right|_p \in T_pM, \quad i = 1, \ldots, d,$$

form a basis for T_pM. If $f : U \to \mathbb{R}$ is any smooth function, then $f \circ \phi^{-1} : \mathbb{R}^d \to \mathbb{R}$ is smooth in a neighborhood of the origin. This function can be expanded as

$$f \circ \phi^{-1}(x) = f \circ \phi^{-1}(0) + \sum_{i=1}^d \frac{\partial(f \circ \phi^{-1})}{\partial x^i}(0)x^i + \sum_{i,j} g_{ij}(x)x^i x^j,$$

where the g_{ij} are some smooth functions in a neighborhood of the origin. It follows that we have the expansion

$$f(q) = f(p) + \sum_{i=1}^d \left.\frac{\partial(f \circ \phi^{-1})}{\partial x^i}\right|_{\phi(p)} x^i(q) + \sum_{i,j} h_{ij}(q)x^i(q)x^j(q),$$

where $h_{ij} \in C^\infty(U)$, valid for any $q \in U$. We conclude that for any tangent vector $\mathbf{v} \in T_pM$

$$\mathbf{v}([f]) = \sum_{i=1}^d \left.\frac{\partial(f \circ \phi^{-1})}{\partial x^i}\right|_{\phi(p)} \mathbf{v}([x^i]).$$

In other words, we have

$$\mathbf{v} = \sum_{i=1}^{d} a^i \left. \frac{\partial}{\partial x^i} \right|_p,$$

where $a^i = \mathbf{v}([x^i])$. This shows that the vectors $(\partial/\partial x^i)|_p \in T_p M$, $i = 1, \ldots, \dim M$ form a generating set. We leave it as an exercise to show that they are linearly independent. $\qquad \square$

From now on, given $\mathbf{v} \in T_p M$ and a smooth function f defined in some neighborhood of $p \in M$ we set

$$\mathbf{v}(f) := \mathbf{v}([f]).$$

Note that $\mathbf{v}(f) = \mathbf{v}(g)$ if f and g coincide in a neighborhood of p and that for $a, b \in \mathbb{R}$

$$\mathbf{v}(af + bg) = a\mathbf{v}(f) + b\mathbf{v}(g), \quad \mathbf{v}(fg) = f(p)\mathbf{v}(g) + \mathbf{v}(f)g(p),$$

where $af + bg$ and fg are defined in the intersection of the domains of f and g.

The proof of Theorem 5.1 shows that if $(U, \phi) = (U, x^1, \ldots, x^d)$ is a coordinate system around p, then any tangent vector $\mathbf{v} \in T_p M$ can be written as

$$\mathbf{v} = \sum_{i=1}^{d} a^i \left. \frac{\partial}{\partial x^i} \right|_p.$$

The numbers $a^i = \mathbf{v}(x^i)$ are called the **components of tangent vector v** in the coordinate system $(U, x^1, \ldots, x^d)$. If we introduce the notation

$$\left. \frac{\partial f}{\partial x^i} \right|_p := \left. \frac{\partial f \circ \phi^{-1}}{\partial x^i} \right|_{\phi(p)},$$

then we can write

$$\mathbf{v}(f) = \sum_{i=1}^{d} a^i \left. \frac{\partial f}{\partial x^i} \right|_p.$$

On the other hand, given another coordinate system $(V, y^1, \ldots, y^d)$, we find that

$$\left. \frac{\partial}{\partial y^j} \right|_p = \sum_{i=1}^{d} \left. \frac{\partial x^i}{\partial y^j} \right|_p \left. \frac{\partial}{\partial x^i} \right|_p.$$

Hence, in this new coordinate system, we have

$$\mathbf{v} = \sum_{j=1}^{d} b^j \left. \frac{\partial}{\partial y^j} \right|_p, \quad \text{with } b^j = \mathbf{v}(y^j),$$

where the new components b^j are related to the old components a^i by the transformation formula

$$a^i = \sum_{j=1}^{d} \left. \frac{\partial x^i}{\partial y^j} \right|_p b^j. \tag{5.1}$$

Let us turn now to the question of how the tangent spaces vary from point to point. We define the **tangent bundle** to M to be union of all tangent spaces:

$$TM := \bigcup_{p \in M} T_p M.$$

Note that we have a natural projection $\pi : TM \to M$ which associates to a tangent vector $\mathbf{v} \in T_p M$ the corresponding base point $\pi(\mathbf{v}) = p$. The term "bundle" comes from the fact that we can picture TM as a set of fibers (the spaces $T_p M$), juxtaposed with each other forming a manifold (see Figure 5.5).

Proposition 5.1. *TM has a natural smooth structure of manifolds of dimension* $2 \dim M$ *such that the projection in the base is a smooth map.*

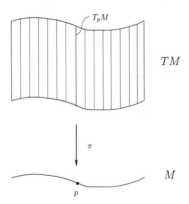

Fig. 5.5.   The tangent bundle to M.

Proof. Let $\{(U_\alpha, \phi_\alpha) : \alpha \in A\}$ be an atlas for M. For each smooth chart $(U_\alpha, \phi_\alpha) = (U_\alpha, x^1, \ldots, x^n)$, we define $\tilde{\phi}_\alpha : \pi^{-1}(U_\alpha) \to \mathbb{R}^{2d}$ by setting

$$\tilde{\phi}_\alpha(\mathbf{v}) := (x^1(\pi(\mathbf{v})), \ldots, x^d(\pi(\mathbf{v})), \mathbf{v}(x^1), \ldots, \mathbf{v}(x^d)).$$

One checks easily that the collection

$$\left\{ \tilde{\phi}_\alpha^{-1}(O) : O \subset \mathbb{R}^{2d} \text{ open}, \alpha \in A \right\}$$

is a basis for a topology of TM, which is Hausdorff and second countable. Now, we have that

(a) TM is a topological manifold with local charts $(\pi^{-1}(U_\alpha), \tilde{\phi}_\alpha)$.
(b) For any pair of charts $(\pi^{-1}(U_\alpha), \tilde{\phi}_\alpha)$ and $(\pi^{-1}(U_\beta), \tilde{\phi}_\beta)$, the transition functions $\tilde{\phi}_\beta \circ \tilde{\phi}_\alpha^{-1}$ are smooth.

We conclude that the collection $\left\{ (\pi^{-1}(U_\alpha), \tilde{\phi}_\alpha) : \alpha \in A \right\}$ is an atlas, and so defines on TM the structure of a smooth manifold of dimension $\dim TM = 2 \dim M$. Finally, the map $\pi : TM \to M$ is smooth because for each α we have that $\phi_\alpha \circ \pi \circ \tilde{\phi}_\alpha^{-1} : \mathbb{R}^{2d} \to \mathbb{R}^d$ is just the projection in the first d components. $\qquad\square$

We say that a d-dimensional manifold M has *trivial tangent bundle* if there is a diffeomorphism $\Psi : TM \to M \times \mathbb{R}^d$ commuting with the projections

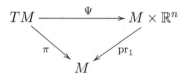

whose restriction to each fiber is linear isomorphism $\Phi|_{T_pM} : T_pM \to \mathbb{R}^d$. For example, $\mathbb{R}^d$ and $\mathbb{T}^d$ have both trivial tangent bundles. However, we will see later that $\mathbb{S}^d$ has trivial tangent bundle if and only if $d = 1, 3$.

Exercises

Exercise 5.1
Show that the three descriptions of tangent vectors are indeed equivalent.

Exercise 5.2

In $\mathbb{R}^3$ consider the usual Cartesian coordinates (x, y, z). One defines **spherical coordinates** in $\mathbb{R}^3$ to be the smooth chart (U, ϕ), where $U = \mathbb{R}^3 \setminus \{(x, 0, z) : x \geq 0\}$ and $\phi = (r, \theta, \varphi)$ is defined as usual by

- $r(x, y, z) := \sqrt{x^2 + y^2 + z^2}$ is the distance to the origin;
- $\theta(x, y, z)$ is the longitude, i.e., the angle in $]0, 2\pi[$ between the vector $(x, y, 0)$ and the x-axis;
- $\varphi(x, y, z)$ is the co-latitude, i.e., the angle in $]0, \pi[$ between the vector (x, y, z) and the z-axis.

Compute:

(a) The components of the tangent vectors to $\mathbb{R}^3$ $\frac{\partial}{\partial r}$, $\frac{\partial}{\partial \theta}$, $\frac{\partial}{\partial \varphi}$ in Cartesian coordinates.

(b) The components of the tangent vectors to $\mathbb{R}^3$ $\frac{\partial}{\partial x}$, $\frac{\partial}{\partial y}$, $\frac{\partial}{\partial z}$ in spherical coordinates.

Exercise 5.3

Let $M \subset \mathbb{R}^n$ be an embedded d-manifold. Show that if $\psi : V \to M \cap U$ is a parameterization of a neighborhood of $p \in M$, then the tangent space $T_p M$ can be identified with the subspace $\psi'(q)(\mathbb{R}^d) \subset \mathbb{R}^n$, where $p = \psi(q)$.

Exercise 5.4

Let $(U, x^1, \ldots, x^d)$ be a local coordinate system in a manifold M. Show that the tangent vectors

$$\frac{\partial}{\partial x^i}\bigg|_p \in T_p M, \quad i = 1, \ldots, d,$$

are linearly independent.

Exercise 5.5

Theorem 5.1 says that the tangent space at $p \in M$ is isomorphic to $(\mathcal{M}_p / \mathcal{M}_p^2)^*$. Can you give a geometric interpretation of $(\mathcal{M}_p / \mathcal{M}_p^k)^*$ for $k > 2$?

Exercise 5.6

Let $(U, \phi) = (U, x^1, \ldots, x^d)$ be a chart for a manifold M. Show that the corresponding chart $\tilde{\phi} : \pi^{-1}(U) \to \mathbb{R}^{2d}$ for TM, given in the proof of Proposition 5.1, is given by

$$v = \sum_{i=1}^{n} v^i \left. \frac{\partial}{\partial x^i} \right|_p \mapsto (x^1(p), \ldots, x^d(p), v^1, \ldots, v^d).$$

Determine the transition functions between two such charts and show that they are smooth.

Exercise 5.7

Show that there is a canonical identification $T(M_1 \times M_2) \simeq TM_1 \times TM_2$ and use this to show that the torus $\mathbb{T}^d$ has a trivial tangent bundle.

Lecture 6

The Differential

A smooth map between two smooth manifolds determines a linear transformation between their tangent spaces.

Definition 6.1. Let $\Psi : M \to N$ be a smooth map. The **differential** of Ψ at $p \in M$ is the linear transformation $d_p\Psi : T_pM \to T_{\Psi(p)}N$ defined by

$$d_p\Psi(\mathbf{v})(f) := \mathbf{v}(f \circ \Psi),$$

where f is any smooth function defined in a neighborhood of $\Psi(p)$.

If $(U, \phi) = (U, x^1, \ldots, x^d)$ is a coordinate system around p and $(V, \psi) = (V, y^1, \ldots, y^e)$ is a coordinate system around $\Psi(p)$, we obtain

$$d_p\Psi \cdot \left.\frac{\partial}{\partial x^i}\right|_p = \sum_{j=1}^{e} \left.\frac{\partial(\psi \circ \Psi \circ \phi^{-1})^j}{\partial x^i}\right|_{\phi(p)} \left.\frac{\partial}{\partial y^j}\right|_{\Psi(p)}.$$

The matrix formed by the partial derivatives $\frac{\partial(\psi \circ \Psi \circ \phi^{-1})^j}{\partial x^i}$ is often abbreviated to $\frac{\partial(y^j \circ \Psi)}{\partial x^i}$, or simply $\frac{\partial \Psi^j}{\partial x^i}$, and is called the **Jacobian matrix** of the smooth map Ψ relative to the specified system of coordinates.

The following result is an immediate consequence of the definitions and the usual chain rule for smooth maps between Euclidean space.

Proposition 6.1 (Chain Rule). *Let* $\Psi : M \to N$ *and* $\Phi : N \to P$ *be smooth maps. Then the composition* $\Phi \circ \Psi$ *is smooth and we have that*

$$\mathrm{d}_p(\Phi \circ \Psi) = \mathrm{d}_{\Psi(p)}\Phi \circ \mathrm{d}_p\Psi.$$

Similarly, it is easy to prove the following proposition that generalizes another well-known result for smooth maps between Euclidean spaces.

Proposition 6.2. *If a smooth map* $\Psi : M \to N$ *has zero differential on a connected open set* $U \subset M$, *then* Ψ *is constant in* U.

An important special case occurs when taking the differential of a smooth function $f : M \to \mathbb{R}$, thought as a smooth map between M and the manifold $\mathbb{R}$, with its canonical smooth structure. In this case, the differential at p is a linear transformation $\mathrm{d}_p f : T_p M \to T_{f(p)}\mathbb{R}$. Since we have a canonical identification $T_x \mathbb{R} \simeq \mathbb{R}$, the differential $\mathrm{d}_p f$ is an element in the dual vector space to $T_p M$. Explicitly, it is given by

$$\mathrm{d}_p f(\mathbf{v}) := \mathbf{v}(f).$$

Definition 6.2. The **cotangent space** to M at a point p is the vector space $T_p^* M$ dual to the tangent space

$$T_p^* M := \{\omega : T_p M \to \mathbb{R}, \text{ with } \omega \text{ linear}\}.$$

Of course, we can define $\mathrm{d}_p f \in T_p^* M$ even if f is a smooth function *defined only* in a neighborhood of p. In particular, if choose a coordinate system $(U, x^1, \ldots, x^d)$ around p, we obtain elements

$$\left\{ \mathrm{d}_p x^1, \ldots, \mathrm{d}_p x^d \right\} \subset T_p^* M.$$

It is then easy to check that

$$\mathrm{d}_p x^i \cdot \left.\frac{\partial}{\partial x^j}\right|_p = \begin{cases} 1 & \text{if } i = j, \\ 0, & \text{if } i \neq j. \end{cases}$$

In other words, we have:

Lemma 6.1. *For a coordinate system* (U, x^i) *of* M *around* p, $\{d_p x^1, \ldots, d_p x^d\}$ *is the basis of* $T_p^* M$ *dual to* $\left\{ \frac{\partial}{\partial x^1}\big|_p, \ldots, \frac{\partial}{\partial x^d}\big|_p \right\}$.

Therefore, once we have fixed a coordinate system $(U, x^1, \ldots, x^d)$ around p, every element $\omega \in T_p^* M$ can be written in the basis $\{d_p x^1, \ldots, d_p x^d\}$ as

$$\omega = \sum_{i=1}^d a_i d_p x^i, \quad \text{with } a_i = \omega(\partial/\partial x^i\big|_p).$$

If $(V, y^1, \ldots, y^d)$ is another coordinate system, we find that

$$\omega = \sum_{j=1}^d b_j d_p y^j, \quad \text{with } b_j = \omega(\partial/\partial y^j\big|_p),$$

and it follows that the components of ω in the two charts are related by

$$a_i = \sum_{j=1}^d \frac{\partial y^j}{\partial x^i}\bigg|_p b_j. \tag{6.1}$$

This transformation formula for the components of elements of $T_p^* M$ should be compared with the corresponding transformation formula (5.1) for the components of elements of $T_p M$.

Similarly to what we did for the tangent bundle, we define the **cotangent bundle** to M as the union of all cotangent spaces

$$T^* M := \bigcup_{p \in M} T_p^* M,$$

with a natural projection $\pi : T^* M \to M$ which is associate to a tangent covector $\omega \in T_p^* M$ the corresponding base point $\pi(\omega) = p$. Again, $T^* M$ has a natural smooth structure of manifold of dimension

$2 \dim M$, such that the projection is a smooth map. The proof is entirely similar to the case of TM, so it is left as an exercise.

Let $\Psi : M \to N$ be a smooth map. We will denote by $\mathrm{d}\Psi : TM \to TN$ the induced map on the tangent bundle obtained by collecting the differentials of Ψ at each point:

$$\mathrm{d}\Psi(\mathbf{v}) := \mathrm{d}_{\pi(\mathbf{v})}\Psi(\mathbf{v}).$$

We call this map the **differential** of Ψ. We leave it as an exercise to check that $\mathrm{d}\Psi : TM \to TN$ is a smooth map between the smooth manifolds TM and TN.

If $f : M \to \mathbb{R}$ is a smooth function, then $\mathrm{d}f : TM \to T\mathbb{R}$. However, $T\mathbb{R} = \mathbb{R} \times \mathbb{R}$ so by projecting in the second factor, we consider $\mathrm{d}f$ as a map:

$$\mathrm{d}f : TM \to \mathbb{R}, \quad \mathrm{d}f(\mathbf{v}) := \mathrm{d}_{\pi(\mathbf{v})}f(\mathbf{v}) = \mathbf{v}(f).$$

If $(U, x^1, \ldots, x^d)$ is a system of coordinates around p, then from the definition we see that $\mathrm{d}_p f \in T_p^* M$ satisfies:

$$\mathrm{d}_p f \cdot \left.\frac{\partial}{\partial x^i}\right|_p = \left.\frac{\partial f}{\partial x^i}\right|_p.$$

It follows that the expression for $\mathrm{d}f$ in local coordinates $(x^1, \ldots, x^d)$ is

$$\mathrm{d}f|_U = \sum_{i=1}^{d} \frac{\partial f}{\partial x^i}\mathrm{d}x^i.$$

Note that in this formula all terms have been precisely defined. This gives some justification for heuristic manipulations with $\mathrm{d}f$ and $\mathrm{d}x^i$ one often finds.

The definitions of tangent space, tangent bundle, and differential, extend to manifolds with boundary.

For example, one can define the **tangent space to a manifold with boundary** of dimension d at some point $p \in M$ exactly as in Definition 5.1 (see Figure 6.1). The tangent space at any point $p \in M$, even at points of the boundary, has dimension d. The tangent bundle TM is now a manifold with boundary of dimension $2 \dim M$. Similarly, the differential of a smooth map $\Psi : M \to N$ between

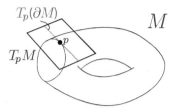

Fig. 6.1. Tangent spaces at boundary point.

manifolds with boundary can be defined as a smooth map between their tangent bundles $\mathrm{d}\Phi : TM \to TN$.

For a manifold with boundary M of dimension $d > 0$, the boundary ∂M is a smooth manifold of dimension $d-1$. Hence, if $p \in \partial M$ we have two tangent spaces: $T_p M$, which has dimension d, and $T_p(\partial M)$, which has dimension $d - 1$. We leave it as an exercise to check that the inclusion $i : \partial M \hookrightarrow M$ is a smooth map and its differential $\mathrm{d}_p i : T_p(\partial M) \to T_p M$ is injective, at any point $p \in \partial M$. It follows that we can identify $T_p(\partial M)$ with its image in $T_p M$, so inside the tangent space to M at points of the boundary we have a well-defined subspace. It is common to denote this subspace also by $T_p(\partial M)$, a practice that we will also adopt.

Exercises

Exercise 6.1
Show that the map $\Psi : \mathbb{RP}^2 \to \mathbb{R}^4$ given by

$$\Psi([x : y : z]) = \frac{1}{x^2 + y^2 + z^2}(xy, xz, y^2 - z^2, 2yz),$$

is smooth, injective and has differential $\mathrm{d}_p \Psi$ injective for all $p \in \mathbb{RP}^2$.

Exercise 6.2
Let $\Psi : \mathbb{CP}^d \to \mathbb{R}^{d+1}$ be the smooth map given by

$$\Psi([z^0 : \cdots : z^d]) = \left(\frac{|z^0|^2}{|z^0|^2 + \cdots + |z^d|^2}, \ldots, \frac{|z^d|^2}{|z^0|^2 + \cdots + |z^d|^2} \right).$$

Find the points $p \in \mathbb{CP}^d$ where the differential $\mathrm{d}_p \Phi$ vanishes.

Exercise 6.3

Let $\pi : \mathbb{S}^d \to \mathbb{RP}^d$ be the map $(x^0, \ldots, x^d) \mapsto [x^0 : \cdots : x^d]$. Show that the differential $\mathrm{d}_p \pi$ is a linear isomorphism for all $p \in \mathbb{S}^2$.

Exercise 6.4

Show that T^*M has a smooth structure of manifold of dimension $2 \dim M$ for which the projection $\pi : T^*M \to M$ is a smooth map.

Exercise 6.5

Check that if M and N are smooth manifolds and $\Psi : M \to N$ is a smooth map, then $\mathrm{d}\Psi : TM \to TN$ is also smooth.

Lecture 7

Immersions, Submersions, and Submanifolds

As we can expect from what we know from calculus in Euclidean space, the properties of the differential of a smooth map between two smooth manifolds reflect the local behavior of the smooth map. In this lecture, we will make this precise.

Definition 7.1. Let $\Psi : M \to N$ be a smooth map:

(a) Ψ is called an **immersion** if $d_p\Psi : T_pM \to T_{\Psi(p)}N$ is injective for all $p \in M$;

(b) Ψ is called a **submersion** if $d_p\Psi : T_pM \to T_{\Psi(p)}N$ is surjective for all $p \in M$;

(c) Ψ is called an **étale**[1] if $d_p\Psi : T_pM \to T_{\Psi(p)}N$ is an isomorphism for all $p \in M$.

Immersions, submersions, and étales have local canonical forms. They are all consequences of the following general result:

Theorem 7.1 (Constant Rank Theorem). *Let $\Psi : M \to N$ be a smooth map and $p \in M$. If $d_q\Psi : T_qM \to T_{\Psi(q)}N$ has constant rank r for all points q in a neighborhood of p, then there are local coordinates $(U, \phi) = (U, x^1, \ldots, x^m)$ for M centered at p and local coordinates $(V, \psi) = (V, y^1, \ldots, y^n)$ for N centered at $\Psi(p)$, such that*

$$\psi \circ \Psi \circ \phi^{-1}(x^1, \ldots, x^m) = (x^1, \ldots, x^r, 0, \ldots, 0).$$

---

[1]We use this term provisionally. We shall see later in Corollary 7.3 that an étale map is the same thing as a local diffeomorphism.

Proof. Let $(\tilde{U}, \tilde{\phi})$ and $(\tilde{V}, \tilde{\psi})$ be local coordinates centered at p and $\Psi(p)$, respectively, with $\Psi(\tilde{U}) \subset \tilde{V}$. Then,

$$\tilde{\psi} \circ \Psi \circ \tilde{\phi}^{-1} : \tilde{\phi}(\tilde{U}) \to \tilde{\psi}(\tilde{V})$$

is a smooth map from a neighborhood of zero in $\mathbb{R}^m$ to a neighborhood of zero in $\mathbb{R}^n$, whose differential has constant rank. Therefore, it is enough to consider the case where $\Psi : \mathbb{R}^m \to \mathbb{R}^n$ is a smooth map

$$(x^1, \ldots, x^m) \mapsto (\Psi^1(x), \ldots, \Psi^n(x)),$$

whose differential has constant rank in a neighborhood of the origin.

Let r be the rank of $d\Psi$. Eventually after some reordering of the coordinates, we can assume that

$$\det \left[\frac{\partial \Psi^j}{\partial x^i} \right]^r_{i,j=1} (0) \neq 0.$$

It follows immediately from the Inverse Function Theorem that the smooth map $\phi : \mathbb{R}^m \to \mathbb{R}^m$ defined by

$$(x^1, \ldots, x^m) \to (\Psi^1(x), \ldots, \Psi^r(x), x^{r+1}, \ldots, x^m),$$

is a diffeomorphism from a neighborhood of the origin. We conclude that

$$\Psi \circ \phi^{-1}(x^1, \ldots, x^m) = (x^1, \ldots, x^r, \Psi^{r+1} \circ \phi^{-1}(x), \ldots, \Psi^n \circ \phi^{-1}(x)).$$

Let q be any point in the domain of $\Psi \circ \phi^{-1}$. We can compute the Jacobian matrix of $\Psi \circ \phi^{-1}$ as:

$$\left[\begin{array}{c|c} I_r & 0 \\ \hline * & \frac{\partial(\Psi^j \circ \phi^{-1})}{\partial x^i}(q) \end{array} \right],$$

where I_r is the $r \times r$ identity matrix and where in the lower right corner $i, j > r$. Since this matrix has exactly rank r, we conclude that

$$\frac{\partial(\Psi^j \circ \phi^{-1})}{\partial x^i}(q) = 0, \quad \text{if } i, j > r.$$

In other words, the components of $\Psi^j \circ \phi^{-1}$, for $j > r$, do not depend on the coordinates $x^{r+1}, \ldots, x^m$

$$\Psi^j \circ \phi^{-1}(x) = \Psi^j \circ \phi^{-1}(x^1, \ldots, x^r), \quad \text{if } j > r.$$

Let us consider now the map $\psi : \mathbb{R}^n \to \mathbb{R}^n$, defined in some neighborhood of the origin, given by

$$\psi(y^1, \ldots, y^n) = (y^1, \ldots, y^r, y^{r+1} - \Psi^{r+1} \circ \phi^{-1}(y), \ldots, y^n - \Psi^n \circ \phi^{-1}(y)).$$

We see that ψ is a diffeomorphism in a neighborhood of the origin since its Jacobian matrix at the origin is

$$\left[\begin{array}{c|c} I_r & 0 \\ \hline * & I_{e-r} \end{array} \right],$$

which is non-singular. But now we compute

$$\psi \circ \Psi \circ \phi^{-1}(x^1, \ldots, x^m) = (x^1, \ldots, x^r, 0, \ldots, 0). \qquad \square$$

An immediate corollary of this result is that an immersion of a m-manifold into a n-manifold, where necessarily $m \leq n$, locally looks like the *inclusion* $\mathbb{R}^m \hookrightarrow \mathbb{R}^n$.

Corollary 7.1. *Let* $\Psi : M \to N$ *be an immersion. Then for each* $p \in M$, *there are local coordinates* $(U, \phi) = (U, x^1, \ldots, x^m)$ *for* M *centered at* p *and local coordinates* $(V, \psi) = (V, y^1, \ldots, y^n)$ *for* N *centered at* $\Psi(p)$, *such that*

$$\psi \circ \Psi \circ \phi^{-1}(x^1, \ldots, x^m) = (x^1, \ldots, x^m, 0, \ldots, 0).$$

Similarly, we conclude that a submersion of a m-manifold into a n-manifold, where necessarily $m \geq n$, locally looks like the *projection* $\mathbb{R}^m \twoheadrightarrow \mathbb{R}^n$.

Corollary 7.2. *Let* $\Psi : M \to N$ *be a submersion. Then for each* $p \in M$, *there are local coordinates* $(U, \phi) = (U, x^1, \ldots, x^m)$ *for* M *centered at* p *and local coordinates* $(V, \psi) = (V, y^1, \ldots, y^n)$ *for* N *centered at* $\Psi(p)$, *such that*

$$\psi \circ \Psi \circ \phi^{-1}(x^1, \ldots, x^m) = (x^1, \ldots, x^n).$$

Since an étale is a smooth map which is simultaneously an immersion and a submersion, we conclude also that an étale is just a local diffeomorphism.

Corollary 7.3. *Let $\Psi : M \to N$ be an étale. Then for each $p \in M$, there are local coordinates $(U, \phi) = (U, x^1, \ldots, x^d)$ for M centered at p and local coordinates $(V, \psi) = (V, y^1, \ldots, y^d)$ for N centered at $\Psi(p)$, such that*

$$\psi \circ \Psi \circ \phi^{-1}(x^1, \ldots, x^d) = (x^1, \ldots, x^d).$$

Let us now turn to the study of sub-objects in the category of smooth manifolds.

Definition 7.2. A **submanifold** of a manifold M is a pair (N, Φ) where N is a manifold and $\Phi : N \to M$ is an injective immersion. When $\Phi : N \to \Phi(N)$ is a homeomorphism for the relative topology on $\Phi(N)$, one calls the pair (N, Φ) an **embedded submanifold** and Φ an **embedding**.

One sometimes uses the term **immersed submanifold** to emphasize that the map $\Phi : N \to M$ is only an immersion and reserves the term *submanifold* for embedded submanifolds. However, in these notes, we will use the term submanifold to denote immersed submanifolds that may fail to be embedded.

Example 7.1. Figure 7.1 illustrates various immersions of $N = \mathbb{R}$ in $M = \mathbb{R}^2$. Note that $(\mathbb{R}, \Phi_1)$ is an embedded submanifold of $\mathbb{R}^2$, while $(\mathbb{R}, \Phi_2)$ is only an immersed submanifold of $\mathbb{R}^2$. On the other hand, Φ_3 is an immersion but it is not injective, so $(\mathbb{R}, \Phi_3)$ is not a submanifold of $\mathbb{R}^2$.

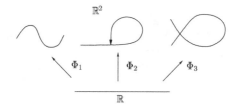

Fig. 7.1.   Immersions of the real line in the plane.

Example 7.2. According to an exercise in the previous lecture, the map $\Psi : \mathbb{RP}^2 \to \mathbb{R}^4$ defined by

$$\Psi([x:y:z]) := \frac{1}{x^2 + y^2 + z^2}(xy, xz, y^2 - z^2, 2yz),$$

is smooth, injective, and has differential $d_p\Psi$ injective for all $p \in \mathbb{RP}^2$. Since $\mathbb{RP}^2$ is compact, this map is an embedding (see the exercises at the end of this lecture). It follows that $\mathbb{RP}^2$ can be realized as an embedded submanifold of $\mathbb{R}^4$.

If (N, Φ) is a submanifold of M, then the linear map $d_p\Phi : T_pN \to T_{\Phi(p)}M$ is injective for each $p \in N$. Hence, we can always identify the tangent space T_pN with its image $d_p\Phi(T_pN)$, which is a subspace of $T_{\Phi(p)}M$. From now on, we will use this identification, so that T_pN will always be interpreted as a subspace of $T_{\Phi(p)}M$.

The local canonical form for immersions (Corollary 7.1) yields the following result.

Proposition 7.1 (Local normal form for immersed submanifolds). *Let (N, Φ) be a submanifold of dimension d of a manifold M. Then for all $p \in N$, there exists a neighborhood U of p and a coordinate system $(V, x^1, \ldots, x^m)$ for M centered at $\Phi(p)$ such that*

$$\Phi(U) = \left\{ q \in V : x^{d+1}(q) = \cdots = x^m(q) = 0 \right\}.$$

Proof. By Corollary 7.1, for any $p \in N$, we can choose coordinates (U, ϕ) for N centered at p and coordinates $(V, \psi) = (V, x^1, \ldots, x^m)$ for M centered at $\Phi(p)$, such that $\psi \circ \Phi \circ \phi^{-1} : \mathbb{R}^d \to \mathbb{R}^m$ is the inclusion. But then $\psi \circ \Phi(U)$ is exactly the set of points in $\psi(V) \subset \mathbb{R}^m$ with the last $m - d$ coordinates equal to 0. $\qquad\square$

As illustrated by Figure 7.2, in the Proposition one may have $\Phi(N) \cap V \neq \Phi(U)$. In other words, there could exist points in $\Phi(N) \cap V$ which *do not belong* to the slice $\{q \in V : x^{d+1}(q) = \cdots = x^m(q) = 0\}$. Whenever (N, Φ) is an embedded submanifold this can be fixed.

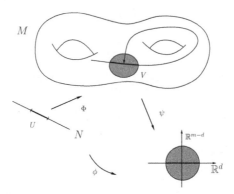

Fig. 7.2.   Immersed submanifold.

Corollary 7.4 (Local normal form for embedded submanifolds). *Let (N, Φ) be an embedded submanifold of dimension d of a manifold M. For each $p \in N$, there exists a chart $(V, x^1, \ldots, x^m)$ of M centered at $\Phi(p)$, such that*

$$\Phi(N) \cap V = \Big\{ q \in V : x^{d+1}(q) = \cdots = x^m(q) = 0 \Big\}.$$

Proof. Fix $p \in N$ and choose a neighborhood U of p and a chart $(V', x^1, \ldots, x^m)$ centered at $\Phi(p)$, as in the proposition. Since (N, Φ) is assumed to be embedded, $\Phi(U)$ is an open subset of $\Phi(N)$ for the relative topology: there exists an open set $V'' \subset M$ such that $\Phi(U) = V'' \cap \Phi(N)$. If we set $V = V' \cap V''$ the restrictions of the x^i to V, yield a coordinate system $(V, x^1, \ldots, x^m)$ such that

$$\Phi(N) \cap V = \Big\{ q \in V : x^{d+1}(q) = \cdots = x^m(q) = 0 \Big\}.$$

$\square$

We would like to think of submanifolds of a manifold M simply as subsets of M. However, this in general is not possible, as illustrated by the following example.

Example 7.3. There are two injective immersions $\Phi_i : \mathbb{R} \to \mathbb{R}^2$, $i = 1, 2$, with image the same infinite symbol (see Figure 7.3).

The previous example shows that one must be careful if we wish to think of a submanifold of M as a subset. In order to deal with the issues that arise in dealing with this idea, it is convenient to

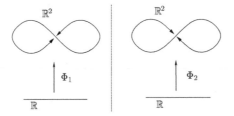

Fig. 7.3. Injective immersions with the same image.

introduce an equivalence relation on the set of submanifolds of a given manifold M, thought of as the set of pairs (N, Φ). For this notion of equivalence, the two submanifolds $(\mathbb{R}, \Phi_1)$ and $(\mathbb{R}, \Phi_2)$ in Example 7.3 will be inequivalent.

Definition 7.3. We say that (N_1, Φ_1) and (N_2, Φ_2) are **equivalent submanifolds** of M if there exists a diffeomorphism $\Psi : N_1 \to N_2$ such that the following diagram commutes

If (N, Φ) is a submanifold of M we can consider the image $\Phi(N) \subset M$ with the unique smooth structure for which $\hat{\Phi} : N \to \Phi(N)$ is a diffeomorphism. Obviously, if we take this smooth structure on $\Phi(N)$, the inclusion $i : \Phi(N) \hookrightarrow M$ is an injective immersion and the following diagram commutes

Therefore, *every* equivalence class of submanifolds of M has a unique representative (A, i), where $A \subset M$ is a subset and $i : A \hookrightarrow M$ is the inclusion. Note, however, that the topology on A is, in general, distinct from the relative topology. The next proposition shows that, once the topology of A is specified, the smooth structure is unique, and this allows one to say "$A \subset M$ is a submanifold".

Theorem 7.2. *Let $A \subset M$ be some subset of a smooth manifold and $i : A \hookrightarrow M$ the inclusion. Then,*

(i) *For each choice of a topology in A there exists at most one smooth structure compatible with this choice and such that (A, i) is a submanifold of M.*

(ii) *If A admits a smooth structure compatible with the relative topology such that (A, i) is a submanifold of M, then this is the only topology in A for which there exists a compatible smooth structure such that (A, i) is a submanifold of M.*

Example 7.4. If $A \subset M$ is an arbitrary subset, in general, there will be no smooth structure on A for which the inclusion $i : A \hookrightarrow M$ is an immersion. For example, the subset

$$A = \{(x, |x|) : x \in \mathbb{R}\} \subset \mathbb{R}^2$$

does not admit such a smooth structure (exercise). On the other hand, if A admits a smooth structure such that the inclusion $i : A \hookrightarrow M$ is an immersion, this smooth structure may not be unique as we saw in Example 7.3.

Example 7.5. The sphere $\mathbb{S}^7 \subset \mathbb{R}^8$ is an embedded submanifold. We have mentioned before that the sphere $\mathbb{S}^7$ has smooth structures compatible with the usual topology but which are not equivalent to the standard smooth structure on the sphere. It follows that for these exotic smooth structures, $\mathbb{S}^7$ is not a submanifold of $\mathbb{R}^8$.

In order to prove Theorem 7.2, we observe that if (N, Φ) is a submanifold of M and $\Psi : P \to M$ is a smooth map such that $\Psi(P) \subset \Phi(N)$, the fact that Φ is 1:1 implies that Ψ factors through a map $\hat{\Psi} : P \to N$, i.e., we have a commutative diagram:

However, the problem is that, in general, the map $\hat{\Psi}$ *is not* smooth, as shown by the example of the infinite symbol.

Example 7.6. Let $\Phi_i : \mathbb{R} \to \mathbb{R}^2$, $i = 1, 2$, be the two injective immersions whose images in $\mathbb{R}^2$ coincide with the infinite symbol, as in Example 7.3. Since $\Phi_1(\mathbb{R}) = \Phi_2(\mathbb{R})$, we have unique maps $\hat{\Phi}_1 : \mathbb{R} \to \mathbb{R}$ and $\hat{\Phi}_2 : \mathbb{R} \to \mathbb{R}$ such that $\Phi_2 \circ \hat{\Phi}_1 = \Phi_1$ and $\Phi_1 \circ \hat{\Phi}_2 = \Phi_2$. It is easy to check that $\hat{\Phi}_1$ and $\hat{\Phi}_2$ are not continuous, hence they are not smooth.

Next we show that what may fail is precisely the continuity of the map $\hat{\Psi}$.

Proposition 7.2. *Let (N, Φ) be a submanifold of M, $\Psi : P \to M$ a smooth map such that $\Psi(P) \subset \Phi(N)$ and $\hat{\Psi} : P \to N$ the induced map.*

(i) *If $\hat{\Psi}$ is continuous, then it is smooth.*
(ii) *If Φ is an embedding, then $\hat{\Psi}$ is continuous (hence smooth).*

Proof. Assume first that $\hat{\Psi}$ is continuous. For each $p \in N$, choose $U \subset N$ and $(V, \phi) = (V, x^1, \dots, x^m)$ as in Proposition 7.1, and consider the smooth map

$$\psi = \pi \circ \phi \circ \Phi : U \to \mathbb{R}^d,$$

where $\pi : \mathbb{R}^m \to \mathbb{R}^d$ is the projection $(x^1, \dots, x^m) \mapsto (x^1, \dots, x^d)$. The pair (U, ψ) is a smooth coordinate system for N centered at p. On the other hand, we see that

$$\psi \circ \hat{\Psi} = \pi \circ \phi \circ \Phi \circ \hat{\Psi} = \pi \circ \phi \circ \Psi,$$

is smooth in the open set $\hat{\Psi}^{-1}(U)$. Since the collection of all such open sets $\hat{\Psi}^{-1}(U)$ covers P, we conclude that $\hat{\Psi}$ is smooth, so (i) holds.

Now if Φ is an embedding, then every open set $U \subset N$ is of the form $\Phi^{-1}(V)$, where $V \subset M$ is open. Hence, $\hat{\Psi}^{-1}(U) = \hat{\Psi}^{-1}(\Phi^{-1}(V)) = \Psi^{-1}(V)$ is also open. We conclude that $\hat{\Psi}$ is continuous, so (ii) also holds. $\square$

Proof of Theorem 7.2. The first item follows immediately from Proposition 7.2(i). As for the second item, let (N, Φ) be a submanifold with $\Phi(N) = A$ and consider the diagram

Since A is assume to have the relative topology, by Proposition 7.2(ii), $\hat{\Phi}$ is smooth. Hence, $\hat{\Phi}$ is an invertible immersion so it is a diffeomorphism (exercise). We conclude that (N, Φ) is equivalent to (A, i), so (ii) holds. $\qquad\square$

The previous discussion justifies considering the following class of submanifolds, which lies in between immersed and embedded submanifolds.

Definition 7.4. A **regularly immersed submanifold** of M is a submanifold (N, Φ) such that every smooth map $\Psi : P \to M$ with $\Psi(P) \subset \Phi(N)$ factors through a smooth map $\hat{\Psi} : P \to N$

The two different immersions of the infinity symbol that we saw above are not regular immersions. On the other hand, Proposition 7.2(ii) shows that embedded submanifolds are always regularly immersed submanifolds. But there are many examples of regularly immersed submanifolds which are not embedded.

Example 7.7. In the 2-torus $\mathbb{T}^2 = \mathbb{S}^1 \times \mathbb{S}^1$ we have a family of submanifolds $(\mathbb{R}, \Phi_a)$, depending on the parameter $a \in \mathbb{R}$, defined by

$$\Phi_a(t) = (e^{it}, e^{iat}).$$

If $a = m/n$ is rational, this is a closed curve, which turns m times in one torus direction and n times in the other torus direction, so this is an embedding. However, if $a \notin \mathbb{Q}$ then the curve is dense in the 2-torus, so this is only an immersed submanifold. Given a map $\hat{\Psi} : P \to \mathbb{R}$ such that the composition $\Phi_a \circ \hat{\Psi}$ is smooth, we see immediately that $\hat{\Psi} : P \to \mathbb{R}$ must be continuous. By Proposition 7.2, we conclude that $\hat{\Psi}$ is smooth. Hence, (N, Φ_a) is a regularly immersed submanifold.

Exercises

Exercise 7.1
Show that a submersion is an open map. What can you say about an immersion?

Exercise 7.2
Show that $\{(x, |x|) : x \in \mathbb{R}\}$ *is not* the image of an immersion $\Phi : \mathbb{R} \to \mathbb{R}^2$.

Exercise 7.3
Show that $\mathbb{S}^3$ has trivial tangent bundle, i.e., there exists a diffeomorphism $\Psi : T\mathbb{S}^3 \to \mathbb{S}^3 \times \mathbb{R}^3$, which makes the following diagram commutative:

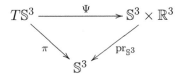

and where the restriction $\Psi : T_p\mathbb{S}^3 \to \mathbb{R}^3$ is linear for every $p \in \mathbb{S}^3$.

Hint: The 3-sphere is the set of quaternions of norm 1.

Exercise 7.4
Let $\{y^1, \ldots, y^e\}$ be some set of smooth functions on a manifold M. Show the following:

(a) If $\{d_p y^1, \ldots, d_p y^e\} \subset T_p^* M$ is a linearly independent set, then the functions $\{y^1, \ldots, y^e\}$ is a part of a coordinate system around p.
(b) If $\{d_p y^1, \ldots, d_p y^e\} \subset T_p^* M$ is a generating set, then a subset of $\{y^1, \ldots, y^e\}$ is a coordinate system around p.
(c) If $\{d_p y^1, \ldots, d_p y^e\} \subset T_p^* M$ is a basis, then the functions $\{y^1, \ldots, y^e\}$ form a coordinate system around p.

Exercise 7.5

Let $\Phi : \mathbb{RP}^2 \to \mathbb{R}^3$ be the map defined by

$$\Phi([x, y, z]) = \frac{1}{x^2 + y^2 + z^2}(yz, xz, xy).$$

Show that Φ is smooth which fails to be an immersion at six points. Sketch its image.

Exercise 7.6

Let M be a manifold, $A \subset M$, and $i : A \hookrightarrow M$ the inclusion. Show that (A, i) is a an embedded submanifold of M of dimension d, if and only if for each $p \in A$ there exists a coordinate system $(U, x^1, \dots, x^m)$ centered at p such that

$$A \cap U = \left\{ p \in U : x^{d+1}(p) = \cdots = x^m(p) = 0 \right\}.$$

Exercise 7.7

Show that a subset $M \subset \mathbb{R}^n$ satisfies Definition 1.1) if and only it is an embedded submanifold (so this justifies us calling M an embedded manifold in $\mathbb{R}^n$).

Exercise 7.8

One says that a subset S of a manifold M has **zero measure** if for every coordinate system (U, ϕ) of M, the set $\phi(S \cap U) \subset \mathbb{R}^d$ has zero measure. Show the following:

(a) A smooth map $\Phi : M \to N$ maps zero measure sets to zero measure sets;
(b) For an immersion $\Phi : N \to M$ with $\dim N < \dim M$, $\Phi(N)$ has zero measure.

Exercise 7.9

Show that for a submanifold (N, Φ) of a manifold M the following are equivalent:

(a) $\Phi(N) \subset M$ is a closed subset and (N, Φ) is embedded.
(b) $\Phi : N \to M$ is a closed map (i.e., $\Phi(A)$ is closed whenever $A \subset N$ is a closed subset).
(c) $\Phi : N \to M$ is a proper map (i.e., $\Phi^{-1}(K) \subset N$ is compact, whenever $K \subset M$ is compact).

Use this to conclude that a submanifold (N, Φ) with N compact, is always an embedded submanifold.

Exercise 7.10

Show that an invertible immersion $\Phi : N \to M$ is a diffeomorphism. Give a counterexample to this statement if N does not have a countable basis.

Exercise 7.11

Let $\pi : \widetilde{M} \to M$ be a covering space of a smooth manifold M, where $\widetilde{M}$ is a second countable topological space. Show that $\widetilde{M}$ has unique smooth structure for which the covering map π is a local diffeomorphism.

Lecture 8

Embeddings and Whitney's Theorem

Definition 8.1. Let $\Psi : M \to N$ be a smooth map.

(i) $p \in M$ is called a **regular point** of Ψ if $\mathrm{d}_p\Psi : T_pM \to T_{\Psi(p)}N$ is surjective. Otherwise, one calls p a **singular point** of Ψ;

(ii) $q \in N$ is called a **regular value** of Ψ if every $p \in \Psi^{-1}(q)$ is a regular point. Otherwise one calls q a **singular value** of Ψ.

The following example gives some evidence for the use of the terms "regular" and "singular".

Example 8.1. Let $\Psi : \mathbb{R}^3 \to \mathbb{R}$ be the map defined by

$$\Psi(x, y, z) := x^2 + y^2 - z^2.$$

This map has Jacobian matrix $[2x \ \ 2y \ \ -2z]$. Therefore, every $(x, y, z) \neq (0, 0, 0)$ is a regular point of Ψ and $(0, 0, 0)$ is a singular point of Ψ. On the other hand, 0 is a singular value of Ψ, while every other value is a regular value of Ψ.

If we consider a regular value c, the level set $\Psi^{-1}(c)$ is a submanifold of $\mathbb{R}^2$ (either a 1 sheet or a 2 sheets hyperboloid). On the other hand, for the singular value 0, we see that $\Psi^{-1}(0)$ is a cone, which is not a manifold at the origin (see Figure 8.1).

In fact, the level sets of regular values are always submanifolds:

Theorem 8.1. *Let $\Psi : M \to N$ be a smooth map and let $q \in N$ be a regular value of Ψ. Then $\Psi^{-1}(q) \subset M$ is an embedded submanifold of dimension $\dim M - \dim N$ and for all $p \in \Psi^{-1}(q)$ we have*

$$T_p(\Psi^{-1}(q)) = \operatorname{Ker} \mathrm{d}_p\Psi.$$

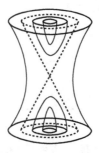

Fig. 8.1.   Regular and singular level sets of $\Psi(x, y, z) = x^2 + y^2 - z^2$.

Proof. If $q \in N$ is a regular value of Ψ there exists an open set $\Psi^{-1}(q) \subset O \subset M$ such that $\Psi|_O$ is a submersion. Therefore, for any $p \in \Psi^{-1}(q)$ we can choose coordinates $(U, x^1, \dots, x^m)$ around p and coordinates $(V, y^1, \dots, y^n)$ around q such that Ψ is represented in these local coordinates by the projection

$$\mathbb{R}^m \to \mathbb{R}^n : (x^1, \dots, x^m) \mapsto (x^1, \dots, x^n).$$

Therefore, we see that

$$\Psi^{-1}(q) \cap U = \left\{ p \in U : x^1(p) = \cdots = x^n(p) = 0 \right\}.$$

It follows that $\Psi^{-1}(q)$ is an embedded submanifold of dimension $m - n = \dim M - \dim N$ (see Exercise 7.9). The statement about the tangent space to $\Psi^{-1}(q)$ is left as an exercise. $\qquad\square$

Example 8.2. Let $M = \mathbb{R}^{d+1}$ and let $\Psi : \mathbb{R}^{d+1} \to \mathbb{R}$ be the smooth map

$$\Psi(x) := \|x\|^2.$$

The Jacobian matrix of Ψ at x is

$$\Psi'(x) = [2x^1, \dots, 2x^{d+1}].$$

Since $\Psi'(x)$ has rank one if $\|x\| > 0$, it follows that any $c = R^2 > 0$ is a regular value of Ψ. The theorem above then asserts that the spheres $\mathbb{S}_R^d = \Psi^{-1}(R^2)$ are embedded submanifolds of $\mathbb{R}^{d+1}$ of codimension 1. Note that for the differential structure on $\mathbb{S}^d$ that we have defined before, $\mathbb{S}^d$ is also an embedded submanifold of $\mathbb{R}^{d+1}$. Hence, the two differential structures coincide.

Not every embedded submanifold $S \subset M$ is of the form $\Psi^{-1}(q)$ for a regular value of some smooth map $\Psi : M \to N$. There are global obstructions that we will study later. Also, singular level sets can be very wild: using a partition of unity argument it is possible to show that for any closed subset $A \subset M$ of a smooth manifold, there exists a smooth function $f : M \to \mathbb{R}$ such that $f^{-1}(0) = A$.

If $N \subset M$ is a submanifold we call the **codimension** of N in M the integer $\dim M - \dim N$. Since a set with a single point is a manifold of dimension 0, the previous result can be restated as saying that if q is a regular value of Ψ, then $\Psi^{-1}(q)$ is an embedded submanifold with $\operatorname{codim} \Psi^{-1}(q) = \operatorname{codim} \{q\}$. In this form, the previous result can be generalized in the following very useful way.

Theorem 8.2. *Let* $\Psi : M \to N$ *be a smooth map and let* $Q \subset N$ *be an embedded submanifold. Assume that for all* $p \in \Psi^{-1}(Q)$ *one has*

$$\operatorname{Im} \mathrm{d}_p\Psi + T_{\Psi(p)}Q = T_{\Psi(p)}N. \tag{8.1}$$

Then $\Psi^{-1}(Q) \subset M$ *is an embedded submanifold with*

$$\operatorname{codim} \Psi^{-1}(Q) = \operatorname{codim} Q$$

and for all $p \in \Psi^{-1}(Q)$ *one has*

$$T_p(\Psi^{-1}(Q)) = (\mathrm{d}_p\Psi)^{-1}(T_{\Psi(p)}Q).$$

Proof. Choose $p_0 \in \Psi^{-1}(Q)$ and set $q_0 = \Psi(p_0)$. Since $Q \subset N$ is assumed to be an embedded submanifold, we can choose a coordinate system $(V, \phi) = (V, y^1, \ldots, y^n)$ for N around q_0, such that

$$Q \cap V = \left\{ q \in V : y^{l+1}(q) = \cdots = y^n(q) = 0 \right\},$$

where $l = \dim Q$. Define a smooth map $\Phi : \Psi^{-1}(V) \to \mathbb{R}^{n-l}$ by

$$\Phi = (y^{l+1} \circ \Psi, \ldots, y^n \circ \Psi).$$

Then we see that $U = \Psi^{-1}(V)$ is an open subset of M which contains p_0 and such that $\Psi^{-1}(Q) \cap U = \Phi^{-1}(0)$. If we can show that 0 is a regular value of Φ, then by Theorem 8.1 it follows that for all $p_0 \in \Psi^{-1}(Q)$, there exists an open set $U \subset M$ such that $\Psi^{-1}(Q) \cap U$ is an embedded submanifold of M of codimension $n - l = \operatorname{codim} Q$. This implies that $\Psi^{-1}(Q)$ is an embedded submanifold of M, as claimed.

To check that 0 is a regular value of Φ, note that $\Phi = \pi \circ \phi \circ \Psi$, where $\pi : \mathbb{R}^n \to \mathbb{R}^{n-l}$ is the projection in the last $n - l$ components. Since π is a submersion, ϕ is a diffeomorphism and $\ker \mathrm{d}_q(\pi \circ \phi) = T_q Q$ for all $q \in Q \cap V$, it follows from (8.1) that $\mathrm{d}_p \Phi = \mathrm{d}_{\Psi(p)}(\pi \circ \phi) \cdot \mathrm{d}_p \Psi$ is surjective for all $p \in \Psi^{-1}(Q) \cap U = \Phi^{-1}(0)$, i.e., 0 is a regular value of Φ.

The statement about the tangent space to $\Psi^{-1}(Q)$ is left as an exercise. $\qquad\square$

Condition (8.1) appearing in the statement of the previous theorem is so important that one has a special name for it.

Definition 8.2. Let $\Psi : M \to N$ be a smooth map. We say that Ψ is **transversal** to a submanifold $Q \subset N$, and we write $\Psi \pitchfork Q$, if

$$\operatorname{Im} \mathrm{d}_p \Psi + T_{\Psi(p)} Q = T_{\Psi(p)} N, \quad \forall p \in \Psi^{-1}(Q).$$

Note that submersions $\Psi : M \to N$ are specially nice: they are transverse to every submanifold $Q \subset N$! So for a submersion the inverse image of any submanifold is a submanifold.

A special case that justifies the use of the term "transversal" is when $M \subset N$ is a submanifold and $\Psi : M \hookrightarrow N$ is the inclusion. In this case, $\Psi^{-1}(Q) = M \cap Q$ and the transversality condition reduces to

$$T_q M + T_q Q = T_q N, \quad \forall q \in M \cap Q.$$

Note that this condition is symmetric in M and Q. So in this case, we simply say that M and Q **intersect transversely** and we write $M \pitchfork Q$.

Corollary 8.1. *If $M, Q \subset N$ are embedded submanifolds such that $M \pitchfork Q$. Then $M \cap Q$ is an embedded submanifold of N with*

$$\dim M \cap Q = \dim M + \dim Q - \dim N.$$

Moreover, for all $q \in M \cap Q$, one has

$$T_q(M \cap Q) = T_q M \cap T_q Q.$$

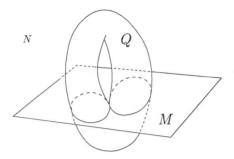

Fig. 8.2.   Non-transverse intersection.

Although Theorem 8.2 and its corollary were stated for embedded submanifolds, you are asked in an exercise in this lecture to check that these results still hold for immersed submanifolds.

Transversality plays an important role because of the following properties:

- *Transversality is a stable property*: If $\Phi : M \to N$ is transverse to Q then any map $\Psi : M \to N$ close enough to Φ is also transverse to Q.
- *Transversality is a generic property*: Any smooth map $\Phi : M \to N$ can be approximated by $\tilde{\Phi} : M \to N$ transverse to Q.

We shall not attempt to make precise these two statements, since we would need to introduce and study appropriate topologies on the space of smooth maps $C^\infty(M, N)$. These types of issues are studied in *Differential Topology* (see, e.g., Hirsch, 1994).

On the other hand, when two submanifolds do not intersect transversally, in general, the intersection is not a manifold as illustrated by Figure 8.2.

Given a manifold M one may wonder if it can be embedded into some Euclidean space. For elementary examples of manifolds, it is not hard to construct explicit embeddings, as we show in the next examples.

Example 8.3. Let $M = \mathbb{S}^1 \times \mathbb{R}$ be a cylinder. We can embed M in $\mathbb{R}^3$ using the map $\Phi : M \to \mathbb{R}^3$ defined by

$$\Phi(\theta, t) = (R\cos\theta, R\sin\theta, t),$$

Fig. 8.3.   Side identification on the 2-torus.

where we identify $\mathbb{S}^1 = [0, 2\pi]/2\pi\mathbb{Z}$. This map is injective and its Jacobian matrix $\Phi'(\theta, t)$ has rank 2, hence Φ is an injective immersion. The image of Φ is the level set of a smooth map, namely

$$\operatorname{Im}\Phi = \left\{ (x, y, z) \in \mathbb{R}^3 : x^2 + y^2 = R^2 \right\} = \Psi^{-1}(c),$$

where $c = R^2$ and $\Psi(x, y, z) := x^2 + y^2$. Since $\Psi'(x, y, z) = [2x, 2y, 0] \neq 0$ if $x^2 + y^2 = c \neq 0$, we conclude that any $c \neq 0$ is a regular value of Ψ, so we have an embedding of $\mathbb{S}^1 \times \mathbb{R}$ in $\mathbb{R}^3$.

Example 8.4. The **2-torus** $M = \mathbb{S}^1 \times \mathbb{S}^1$ can also be embedded in $\mathbb{R}^3$ as follows. First, we can think of the two torus as $\mathbb{S}^1 \times \mathbb{S}^1 = [0, 2\pi]/2\pi\mathbb{Z} \times [0, 2\pi]/2\pi\mathbb{Z}$. Note that this amounts to think of the torus as a square of side 2π where we identify the sides of the square (see Figure 8.3).

Now define $\Phi : M \to \mathbb{R}^3$ by

$$\Phi(\theta, \phi) = ((R + r\cos\phi)\cos\theta, (R + r\cos\phi)\sin\theta, r\sin\phi).$$

It is easy to check that if $R > r > 0$, then Φ is an injective immersion with image:

$$\operatorname{Im}\Phi = \big\{ (x, y, z) \in \mathbb{R}^3 : (x^2 + y^2 + z^2 - R^2 - r^2)^2 + 4R^2 z^2$$
$$= 4R^2 r^2 \big\} = \Psi^{-1}(c),$$

where $c = 4R^2 r^2$ and $\Psi : \mathbb{R}^3 \to \mathbb{R}$ is the smooth map

$$\Psi(x, y, z) = (x^2 + y^2 + z^2 - R^2 - r^2)^2 + 4R^2 z^2.$$

We leave it as an exercise to check that every $c \neq 0$ is a regular value of Ψ, so this gives an embedding of $\mathbb{S}^1 \times \mathbb{S}^1$ in $\mathbb{R}^3$.

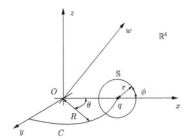

Fig. 8.4. Coordinates for the Klein bottle.

Example 8.5. The **Klein bottle** is the subset $K \subset \mathbb{R}^4$ defined as follows: Let Ox, Oy, Oz, and Ow, be the coordinate axes in $\mathbb{R}^4$ and denote by C a circle of radius R in the plane xOy. Let θ be the angle coordinate on this circle (say, measured from the Ox-axis). If $\mathbb{S}^1$ is a circle of radius r in the plane xOz, with center at $q \in C$, then K is the figure obtained by rotating this circle around the zOw plane so that when its center $q \in C$ is rotated an angle θ, the plane where $\mathbb{S}^1$ lies has rotated an angle $\theta/2$ around the Oq-axis in the 3-space $OqOzOw$. Let ϕ be the angle coordinate in the circle $\mathbb{S}^1$ (say, measured from the Oq-axis). See Figure 8.4.

The points of K with $\theta \neq 0$ and $\phi \neq 0$ can be parameterized by the map $\Phi_1 :]0, 2\pi[\times]0, 2\pi[\to \mathbb{R}^4$ given by

$$\Phi_1(\theta, \phi) = ((R + r\cos\phi)\cos\theta, (R + r\cos\phi)\sin\theta, r\sin\phi\cos\theta/2,$$
$$r\sin\phi\sin\theta/2).$$

We can change the origin of θ and ϕ, obtaining new parameterizations, which all together cover K. It is easy to check that three parameterizations Φ_1, Φ_2, and Φ_3 are enough to cover K. For these parameterizations, the transitions $\Phi_i \circ \Phi_j^{-1}$ are C^∞, so K is an embedded manifold in $\mathbb{R}^4$. Any of these parameterizations amount to think of K as a square of side 2π where we identify the sides as in Figure 8.5.

One can also express the Klein bottle as a level set $K = \Psi^{-1}(c, 0)$, where $c = 4R^2r^2$ and $\Psi : \mathbb{R}^4 \to \mathbb{R}^2$ is the smooth map

$$\Psi(x, y, z) = ((x^2 + y^2 + z^2 + w^2 - R^2 - r^2)^2$$
$$+ 4R^2(z^2 + w^2), y(z^2 - w^2) - 2xzw).$$

Fig. 8.5.   Side identification on the Klein bottle.

For $c \neq 0$, one check that $(c, 0)$ is a regular value of Ψ, so that K is an embedded submanifold of $\mathbb{R}^4$.

Actually, any compact manifold can be embedded in an Euclidean space of large enough dimension.

Theorem 8.3 (Whitney). *Let M be a compact manifold. There exists an embedding $\Phi : M \to \mathbb{R}^m$ for some integer m.*

Proof. Since M is compact, we can find a finite collection of coordinate systems $\{(U_i, \phi_i) : i = 1, \ldots, N\}$ such that

(a) $\overline{B_1(0)} \subset \phi_i(U_i)$;

(b) $\bigcup_{i=1}^{N} \phi_i^{-1}(B_1(0)) = M$.

Let $\lambda_i : M \to \mathbb{R}$, $i = 1, \ldots, N$, be smooth functions such that

$$\lambda_i(p) = \begin{cases} 1, & \text{if } p \in \phi_i^{-1}(B_1(0)), \\ 0, & \text{if } p \notin U_i. \end{cases}$$

Also, let $\psi_i : M \to \mathbb{R}^d$, $i = 1, \ldots, N$, be smooth maps defined by

$$\psi_i(p) = \begin{cases} \lambda_i \phi_i(p), & \text{if } p \in U_i, \\ 0, & \text{if } p \notin U_i. \end{cases}$$

We claim that the smooth map $\Phi : M \to \mathbb{R}^{Nd+N}$ defined by

$$\Phi(p) = (\psi_1(p), \lambda_1(p), \ldots, \psi_N(p), \lambda_N(p))$$

is the desired embedding. In fact, we have that

(i) Φ is an immersion: if $p \in M$ then $p \in \phi_i^{-1}(B_1(0))$, for some i. Hence, we have that $\psi_i = \phi_i$ in a neighborhood p. We conclude that $d_p \psi_i = d_p \phi_i$ is injective. This shows that $d_p \Phi$ is injective.

(ii) Φ is injective: Let $p, q \in M$, $p \neq q$, and choose i such that $p \in \lambda_i^{-1}(1)$. If $q \notin \lambda_i^{-1}(1)$, then $\lambda_i(p) \neq \lambda_i(q)$ so that $\Phi(p) \neq \Phi(q)$. On the other hand, if $q \in \lambda_i^{-1}(1)$, then $\psi_i(p) = \phi_i(p) \neq \phi_i(q) = \psi_i(q)$, since ϕ_i is injective. In any case, $\Phi(p) \neq \Phi(q)$, so Φ is injective.

Since M is compact, we conclude that Φ is an embedding. $\square$

The previous result also holds for non-compact manifolds (see the exercises in this lecture) and is valid also for manifolds with boundary. It is a weaker version of the following result:

Theorem 8.4 (Whitney, 1944). *Any smooth manifold (compact or not) of dimension d can be embedded in $\mathbb{R}^{2d}$.*

As the example of the Klein bottle shows, there are smooth manifolds of dimension d which cannot be embedded in $\mathbb{R}^{2d-1}$. On the other hand, for $d > 1$, Whitney also showed that any manifold of dimension d can be *immersed* in $\mathbb{R}^{2d-1}$.

Whitney's results are not the best possible. Ralph Cohen (1985) showed that a compact manifold of dimension d can be *immersed* in $\mathbb{R}^{2d-a(d)}$ where $a(d)$ is the number of 1's in the binary expression of d, and this is the best possible!! (e.g., every compact 5-manifold can immersed in $\mathbb{R}^8$, but there are compact 5-manifolds which cannot be immersed in $\mathbb{R}^7$). On the other hand, the best optimal *embedding* dimension is only known for a few dimensions.

Exercises

Exercise 8.1
Consider the sets of orthogonal and symmetric $n \times n$ matrices:

$$O(n) = \{A : AA^T = I\},$$
$$S(n) = \{A : A = A^T\}.$$

Show that $O(n)$ and $S(n)$ are embedded submanifolds of the space $\mathbb{R}^{n^2}$ of all $n \times n$ matrices and check that they intersect transversely at I. Use this to conclude that there is a neighborhood of I where the only $n \times n$-matrix which is both orthogonal and symmetric is I itself.

Exercise 8.2

Furnish the details of the example of the Klein bottle K and show that K is an embedded manifold in $\mathbb{R}^4$.

Exercise 8.3

Prove that the map $\Phi : \mathbb{R}^3 \to \mathbb{R}^4$, $\Phi(x, y, z) := (x^2 - y^2, xy, xz, yz)$, induces an embedding of $\mathbb{RP}^2$ in $\mathbb{R}^4$.

Exercise 8.4

If $\Psi : M \to N$ is a smooth map with a regular value $q \in N$, show that

$$T_p \Psi^{-1}(q) = \{\mathbf{v} \in T_p M : \mathrm{d}_p \Psi \cdot \mathbf{v} = 0\}.$$

Exercise 8.5

Let $\Psi : M \to N$ be a smooth map which is transversal to a submanifold $Q \subset N$ (not necessarily embedded). Show that $\Psi^{-1}(Q)$ is a submanifold of M (not necessarily embedded) and that

$$T_p \Psi^{-1}(Q) = \{\mathbf{v} \in T_p M : \mathrm{d}_p \Psi \cdot \mathbf{v} \in T_{\Psi(p)} Q\}.$$

Exercise 8.6

Extend Theorem 8.2 to the case where $\Psi : M \to N$ is a smooth map between manifolds with boundary such that $\Psi(\partial M) = \partial N$. Show that the conclusion of the theorem may fail if this last condition is omitted.

Exercise 8.7

Let M and N be smooth manifolds and let $S \subset M \times N$ be a submanifold. Denote by $\pi_M : M \times N \to M$ and $\pi_N : M \times N \to N$ the projections on each factor. Show that the following are equivalent:

(a) S is the graph of a smooth map $\Phi : M \to N$;
(b) $\pi_M|_S$ is a diffeomorphism from S onto M;
(c) For each $p \in M$, the submanifolds S and $\{p\} \times N = \pi_M^{-1}(p)$ intersect transversely and the intersection consists of a single point.

Moreover, if any of these hold then S is an embedded submanifold.

The next sequence of exercises give a sketch of the proof of the weak Whitney's Embedding Theorem for non-compact manifolds. It uses the following result which we will not discuss in these lectures (see, e.g., Hirsch, 1994).

Theorem 8.5 (Sard's Theorem). *The set of singular values of any smooth map* $\Psi : M \to N$ *has zero measure.*

Exercise 8.8
Using Sard's Theorem, show that if $\Phi : M \to N$ is a smooth map between smooth manifolds and $\dim M < \dim N$ then $\Phi(M)$ has zero measure.

Exercise 8.9
Let $M \subset \mathbb{R}^n$ be a smooth submanifold of dimension d. Given $v \in \mathbb{R}^n \setminus \mathbb{R}^{n-1}$ denote by $\pi_v : \mathbb{R}^n \to \mathbb{R}^{n-1}$ the linear projection with kernel $\mathbb{R}v$. Show that if $n > 2d + 1$, there is a dense set of vectors $v \in \mathbb{R}^n \setminus \mathbb{R}^{n-1}$ for which $\pi_v|_M$ is an injective immersion of M in $\mathbb{R}^{n-1}$. Conclude that any compact manifold with boundary of dimension d can be embedded in $\mathbb{R}^{2d+1}$.

Hint: Check that the proof given in the text of Whitney's embedding theorem is valid for compact manifolds with boundary. Then apply Sard's theorem in a clever way.

Exercise 8.10
Using a smooth exhaustion function, show that any smooth manifold M of dimension d can be embedded in $\mathbb{R}^{2d+1}$.

Hint: If $f : M \to \mathbb{R}$ is a smooth exhaustion function, then by Sard's Theorem, in each interval $[i, i+1[$, the function f has a regular value a_i. It follows that the sets $E_0 = f^{-1}(] - \infty, a_2]$, $E_i = f^{-1}([a_{i-1}, a_{i+1}]$ $(i = 1, 2, \dots)$, are all compact submanifolds of M of dimension d to which the previous result can be applied. Now use a partition of unity to build an embedding of M in $\mathbb{R}^{2d+1}$.

Lecture 9

Foliations

A foliation is a nice decomposition of a manifold into submanifolds.

Definition 9.1. Let M be a manifold of dimension d. A **foliation** of dimension k of M is a decomposition $\{L_\alpha : \alpha \in A\}$ of M into disjoint path-connected subsets satisfying the following property: for any $p \in M$ there exists a smooth chart $\phi = (x^1, \ldots, x^k, y^1, \ldots, y^{d-k})$: $U \to \mathbb{R}^d = \mathbb{R}^k \times \mathbb{R}^{d-k}$, such the connected components of $L_\alpha \cap U$ are the sets

$$\{p \in U : y^1(p) = \text{const.}, \ldots, y^{d-k}(p) = \text{const.}\}.$$

We will denote a foliation by $\mathcal{F} = \{L_\alpha : \alpha \in A\}$. The connected sets L_α are called **leaves** leaf of a foliation of $\mathcal{F}$ and a chart (U, ϕ) as in the definition is called a **foliated chart** (see Figure 9.1). The connected components of $U \cap L_\alpha$ are called **plaques** of L_α.

A **path of plaques** is a collection of plaques $P_1, \ldots, P_l$ such that $P_i \cap P_{i+1} \neq \emptyset$, for all $i = 1, \ldots, l - 1$. The integer l is called the **length of the path of plaques**. Two points $p, q \in M$ belong to the same leaf if and only if there exists a path of plaques $P_1, \ldots, P_l$, with $p \in P_1$ and $q \in P_l$.

Each leaf of a k-dimensional foliation of M is a submanifold of M of dimension k. In general, these are only immersed submanifolds: a leaf can intersect a foliated coordinate chart an infinite number of times and accumulate overt itself. Before we check that leaves are submanifolds, let us look at some examples.

Example 9.1. Let $\Phi : M \to N$ be a submersion. By the local normal form for submersions, the connected components of the fibers

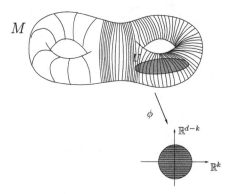

Fig. 9.1. A foliated chart.

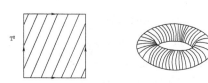

Fig. 9.2. A foliated torus.

$\Phi^{-1}(q)$, where $q \in N$, form a foliation of M of codimension equal to the dimension of N. In this case, all leaves are actually embedded submanifolds.

Example 9.2. In $\mathbb{R}^2$, take the foliation by straight lines with a fixed slope $a \in \mathbb{R}$. This is just a special case of the previous example, where $\Phi : \mathbb{R}^2 \to \mathbb{R}$ is given by

$$\Phi(x, y) = y - ax.$$

Now let $\mathbb{T}^2 = \mathbb{R}^2 / \mathbb{Z}^2$ be the torus. Then we have an induced foliation on $\mathbb{T}^2$, and there are two possibilities. If $a \in \mathbb{Q}$, the leaves are closed curves, hence they are embedded submanifolds. However, if $a \notin \mathbb{Q}$, then the leaves are dense in the torus, so they are only immersed submanifolds. One can also use the model for a torus as a square with sides identified to picture this foliation (see Figure 9.2).

Example 9.3. Let $\Phi : \mathbb{R}^3 \to \mathbb{R}$ be the smooth map defined by

$$\Phi(x, y, z) = f(x^2 + y^2)e^{-z},$$

where $f \in C^\infty(\mathbb{R})$ is a smooth function with $f(0) = 1$, $f(1) = 0$ and $f'(t) < 0$. It is easy to check that Φ is a submersion and so determines

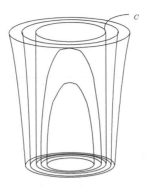

Fig. 9.3. A foliation of $\mathbb{R}^3$.

a foliation $\mathcal{F}$ of $\mathbb{R}^3$ whose leaves are the pre-images $\{\Phi^{-1}(c)\}_{c \in \mathbb{R}}$. When $c = 0$, we obtain as leaf the cylinder $C = \{(x, y, z) : x^2 + y^2 = 1\}$. This cylinder splits the leaves into two classes (see Figure 9.3):

- The leaves with $c > 0$ lying in the interior of the cylinder C, which are all diffeomorphic to $\mathbb{R}^2$.
- The leaves with $c < 0$ lying in the exterior of the cylinder C, which are all diffeomorphic to C.

An explicit parameterization of the leaves with $c \neq 0$ is given by

$$(x, y) \mapsto (x, y, \log(c/f(x^2 + y^2))).$$

For the first type of leaves, $c > 0$ and $x^2 + y^2 < 1$, while for the second type of leaves $c < 0$ and $x^2 + y^2 > 1$.

Example 9.4. The foliation of $\mathbb{R}^3$ in the previous example is invariant under translations in the Oz-axis direction. If we identify $\mathbb{R}^3 = \mathbb{R}^2 \times \mathbb{R}$, we obtain a foliation in the quotient $\mathbb{R}^2 \times \mathbb{S}^1 = \mathbb{R}^2 \times \mathbb{R}/\mathbb{Z}$. If we restrict this foliation to $\operatorname{Int} D^2 \times \mathbb{S}^1$, where $D^2 = \{(x, y) : x^2 + y^2 \leq 1\}$, we obtain a foliation of the solid 2-torus called the **Reeb foliation**. Note that the boundary 2-torus is a leaf of this foliation (see Figure 9.4).

This example suggests that foliations of manifolds with boundary are also interesting. We will not pursue this topic, but you should be aware of the existence of foliations on manifolds with boundary (see Candel and Conlon, 2000).

Fig. 9.4.   The Reeb foliation of the solid 3-torus.

Example 9.5. The 3-sphere $\mathbb{S}^3$ can be obtained by "gluing" two solid 2-torus along its boundary:

$$\mathbb{S}^3 = T_1 \cup_\Phi T_2,$$

where $\Phi : \partial T_1 \to \partial T_2$ is a diffeomorphism that takes the meridians of ∂T_1 in the circles of latitude of ∂T_2, and vice-versa. Explicitly, if we write

$$\mathbb{S}^3 = \{(x, y, z, w) : x^2 + y^2 + z^2 + w^2 = 1\},$$

then we can take

$$T_1 := \{(x, y, z, w) \in \mathbb{S}^3 : x^2 + y^2 \leq 1/2\},$$
$$T_2 := \{(x, y, z, w) \in \mathbb{S}^3 : x^2 + y^2 \geq 1/2\}.$$

Each of these solid 2-torus admits a 2-dimensional foliation as in the previous example. One then obtains a famous 2-dimensional foliation of the sphere $\mathbb{S}^3$, called the **Reeb foliation** of $\mathbb{S}^3$. Note that it is not at all obvious that one can glue these foliations to obtain a smooth foliation. Details on this construction can be found, e.g., in Candel and Conlon (2000).

Proposition 9.1. *Let $\mathcal{F}$ be a k-dimensional foliation of a smooth manifold M. Every leaf $L \in \mathcal{F}$ is a regularly immersed submanifold of dimension k.*

Proof. Let L be a leaf of $\mathcal{F}$. On each plaque of L we consider the relative topology, and we furnish L with the topology generated by the open sets in the plaques of L. For each plaque P, associated with a foliated chart $(U, \phi) = (U, x^1, \ldots, x^k, y^1, \ldots, y^{d-k})$, we consider the map $\psi : P \to \mathbb{R}^k$ obtained by choosing the first k-components so

$$\psi(p) = (x^1(p), \ldots, x^k(p)).$$

The pairs (P, ψ) give charts for L, which turn L into a Hausdorff topological manifold. The transition functions for these charts are clearly smooth, so we can consider the maximal atlas that contains all the charts (U, ψ). To check that L is a manifold, we only need to check that the topology admits a countable basis. For that we apply the following lemma.

Lemma 9.1. *Let L be a leaf of $\mathcal{F}$ and $\{U_n : n \in \mathbb{Z}\}$ a covering of M by domains of foliated charts. The number of plaques of L in this covering, i.e., the number of connected components of $L \cap U_n$, $n \in \mathbb{Z}$, is countable.*

Fix a plaque P_0 of L in the covering $\{U_n : n \in \mathbb{Z}\}$. If a plaque P' belongs to L then there exists a **path of plaques** $P_1, \ldots, P_l$ in the covering, with $P_i \cap P_{i+1} \neq \emptyset$ which connects P' to P_0. Therefore, it is enough to check that the collection of such paths is countable.

For each path of plaques $P_1, \ldots, P_l$ let us call l the length of the path. Using induction on n, we show that the collection of paths of length less or equal to n is countable:

- The collection of paths of length 1 has only one element hence is countable.
- Assume that the collection of paths of length $n-1$ is countable. Let $P_1, \ldots, P_{n-1}$ be a path of length $n-1$, corresponding to domains of foliated charts $U_1, \ldots, U_{n-1}$. In order to obtain a path of plaques of length n, we choose a domain of a foliated chart $U_n \neq U_{n-1}$ and we consider the plaques P', which are connected components of $L \cap U_n$, such that the intersection with P_{n-1} is non-empty. Now observe that

$$(L \cap U_n) \cap P_{n-1} = U_n \cap P_{n-1}.$$

Hence, these intersections form an open cover of the plaque P_{n-1}. This cover has a countable subcover, so the collection of all such P' is countable. It follows that the collection of paths of length less or equal than n is countable.

We leave it as an exercise to check that the leaves are actually regularly immersed submanifolds. □

Corollary 9.1. *Each leaf of a foliation intersects the domain of a foliated chart at most a countable number of times.*

Let us describe two constructions which allows one to obtain new foliations out of other foliations.

Product of Foliations

Let $\mathcal{F}_1$ and $\mathcal{F}_2$ be foliations of M_1 and M_2, respectively. Then the **product foliation** $\mathcal{F}_1 \times \mathcal{F}_2$ is a foliation of $M_1 \times M_2$ defined as follows: if $\mathcal{F}_1 = \{L_\alpha^{(1)}\}_{\alpha \in A}$ and $\mathcal{F}_2 = \{L_\beta^{(2)}\}_{\beta \in B}$, then

$$\mathcal{F}_1 \times \mathcal{F}_2 = \{L_\alpha^{(1)} \times L_\beta^{(2)}\}_{(\alpha,\beta) \in A \times B}.$$

It should be clear that $\dim(\mathcal{F}_1 \times \mathcal{F}_2) = \dim \mathcal{F}_1 + \dim \mathcal{F}_2$ and, hence, that $\mathrm{codim}\,(\mathcal{F}_1 \times \mathcal{F}_2) = \mathrm{codim}\,\mathcal{F}_1 + \mathrm{codim}\,\mathcal{F}_2$

Pull-Back of a Foliation

Let $\Phi : M \to N$ be a smooth map between smooth manifolds. If $\mathcal{F}$ is a foliation of N we will say that Φ **is transversal to** $\mathcal{F}$ and write $\Phi \pitchfork \mathcal{F}$ if Φ is transversal to every leaf L of $\mathcal{F}$:

$$\mathrm{d}_p\Phi(T_pM) + T_{\Phi(p)}L = T_{\Phi(p)}N, \quad \forall p \in M.$$

Whenever $\Phi \pitchfork \mathcal{F}$ one defines the **pull-back foliation** $\Phi^*(\mathcal{F})$ to be the foliation of M whose leaves are the connected components of $\Phi^{-1}(L)$, where $L \in \mathcal{F}$. It should be clear that $\mathrm{codim}\,\Phi^*(\mathcal{F}) = \mathrm{codim}\,\mathcal{F}$.

The definition of a foliation is not very practical and it is convenient to have alternative characterizations of foliations, which we discuss next.

Foliations via Smooth $\mathcal{G}_d^k$-Structures

Let $\mathcal{F} = \{L_\alpha : \alpha \in A\}$ be a k-dimensional foliation of M. If (U, ϕ) and (V, ψ) are foliated charts then the change of coordinates $\psi \circ \phi^{-1} : \phi(U \cap V) \to \psi(U \cap V)$ is of the form (see Figure 9.5):

$$\mathbb{R}^k \times \mathbb{R}^{d-k} \ni (x, y) \mapsto (h_1(x, y), h_2(y)) \in \mathbb{R}^k \times \mathbb{R}^{d-k}.$$

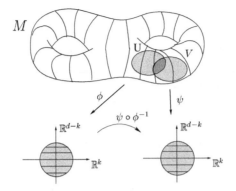

Fig. 9.5. Transitions between foliation charts.

In other words, we have that the transition functions satisfy:

$$\frac{\partial(\psi \circ \phi^{-1})^j}{\partial x^i} = 0, \quad (i = 1, \ldots, k, \ j = k + 1, \ldots, d). \tag{9.1}$$

Conversely, denote by $\mathcal{G}_d^k$ the diffeomorphisms $\mathbb{R}^d \to \mathbb{R}^d$ defined on some open set that satisfies condition (9.1). We can refine the notion of smooth structure by requiring that in Definition 2.2, the transition functions belong to $\mathcal{G}_d^k$, and we then speak of a **smooth $\mathcal{G}_d^k$-structure**. An ordinary smooth structure on M is just a $\mathcal{G}_d^d$-structure: the leaves are the connected components of M.

We have the following alternative description of a foliation.

Proposition 9.2. *Let M be a smooth d-dimensional manifold. Given a foliation $\mathcal{F} = \{L_\alpha : \alpha \in A\}$ of M of dimension k the collection of all foliated charts $\mathcal{C} = \{(U, \phi)\}$ defines a smooth $\mathcal{G}_d^k$-structure. Conversely, for every smooth $\mathcal{G}_d^k$-structure $\mathcal{C}$ on a topological space M, there is smooth structure that makes M into a d-dimensional manifold and there exists a foliation $\mathcal{F}$ of M of dimension k, for which the foliated charts are the elements of $\mathcal{C}$.*

Proof. We have shown above that every k-dimensional foliation of a d-dimensional manifold determines a smooth $\mathcal{G}_d^k$-structure. We will show that, conversely, given a smooth $\mathcal{G}_d^k$-structure $\mathcal{C} = \{(U, \phi)\}$ we can associate to it a smooth structure on M of dimension d and a k-dimensional foliation $\mathcal{F}$ of M.

It should be clear that a smooth $\mathcal{G}_d^k$-structure $\mathcal{C} = \{(U, \phi)\}$ determines a smooth structure on M of dimension d, since it is in particular an atlas. In order to build $\mathcal{F}$, we first observe that we can choose an atlas defining $\mathcal{C}$ with the property that the slices $\phi^{-1}(\mathbb{R}^k \times \{c\})$, for $c \in \mathbb{R}^{d-k}$, are connected. We call these slices *plaques* and note that M is covered by all such plaques. Hence, we can define an equivalence relation in M by

- $p \sim q$ if there exists a path of plaques $P_1, \ldots, P_l$ with $p \in P_1$ and $q \in P_l$.

Let $\mathcal{F}$ be the set of equivalence classes of $\sim$. We show that $\mathcal{F}$ is a foliation of M, i.e., that the condition in Definition 9.1 is satisfied.

Let $p_0 \in M$ and consider a plaque P_0 which contains p_0. Then,

$$P_0 = \phi^{-1}(\mathbb{R}^k \times \{c_0\}),$$

for some smooth chart $(U, \phi) \in \mathcal{C}$ with $\phi(p_0) = (a_0, c_0) \in \mathbb{R}^k \times \mathbb{R}^{d-k}$. We claim that (U, ϕ) is a foliated chart: let $L \in \mathcal{F}$ be an equivalence class that intersects U. If $p \in U \cap L$, then $\phi(p) = (a, c) \in \mathbb{R}^k \times \mathbb{R}^{d-k}$, so we see that that the plaque

$$P = \phi^{-1}(\mathbb{R}^k \times \{c\}),$$

is contained in L. Since P is connected, it is clear that P is contained in the connected component of $L \cap U$ that contains p. If we can show that this connected component is actually P it will follow that (U, ϕ) is a foliated chart.

Let $q \in L \cap U$ be some point in the connected component of $L \cap U$ containing p. We claim that $q \in P$. By the definition of $\sim$, there exists a path of plaques $P_1, \ldots, P_l$, with $p \in P_1$ and $q \in P_l$, and such that $P_i \subset U$. Each plaque P_i is associated to a smooth chart $(U_i, \phi_i) \in \mathcal{C}$ such that

$$P_i = \phi_i^{-1}(\mathbb{R}^k \times \{c_i\}).$$

We can assume also that $U_1 = U$, $\phi_1 = \phi$, $P_1 = P$ and $c_1 = c$. Since $\phi_2 \circ \phi^{-1} \in \mathcal{G}_d^k$, we have that

$$\phi_2^{-1}(\mathbb{R}^k \times \{c_2\}) \subset \phi_2^{-1} \circ \phi_2 \circ \phi^{-1} \circ (\mathbb{R}^k \times \{\bar{c}_2\}) = \phi^{-1}(\mathbb{R}^k \times \{\bar{c}_2\}),$$

for some $\bar{c}_2 \in \mathbb{R}^{d-k}$. Since $P_2 \cap P_1 \neq \emptyset$ and the plaques $\phi^{-1} \circ (\mathbb{R}^k \times \{c\})$ are disjoint, we conclude that $\bar{c}_2 = c_1$ and $P_2 \subset P_1 = P$. By induction, $P_i \subset P$ so $q \in P$, as claimed. $\qquad\square$

Foliations via Submersions

We saw before that the connected components of the fibers of a submersion is an example of a foliation. Actually, every foliation is locally of this form. If $\mathcal{F} = \{L_\alpha\}_{\alpha \in A}$ is a foliation of M of dimension k, for any foliated chart

$$\phi = (x^1, \ldots, x^k, y^1, \ldots, y^{d-k}) : U \to \mathbb{R}^d,$$

the projection in the last $(d-k)$-components gives a submersion

$$\psi := (y^1, \ldots, y^{d-k}) : U \to \mathbb{R}^{d-k},$$

whose fibers are the connected components of $L_\alpha \cap U$. Given another foliated chart

$$\bar{\phi} = (\bar{x}^1, \ldots, \bar{x}^k, \bar{y}^1, \ldots, \bar{y}^{d-k}) : \bar{U} \to \mathbb{R}^d,$$

with $U \cap \bar{U} \neq \emptyset$, the corresponding submersion

$$\bar{\psi} := (\bar{y}^1, \ldots, \bar{y}^{d-k}) : \bar{U} \to \mathbb{R}^{d-k},$$

we have a change of charts of the form

$$\bar{\phi} \circ \phi^{-1}(x, y) = (h_1(x, y), h_2(y)),$$

where h_2 has Jacobian matrix

$$\left[\frac{\partial h_2^j}{\partial y^i} \right]_{i,j=1}^{d-k}$$

with rank $d - k$. We conclude that the two submersions ψ and $\bar{\psi}$ differ by a local diffeomorphism: for every $p \in U \cap \bar{U}$ there exists an open neighborhood $p \in U_p \subset U \cap \bar{U}$ and a local diffeomorphism $\Psi : \mathbb{R}^{d-k} \to \mathbb{R}^{d-k}$, such that

$$\bar{\psi}|_{U_p} = \Psi \circ \psi|_{U_p}.$$

This suggests another way of defining foliations which can be made precise as follows.

Proposition 9.3. *Let M be a d-dimensional manifold. Every k-dimensional foliation $\mathcal{F}$ of M determines a collection $\{\psi_i\}_{i \in I}$ of submersions $\psi_i : U_i \to \mathbb{R}^{d-k}$, where $\{U_i\}_{i \in I}$ is an open cover of M, which satisfies the following property: for every $i, j \in I$ and $p \in U_i \cap U_j$, there exists a local diffeomorphism ψ_{ji}^p of $\mathbb{R}^{d-k}$, such that*

$$\psi_j = \psi_{ji}^p \circ \psi_i,$$

in an open neighborhood U_p of p. Conversely, every such collection determines a foliation of M.

We have already seen how to a foliation we can associate a collection of submersions. We leave it as an exercise to prove the converse.

Foliations appear naturally in many problems in differential geometry and we shall see many other examples of foliations during the course of these lectures. Our discussion here only touches upon very elementary aspects of the theory of foliations. For more in-depth discussions,d we refer to standard textbooks such as Candel and Conlon (2000) and Moerdijk and Mrcun (2003).

Exercises

Exercise 9.1
Show that the leaves of a foliation are regularly immersed submanifolds.

Exercise 9.2
Let $\mathcal{F}$ be the Reeb foliation of S^3 and let $\Phi : S^3 \to N$ be a continuous map whose restriction to each leaf of $\mathcal{F}$ is constant. Show that Φ is constant.

Exercise 9.3
Prove Proposition 9.3.

Exercise 9.4
Let $\mathcal{F}_1 = \{L_\alpha^{(1)}\}_{\alpha \in A}$ and $\mathcal{F}_2 = \{L_\beta^{(2)}\}_{\beta \in B}$ be foliations. Using your favorite definition of a foliation, show that the product $\mathcal{F}_1 \times \mathcal{F}_2$ is a foliation:

$$\mathcal{F}_1 \times \mathcal{F}_2 := \{L_\alpha^{(1)} \times L_\beta^{(2)}\}_{(\alpha,\beta) \in A \times B}.$$

Exercise 9.5

Let $\Phi : M \to N$ be a smooth map and $\mathcal{F} = \{L_\alpha\}_{\alpha \in A}$ a foliation of N such that $\Phi \pitchfork \mathcal{F}$. Using your favorite definition of a foliation, show that the pull-back $\Phi^*(\mathcal{F})$ is a foliation:

$$\Phi^*(\mathcal{F}) := \{\text{connected components of } \Phi^{-1}(L_\alpha)\}_{\alpha \in A}.$$

Exercise 9.6

Let $\mathcal{F}_1$ and $\mathcal{F}_2$ be two foliations of a smooth manifold M such that $\mathcal{F}_1 \pitchfork \mathcal{F}_2$, i.e., such that

$$T_p M = T_p L^{(1)} + T_p L^{(2)}, \quad \forall p \in M,$$

where $L^{(1)}$ and $L^{(2)}$ are the leaves of $\mathcal{F}_1$ and $\mathcal{F}_2$ through p. Show that there exists a foliation $\mathcal{F}_1 \cap \mathcal{F}_2$ of M whose leaves are the connected components of $L_\alpha^{(1)} \cap L_\beta^{(2)}$, and which satisfies $\operatorname{codim} \mathcal{F} = \operatorname{codim} \mathcal{F}_1 + \operatorname{codim} \mathcal{F}_2$.

Exercise 9.7

Given a foliation $\mathcal{F}$ of M, one denotes by $M/\mathcal{F}$ the space of leaves of $\mathcal{F}$ with the quotient topology. Describe for each of the examples given in the text their space of leaves.

Exercise 9.8

Given a foliation $\mathcal{F}$ of M, one denotes by $M/\mathcal{F}$ the space of leaves of $\mathcal{F}$ with the quotient topology. Describe for each of the examples given in the text their space of leaves.

Lecture 10

Quotients

We have seen several constructions that produce new manifolds out of old manifolds, such as the product of manifolds or the pullback of submanifolds under transversal maps. We will now study another important, but more delicate, such construction: forming quotients of manifolds.

Let X be a topological space. If $\sim$ is an equivalence relation on X, we will denote by $X/\sim$ the set of equivalence classes of $\sim$ and by $\pi : X \to X/\sim$ the *quotient map* which associates to each $x \in X$ its equivalence class $\pi(x) = [x]$. In $X/\sim$ we consider the *quotient topology*: a subset $V \subset X/\sim$ is open if and only if $\pi^{-1}(V)$ is open. This is the largest topology in $X/\sim$ for which the quotient map $\pi : X \to X/\sim$ is continuous. We have the following basic result about the quotient topology which we leave as an exercise.

Lemma 10.1. *Let X be a Hausdorff topological space and let $\sim$ be an equivalence relation on X such that $\pi : X \to X/\sim$ is an open map. Then $X/\sim$ is Hausdorff if and only if the graph of $\sim$*

$$R := \{(x,y) \in X \times X : x \sim y\},$$

is a closed subset of $X \times X$.

Let M be a smooth manifold and let $\sim$ be an equivalence relation on M. We would like to know when there exists a smooth structure on $M/\sim$, compatible with the quotient topology, such that $\pi : M \to M/\sim$ becomes a submersion. Before we can state a result that gives a complete answer to this question, we need one definition. For that, recall that a continuous map $\Phi : X \to Y$, between two Hausdorff topological spaces is called a **proper map** if $\Phi^{-1}(K) \subset X$ is compact whenever $K \subset Y$ is compact. A proper map is always a closed map.

Definition 10.1. A **proper submanifold** of M is a submanifold (N, Φ) such that $\Phi : N \to M$ is a proper map.

By Exercise 7.9, any proper submanifold is an embedded submanifold. Also, if $\Phi : N \to M$ is proper, then its image $\Phi(N)$ is a closed subset of M. Conversely, every embedded closed submanifold of M is a proper submanifold.

Theorem 10.1 (Godement's Criterion). *Let M be a smooth manifold and let $\sim$ be an equivalence relation on M. The following statements are equivalent:*

(i) *There exists a smooth structure on $M/\sim$, compatible with the quotient topology, such that $\pi : M \to M/\sim$ is a submersion.*

(ii) *The graph R of $\sim$ is a proper submanifold of $M \times M$ and the restriction of the projection $p_1 : M \times M \to M$ to R is a submersion.*

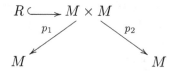

Proof. We must show both implications:

(i) $\Rightarrow$ (ii). The graph of the quotient map, as for every smooth map, is a closed embedded submanifold

$$\mathcal{G}(\pi) = \{(p, \pi(p)) : p \in M\} \subset M \times M/\sim.$$

Since $\mathrm{Id} \times \pi : M \times M \to M \times M/\sim$ is a submersion and

$$R = (\mathrm{Id} \times \pi)^{-1}(\mathcal{G}(\pi)),$$

we conclude that $R \subset M \times M$ is an embedded closed submanifold, i.e., is a proper submanifold.

On the other hand, the map $(\mathrm{Id} \times \pi)|_R : R \to \mathcal{G}(\pi)$ is a submersion while $\mathcal{G}(\pi) \to M$, $(p, \pi(p)) \mapsto p$ is a diffeomorphism, hence their composition $p_1|R$ is a submersion.

(ii) $\Rightarrow$ (i). We split the proof into several lemmas. The first of these lemmas states that we can "straighten out" $\sim$.

Lemma 10.2. *For every $p \in M$, there exists a local chart $(U, (x^1, \ldots, x^d))$ centered at p, such that*

$$\forall q, q' \in U, \quad q \sim q' \text{ if and only if } x^{k+1}(q)$$
$$= x^{k+1}(q'), \ldots, x^d(q) = x^d(q'),$$

where k is an integer independent of p.

To prove this lemma, let $\Delta \subset M \times M$ be the diagonal. Note that $\Delta \subset R \subset M \times M$, and since Δ and R are both embedded submanifolds of $M \times M$, we have that Δ is an embedded submanifold of R. Therefore, for each $p \in M$, there exists a neighborhood O of (p, p) in $M \times M$ and a submersion $\Phi : O \to \mathbb{R}^{d-k}$, where $d - k = \operatorname{codim} R$, such that

$$(q, q') \in O \cap R \quad \text{if and only if } \Phi(q, q') = 0.$$

We have that $k \geq 0$, since $\Delta \subset R$ and $\operatorname{codim} \Delta = d$.

Next, we observe that the differential of the map $q \mapsto \Phi(q, p)$ has maximal rank at $q = p$. In fact, after identifying $T_{(p,p)}(M \times M) = T_p M \times T_p M$, we see that $\mathrm{d}_{(p,p)} \Phi$ is zero precisely in the subspace formed by pairs $(\mathbf{v}, \mathbf{v}) \in T_p M \times T_p M$, and this subspace is complementary to the subspace formed by elements of the form $(\mathbf{v}, 0) \in T_p M \times T_p M$. We conclude that there exists a neighborhood V' of p such that $V' \times V' \subset O$, and the map $q \mapsto \Phi(q, p)$ is a submersion in V'. By the local canonical form for submersions, there exist a chart $(V, \phi) = (V, (u^1, \ldots, u^k, v^1, \ldots, v^{d-k}))$ centered at p, with $V \subset V'$, such that

$$\Phi \circ (\phi^{-1} \times \phi^{-1})(u^1, \ldots, u^k, v^1, \ldots, v^{d-k}, 0, \ldots, 0) = (v^1, \ldots, v^{d-k}).$$

In the domain of this chart, the points $q \in V$ such that $q \sim p$ are precisely the points satisfying $v^1(q) = 0, \ldots, v^{d-k}(q) = 0$.

Now set $\widehat{\Phi} = \Phi \circ (\phi^{-1} \times \phi^{-1})$. The smooth map

$$\mathbb{R}^d \times \mathbb{R}^{d-k} \to \mathbb{R}^{d-k}, \quad (u, v, w) \mapsto \widehat{\Phi}((u, v), (0, w)),$$

satisfies

$$\widehat{\Phi}((u,v),(0,0)) = v.$$

so the matrix of partial derivatives $\partial \widehat{\Phi}^i / \partial v^j$, $(i,j = 1,\ldots,d-k)$ is non-degenerate. We can apply the Implicit Function Theorem to conclude that there exists a local defined smooth function $\mathbb{R}^k \times \mathbb{R}^{d-k} \to \mathbb{R}^{d-k}$, $(u,w) \mapsto v(u,w)$, such that

$$\widehat{\Phi}((u,v),(0,w)) = 0 \ \text{ if and only if } v = v(u,w).$$

Since $v(0,w) = w$ is a solution, uniqueness implies that

$$\phi^{-1}(0,w) \sim \phi^{-1}(0,w') \ \text{ if and only if } w = w'.$$

This shows that the map $(u,w) \mapsto (u,v(u,w))$ is a local diffeomorphism. Hence, there exists an open set U where

$$(x^1,\ldots,x^d) = (u^1,\ldots,u^k,w^1,\ldots,w^{d-k})$$

are local coordinates and in these coordinates, we have

$$\forall q, q' \in U, \ q \sim q' \ \text{ if and only if } \ x^{k+1}(q)$$
$$= x^{k+1}(q'),\ldots, x^d(q) = x^d(q'),$$

so the lemma follows.

Since the functions $x^{k+1},\ldots,x^d$ given by this lemma induce well-defined functions $\bar{x}^{k+1},\ldots,\bar{x}^d$ on the quotient $M/\sim$, we consider the pairs of the form $(\pi(U),\bar{x}^{k+1},\ldots,\bar{x}^d)$ and prove they furnish an atlas.

Lemma 10.3. *The collection* $\mathcal{C} = \{(\pi(U),\bar{x}^{k+1},\ldots,\bar{x}^d)\}$ *gives* $M/\sim$, *with the quotient topology, the structure of a topological manifold of dimension* $d-k$.

To prove this, note that $\pi : M \to M/\sim$ is an open map. In fact, for any $V \subset M$, we have that

$$\pi^{-1}(\pi(V)) = p_1|_R((p_2|_R)^{-1}(V)).$$

By assumption, $p_1|_R$ is a submersion hence it is an open map. Therefore, if $V \subset M$ is open then $\pi^{-1}(\pi(V))$ is also open, so $\pi(V) \subset M/\sim$

is open. In particular, we can conclude that $\pi(U)$ is open. Since the map

$$(x^{k+1}, \ldots, x^d) : U \to \mathbb{R}^{d-k}$$

is both continuous and open, it follows that the induced map

$$(\bar{x}^{k+1}, \ldots, \bar{x}^d) : \pi(U) \to \mathbb{R}^{d-k}$$

is continuous, open, and injective, hence a homeomorphism onto its image. This proves the lemma.

Lemma 10.4. *The family* $\mathcal{C} = \{(\pi(U), \bar{x}^{k+1}, \ldots, \bar{x}^d)\}$ *is an atlas generating a smooth structure for* $M/\sim$ *such that* $\pi : M \to M/\sim$ *is a submersion.*

To see this, take two pairs of charts in $\mathcal{C}$

$$(\pi(U), \bar{\phi}) := (\pi(U), \bar{x}^{k+1}, \ldots, \bar{x}^d),$$
$$(\pi(V), \bar{\psi}) := (\pi(V), \bar{y}^{k+1}, \ldots, \bar{y}^d),$$

which correspond to two charts in M

$$(U, \phi) := (U, x^1, \ldots, x^d), \quad (V, \psi) := (V, y^1, \ldots, y^d).$$

The corresponding transition function

$$\bar{\psi} \circ \bar{\phi}^{-1} : \mathbb{R}^{d-k} \to \mathbb{R}^{d-k},$$

composed with the projection $p : \mathbb{R}^d \to \mathbb{R}^{d-k}$ in the last $d - k$ components equals

$$\bar{\psi} \circ \bar{\phi}^{-1} \circ p = p \circ \psi \circ \phi^{-1}.$$

Since the right-hand side is a smooth map $\mathbb{R}^d \to \mathbb{R}^{d-k}$, it follows that $\bar{\psi} \circ \bar{\phi}^{-1}$ is smooth.

In order to check that $\pi : M \to M/\sim$ is a submersion, it is enough to observe that in the charts $(U, x^1, \ldots, x^d)$ for M and $(\pi(U), \bar{x}^{k+1}, \ldots, \bar{x}^d)$ for $M/\sim$, this map corresponds to the projection $p : \mathbb{R}^d \to \mathbb{R}^{d-k}$. This completes the proof of the lemma.

To finish the proof of Theorem 10.1, we check that

Lemma 10.5. *The quotient topology* $M/\sim$ *is Hausdorff and second countable.*

It is obvious that if M has a countable basis, then the quotient topology also has a countable basis. Since the graph R of $\sim$ is closed in $M \times M$, M is Hausdorff and π is an open map, it follows from Lemma 10.1 that $M/\sim$ is Hausdorff. $\qquad\square$

Remark 10.1. The proof shows that if we assume that R is embedded, not closed, and $p_1|_R : R \to M$ is a submersion, then the quotient $M/\sim$ is a smooth manifold, second countable, but not Hausdorff (see Exercise 10.4 for an example).

We will now study two important examples of quotients.

Leaf Spaces of Foliations

Let $\mathcal{F}$ be a foliation of a smooth manifold M. Since $\mathcal{F}$ is a partition of M, it determines an equivalence relation on M, namely:

$$p \sim q \quad \text{if and only if } p \text{ and } q \text{ belong to the same leaf.}$$

The set of equivalence classes

$$M/\mathcal{F} := M/\sim$$

is the collection of all leaves of $\mathcal{F}$ and hence is called the *leaf space* of the foliation.

In general, the leaf space of a foliation does not carry a smooth structure compatible with the quotient topology. One can use Godement's Criterion to find when this happens.

Corollary 10.1. *Let $\mathcal{F}$ be a foliation of a smooth manifold M. The following statements are equivalent:*

(i) *There exists a smooth structure on $M/\mathcal{F}$, compatible with the quotient topology, such that $\pi : M \to M/\mathcal{F}$ is a submersion.*

(ii) *The leaf space $M/\mathcal{F}$ is Hausdorff and there is a cover of M by foliated charts with the property that each leaf of $\mathcal{F}$ intersects each chart at most once.*

A foliation satisfying either of the equivalent conditions in this corollary is called a **simple foliation**. We leave the proof as an exercise.

As a side remark, note that the proof of Godement's Criterion actually amounts to show that the equivalence classes of R form a simple foliation of M.

Orbit Spaces of Discrete Group Actions

A very important class of equivalence relations on manifolds is given by actions of groups of diffeomorphisms. If G is a group, we recall that an **action** of G on a set M is a group homomorphism $\widehat{\Psi}$ from G to the group of bijections of M. One can also view an action as a map $\Psi : G \times M \to M$, which we write as $(g, p) \mapsto g \cdot p$, if one sets

$$g \cdot p := \widehat{\Psi}(g)(p).$$

Since $\widehat{\Psi}$ is a group homomorphism, it follows that

(a) $e \cdot p = p$, for all $p \in M$;
(b) $g \cdot (h \cdot p) = (gh) \cdot p$, for all $g, h \in G$ and $p \in M$.

Conversely, any map $\Psi : G \times M \to M$ satisfying (a) and (b), determines a homomorphism $\widehat{\Psi}$. From now on, we will denote an action by $\Psi : G \times M \to M$, and for each $g \in G$ we denote by Ψ_g the bijection

$$\Psi_g : M \to M, \quad p \mapsto g \cdot p$$

Given an action of G on M the quotient $G \backslash M$ is, by definition, the set of equivalence classes determined by the **orbit equivalence relation**:

$$p \sim q \iff \exists g \in G : q = g \cdot p.$$

Assume now that M is a manifold. We say that that a group G *acts on M by diffeomorphisms* if, for each $g \in G$, $\Psi_g : M \to M$ is a diffeomorphism. This means that we have a group homomorphism $\widehat{\Psi} : G \to \mathrm{Diff}(M)$, where $\mathrm{Diff}(M)$ denotes the group of all diffeomorphisms of M. We can also express this condition by saying that the map $\Psi : G \times M \to M$ is smooth, where G is viewed as a smooth 0-dimensional manifold with the discrete topology. So we will also say in this case that the *discrete group G acts smoothly on M*.

We will now discuss conditions on an action by diffeomorphisms for the quotient $G \backslash M$ to be a manifold. We recall that a **free action** is an action $G \times M \to M$ such that each $g \neq e$ acts without fixed points, i.e.,

$$g \cdot p = p \quad \text{for some } p \in M \quad \implies \quad g - e.$$

Denoting by G_p the **isotropy subgroup** of $p \in M$, i.e.,

$$G_p = \{g \in G : g \cdot p = p\},$$

an action is free if and only if $G_p = \{e\}$, for all $p \in M$.

Definition 10.2. A smooth action $\Psi : G \times M \to M$ of a discrete group G on a smooth manifold M is said to be **proper** if the map:

$$G \times M \to M \times M, \quad (g, p) \mapsto (g \cdot p, p),$$

is a proper map.

Example 10.1. Actions of finite groups are always proper (exercise). For example, the $\mathbb{Z}_2$-action on $M = \mathbb{S}^d$, defined by

$$\pm 1 \cdot (x^1, \dots, x^{d+1}) := \pm(x^1, \dots, x^{d+1}).$$

is a free and proper action.

Example 10.2. Let $\mathbb{Z}^d$ act on $\mathbb{R}^d$ by translations:

$$(n_1, \dots, n_d) \cdot (x^1, \dots, x^d) := (x^1 + n_1, \dots, x^d + n_d).$$

It is easy to see that his action is also free and proper. So there are proper actions of infinite discrete groups G.

Example 10.3. The orthogonal group $O(d)$ consisting of orthogonal matrices of size d acts smoothly on the sphere $\mathbb{S}^d$ by matrix multiplication:

$$A \cdot x := Ax.$$

The isotropy group of x_0 consists of those orthogonal matrices fixing x_0 and contains, e.g., the rotations with the axis of the line through x_0 and the origin. Since a proper smooth action $G \times M \to M$ of a discrete group must have finite isotropy groups (exercise) this action is not proper. Note that $O(d)$ is considered here as a discrete group.

Godement's Criterion yields conditions for an orbit space to be smooth.

Corollary 10.2. *Let* $\Psi : G \times M \to M$ *be a free and proper smooth action of a discrete group G on M. There exists a unique* *smooth structure on $G\backslash M$ such that $\pi : M \to G\backslash M$ is a local diffeomorphism.*

Proof. We check that condition (ii) of Theorem 10.1 holds.

We claim that $R \subset M \times M$ is a proper submanifold. Since the action if free and proper, one finds (see Exercise 10.6) for each $p_0 \in M$ an open set $p_0 \in U$, such that

$$g \cdot U \cap U = \emptyset, \quad \forall g \in G \setminus \{e\}.$$

For such an open set, if $g_0 \in G$, we have

$$(U \times g_0 \cdot U) \cap R = \{(q, g_0 \cdot q) : q \in U\}.$$

Hence, the map

$$U \to (U \times g_0 \cdot U) \cap R, \quad q \mapsto (q, g_0 \cdot q),$$

is a parameterization of $O \cap R$, with $O \subset M \times M$ open. It follows that R can be covered by open sets $O \cap R$ embedded in $M \times M$, so R is an embedded submanifold. Also, the action being proper, the inclusion

$$R = \{(p, g \cdot p) : p \in M, g \in G\} \hookrightarrow M \times M$$

is a proper map.

Finally, we observe that the projection $p_1 : M \times M \to M$ restricted to R is an inverse to the parameterizations of R constructed above, hence $p_1|_R$ is a local diffeomorphism. $\qquad\square$

Under the conditions of this corollary, it is easy to check that the projection $\pi : M \to G\backslash M$ is in fact a covering map. Therefore, if M is simply connected, then M is the universal covering space of $G\backslash M$ and we conclude that $\pi_1(G\backslash M) \simeq G$.

Example 10.4. The action $\mathbb{Z}_2 \times \mathbb{S}^d \to \mathbb{S}^d$ defined in Example 10.1 is free and proper so the orbit space $\mathbb{Z}_2 \backslash \mathbb{S}^d$ is a manifold. We claim that this manifold is diffeomorphic to $\mathbb{RP}^d$: the map $\mathbb{S}^d \to \mathbb{RP}^d$ given by $(x^1, \ldots, x^d) \mapsto [x^1 : \cdots : x^d]$ induces a diffeomorphism $\mathbb{Z}_2 \backslash \mathbb{S}^d \to \mathbb{RP}^d$ such that the following diagram commutes:

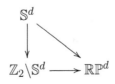

For $d > 1$, $\mathbb{S}^d$ is simply connected, so we conclude also that the quotient map is a covering map and that $\pi_1(\mathbb{RP}^d) = \mathbb{Z}_2$.

Example 10.5. The action $\mathbb{Z}^d \times \mathbb{R}^d \to \mathbb{R}^d$ defined in Example 10.2 is also free and proper, so the orbit space $\mathbb{Z}^d \backslash \mathbb{R}^d$ is a smooth manifold. This manifold is diffeomorphic to d-torus $\mathbb{T}^d$: the map $\mathbb{R}^d \to \mathbb{T}^d$ given by $(x^1, \ldots, x^d) \to (e^{2\pi i x^1}, \ldots, e^{2\pi i x^1})$ induces a diffeomorphism $\mathbb{Z}^d \backslash \mathbb{R}^d \to \mathbb{T}^d$ such that we have a commutative diagram

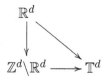

Since $\mathbb{R}^d$ is simply connected, we conclude also that the quotient map is a covering map and that $\pi_1(\mathbb{T}^d) = \mathbb{Z}^n$.

Example 10.6. Let $(\mathbb{R}, +)$ act on $\mathbb{R}^2$ by translations in the x-direction defined by

$$\lambda \cdot (x^1, x^2) = (x^1 + \lambda, x^2).$$

This is a free but non-proper action of a discrete group. However, the orbits of this action form a simple foliation of $\mathbb{R}^2$ so that $\mathbb{R} \backslash \mathbb{R}^2$ inherits a smooth structure. The quotient $\mathbb{R} \backslash \mathbb{R}^2$ is diffeomorphic to $\mathbb{R}$.

The issue in the last example is that one should consider on the group $(\mathbb{R}, +)$ the usual topology, instead of the discrete topology. Later we will study *Lie groups*, which are groups carrying a compatible smooth structure (of positive dimension). Quotients for Lie group actions give rise to many other examples of manifolds.

Exercises

Exercise 10.1
Let X be a Hausdorff topological space and $\sim$ an equivalence relation in X such that $\pi : X \to X/\sim$ is an open map, for the quotient topology. Show that $X/\sim$ with the quotient topology is Hausdorff if and only if the graph of $\sim$ is closed in $X \times X$.

Exercise 10.2

Let $\pi : M \to Q$ be a surjective submersion, $\Phi : M \to N$ and $\Psi : Q \to N$ any maps into a smooth manifold N such that the following diagram commutes:

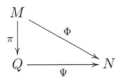

Show that Φ is smooth if and only if Ψ is smooth. Use this to conclude that if M is a manifold, $\sim$ is an equivalence relation satisfying any of the conditions of Theorem 10.1, and $\Phi : M \to N$ is a smooth map such that $\Phi(x) = \Phi(y)$ whenever $x \sim y$, then there is an induced smooth map $\overline{\Phi} : M/\sim \to N$ such that the following diagram commutes

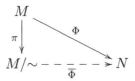

Exercise 10.3

Use Godement's Criterion to prove Corollary 10.1 characterizing simple foliations.

Exercise 10.4

Let $\mathcal{F}$ be the foliation of $M = \mathbb{R}^2 \setminus \{0\}$ whose leaves are the connected components of the horizontal lines $y = \text{const}$. Show that the leaf space $M/\mathcal{F}$ has a non-Hausdorff smooth structure (this non-Hausdorff manifold is sometimes called the **line with two origins**).

Exercise 10.5

Show that any smooth action $G \times M \to M$ of a finite group G on a manifold M is proper.

Exercise 10.6

A smooth action $\Psi : G \times M \to M$ of a discrete group G is said to be **properly discontinuous** if the following two conditions are satisfied:

(a) For every $p \in M$, there exists a neighborhood U of p, such that

$$g \cdot U \cap U = \emptyset, \quad \forall g \in G \setminus G_p.$$

(b) If $p, q \in M$ do not belong to the same orbit, then there are open neighborhoods U of p and V of q, such that

$$g \cdot U \cap V = \emptyset, \quad \forall g \in G.$$

Show that a free action of a discrete group is proper if and only if it is properly discontinuous.

Exercise 10.7

Show that for a proper and free action of a discrete group $G \times M \to M$, the projection $\pi : M \to G \backslash M$ is a covering map.

PART 2
Lie Theory

Lecture 11

Vector Fields and Flows

Definition 11.1. A **vector field** on a manifold M is a section of the tangent bundle $\pi : TM \to M$, i.e., a map $X : M \to TM$ such that $\pi \circ X = \mathrm{Id}$. We say that the vector field X is **smooth** or C^∞, if the map $X : M \to TM$ is smooth. We will denote by $\mathfrak{X}(M)$ the set of smooth vector fields on a manifold M.

If X is a vector field on M, we denote by X_p, rather than $X(p)$, the value of X at $p \in M$. For each $p \in M$, X_p is a derivation, hence, given any $f \in C^\infty(M)$ we can define a new function $X(f) : M \to \mathbb{R}$ by setting

$$X(f)(p) := X_p(f).$$

Recalling the definition of the differential of a function, ones sees immediately that this definition is equivalent to

$$X(f) := \mathrm{d}f(X).$$

Also, from Definition 5.1 of a tangent vector, we see that $f \mapsto X(f)$ satisfies for any $a, b \in \mathbb{R}$ and smooth functions f, g:

(a) $X(af + bg) = aX(f) + bX(g)$;
(b) $X(fg) = X(f)g + fX(g)$.

Fix a chart (U, x^i) for M. Then the vector fields $\frac{\partial}{\partial x^i} \in \mathfrak{X}(U)$, $i = 1, \ldots, d$, defined by

$$p \mapsto \left. \frac{\partial}{\partial x^i} \right|_p ,$$

yield a basis for T_pM at each $p \in U$. Therefore, if $X \in \mathfrak{X}(M)$ is any vector field on M, its restriction to the open set U, denoted by $X|_U$, can be written in the form

$$X|_U = \sum_{i=1}^{d} X^i \frac{\partial}{\partial x^i},$$

for unique functions $X^i : U \to \mathbb{R}$, $i = 1, \ldots, d$. The functions X^i are called the **components of the vector field** X with respect to the chart (U, x^i). We have the following equivalent characterizations of smooth vector fields.

Lemma 11.1. *If X is a vector field on M, the following statements are equivalent:*

(i) *The vector field X is C^∞.*
(ii) *The components X^i of X with respect to any chart (U, x^i) are C^∞.*
(iii) *For any $f \in C^\infty(M)$, the function $X(f)$ is C^∞.*

Proof. We show that (i) $\Rightarrow$ (ii) $\Rightarrow$ (iii) $\Rightarrow$ (i).

To show that (i) $\Rightarrow$ (ii), note that if X is C^∞ and U is an open set, the restriction $X|_U$ is also C^∞. Hence, if (U, x^i) is any chart, we have that $dx^i(X|_U) := dx^i \circ X|_U$ is C^∞. But

$$dx^i(X|_U) = dx^i \left(\sum_{j=1}^{d} X^j \frac{\partial}{\partial x^j} \right) = X^i.$$

To show that (ii) $\Rightarrow$ (iii), note that $f \in C^\infty(M)$ if and only if $f|_U \in C^\infty(U)$, for every domain U of a chart. But

$$X(f)|_U = \sum_{i=1}^{d} X^i \frac{\partial f}{\partial x^i} \in C^\infty(U).$$

To show that (iii) $\Rightarrow$ (i), it is enough to show that $X|_U$ is C^∞, for every domain U of a chart. Recall that, if $(U, x^1, \ldots, x^d)$ is a chart for M, then

$$\left(\pi^{-1}(U), x^1 \circ \pi, \ldots, x^d \circ \pi, dx^1, \ldots, dx^d \right)$$

is a chart for TM. Since

$$x^i \circ \pi \circ X|_U = x^i \in C^\infty(U), \quad dx^i \circ X|_U = X(x^i) \in C^\infty(U),$$

we conclude that $X|_U$ is C^∞. □

The previous lemma shows that a smooth vector field $X \in \mathfrak{X}(M)$ defines a map $D_X : C^\infty(M) \to C^\infty(M)$, $f \mapsto X(f)$, satisfying properties (a) and (b) above. We call D_X is a **linear derivation** of the algebra $C^\infty(M)$. The converse is also true.

Lemma 11.2. *Every linear derivation* $D : C^\infty(M) \to C^\infty(M)$ *determines a smooth vector field* $X \in \mathfrak{X}(M)$ *through the formula*

$$X_p(f) := D(f)(p).$$

Proof. Recalling Definition 5.1, we only need to show that $X_p(f)$ only depends on the germ $[f] \in \mathcal{G}_p$, i.e., if $f, g \in C^\infty(M)$ are two functions which agree in some neighborhood U of p, then $D(f)(p) = D(g)(p)$. This follows from the fact that derivations are local: if D is a derivation and $f \in C^\infty(M)$ is zero on some open set $U \subset M$, then $D(f)$ is also zero in U. To see this, let $p \in U$ and choose $g \in C^\infty(M)$ such that $g(p) > 0$ and $\operatorname{supp} g \subset U$. Since $gf := 0$, we have that

$$0 = D(gf) = D(g)f + gD(f).$$

If we evaluate both sides at p, we obtain $D(f)(p) = 0$. Hence, $D(f)|_U = 0$ as claimed. □

From now on, we will not distinguish between a vector field and the associated derivation of $C^\infty(M)$, so we will use the same letter to denote them.

Recall that a **path** in a manifold M is a continuous map $\gamma : I \to M$, where $I \subset \mathbb{R}$ is an interval. A **smooth path** is a path for which γ is C^∞. Note that if $\partial I \neq \emptyset$, i.e., is not an open interval, then γ is smooth if and only if it has a smooth extension to a smooth path defined in an open interval $J \supset I$. If $\gamma : I \to M$ is a smooth path, its **derivative** is

$$\frac{d\gamma}{dt}(t) := d\gamma \cdot \left.\frac{\partial}{\partial t}\right|_t \in T_{\gamma(t)}M, \quad (t \in I).$$

We often abbreviate writing $\dot{\gamma}(t)$ instead of $\frac{d\gamma}{dt}(t)$. The derivative $t \mapsto \dot{\gamma}(t)$ is a smooth path in the manifold TM.

Definition 11.2. Let $X \in \mathfrak{X}(M)$ be a vector field. A smooth path $\gamma : I \to M$ is called an **integral curve** of X if

$$\dot{\gamma}(t) = X_{\gamma(t)}, \quad \forall t \in I. \tag{11.1}$$

In a chart (U, x^i) a path $\gamma(t)$ is determined by its components $\gamma^i(t) = x^i(\gamma(t))$. Its derivative is then given by

$$\dot{\gamma}(t) = d\gamma \cdot \frac{\partial}{\partial t} = \sum_{i=1}^{d} \frac{d\gamma^i}{dt} \frac{\partial}{\partial x^i}.$$

It follows that the integral curves of a vector field X with components X^i relative to a chart (U, x^i) are the solutions of the system of o.d.e.'s

$$\frac{d\gamma^i}{dt} = X^i(\gamma^1(t), \dots, \gamma^d(t)), \quad (i = 1, \dots, d). \tag{11.2}$$

This system is the local form of the equation (11.1). Note that it is common to write $x^i(t)$ for the components $\gamma^i(t) = x^i(\gamma(t))$ so that this system of equations becomes

$$\frac{dx^i}{dt} = X^i(x^1(t), \dots, x^d(t)), \quad (i = 1, \dots, d).$$

Example 11.1. In $\mathbb{R}^2$ consider the vector field $X = x\frac{\partial}{\partial y} - y\frac{\partial}{\partial x}$. The equations for the integral curves (11.2) are

$$\begin{cases} \dot{x}(t) = -y(t), \\ \dot{y}(t) = x(t). \end{cases}$$

Hence, the curves $\gamma(t) = (R\cos t, R\sin t)$ are integral curves of this vector field.

This vector field is tangent to the submanifold $\mathbb{S}^1 = \{(x, y) : x^2 + y^2 = 1\}$, so defines a vector field on the circle: $Y = X|_{\mathbb{S}^1}$. If we consider the angle coordinate θ on the circle, the smooth functions $C^\infty(\mathbb{S}^1)$ can be identified with the 2π-periodic smooth functions $f(\theta) = f(\theta + 2\pi)$. It is easy to see that the vector field Y, thought of as a derivation, is given by

$$Y(f)(\theta) = f'(\theta).$$

Hence, we will write this vector field as

$$Y = \frac{\partial}{\partial \theta},$$

although the function θ is not a globally defined smooth coordinate on $\mathbb{S}^1$.

On the cylinder $M = \mathbb{S}^1 \times \mathbb{R}$, with coordinates (θ, x), consider the vector field

$$Z := \frac{\partial}{\partial \theta} + x \frac{\partial}{\partial x}.$$

The integral curve of Z through a point (θ_0, x_0) is given by

$$\gamma(t) = (\theta_0 + t, x_0 e^t).$$

If $x_0 = 0$, this is just a circle around the cylinder. If $x_0 \neq 0$ this is a spiral that approaches the circle when $t \to -\infty$ and goes to infinity when $t \to +\infty$.

Standard results about existence, uniqueness, and maximal interval of definition of solutions a system of o.d.e.'s lead to the following proposition.

Proposition 11.1. *Let $X \in \mathfrak{X}(M)$ be a vector field. For each $p \in M$, there exist real numbers $a_p, b_p \in \mathbb{R} \cup \{\pm\infty\}$ and a smooth path $\gamma_p :]a_p, b_p[\to M$, such that*

(i) *$0 \in]a_p, b_p[$ and $\gamma_p(0) = p$;*
(ii) *γ_p is an integral curve of X;*
(iii) *If $\eta :]c, d[\to M$ is any integral curve of X with $\eta(0) = p$, then $]c, d[\subset]a_p, b_p[$ and $\gamma_p|_{]c,d[} = \eta$.*

We call the integral curve γ_p given by this proposition the **maximal integral curve** of X through p. For each $t \in \mathbb{R}$, we define the domain $D_t(X)$ consisting of those points for which the integral curve through p exists at least until time t

$$D_t(X) := \{p \in M : t \in]a_p, b_p[\}.$$

If it is clear the vector field we are referring to we will write D_t instead of $D_t(X)$. The **flow of the vector field** $X \in \mathfrak{X}(M)$ is the map $\phi_X^t : D_t \to M$ given by

$$\phi_X^t(p) := \gamma_p(t).$$

The next result gives the basic properties of the flow of a vector field. The proof is left as an exercise.

Proposition 11.2. *Let $X \in \mathfrak{X}(M)$ be a vector field with flow ϕ_X^t. Then,*

(i) *For each $p \in M$, there exists a neighborhood U of p and $\varepsilon > 0$, such that the map $(-\varepsilon, \varepsilon) \times U \to M$, $(t, q) \mapsto \phi_X^t(q)$, is well defined and smooth.*

(ii) *For each $t \in \mathbb{R}$, D_t is open and $\bigcup_{t>0} D_t = M$.*

(iii) *For each $t \in \mathbb{R}$, $\phi_X^t : D_t \to D_{-t}$ is a diffeomorphism and*

$$(\phi_X^t)^{-1} = \phi_X^{-t};$$

(iv) *For each $s, t \in \mathbb{R}$, the domain of $\phi_X^t \circ \phi_X^s$ is contained in D_{t+s} and*

$$\phi_X^{t+s} = \phi_X^t \circ \phi_X^s.$$

One calls a vector field X **complete** if $D_t(X) = M$, for every $t \in \mathbb{R}$, i.e., if the maximal integral curve through any $p \in M$ is defined for all $t \in]-\infty, +\infty[$. In this case, the flow of X is a map:

$$\mathbb{R} \times M \to M, \quad (t, p) \mapsto \phi_X^t(p).$$

The properties above then say that this map defines an action of the group $(\mathbb{R}, +)$ on M. In other words, the map

$$\mathbb{R} \to \mathrm{Diff}(M), \quad t \mapsto \phi_X^t,$$

is a group homomorphism from $(\mathbb{R}, +)$ to the group $(\mathrm{Diff}(M), \circ)$ of diffeomorphisms of M. One often says that ϕ_X^t is a 1-*parameter group of transformations* of M. In the non-complete case, one also says that ϕ_X^t is a 1-parameter group of *local* transformations of M.

Example 11.2. The vector field $X = x\frac{\partial}{\partial y} - y\frac{\partial}{\partial x}$ in $\mathbb{R}^2$ is complete (see Example 11.1) and is flow is given by

$$\phi_X^t(x, y) = (x \cos t - y \sin t, x \sin t + y \cos t).$$

Example 11.3. The vector field $Y = -x^2\frac{\partial}{\partial x} - y\frac{\partial}{\partial y}$ in $\mathbb{R}^2$ is not complete: the integral curve through a point (x_0, y_0) is the solution to the system of o.d.e.'s

$$\begin{cases} \dot{x}(t) = -x^2, & x(0) = x_0, \\ \dot{y}(t) = -y, & y(0) = y_0. \end{cases}$$

After solving this system, ones obtains the flow of Y

$$\phi_X^t(x, y) = \left(\frac{x}{xt + 1}, ye^{-t} \right).$$

It follows that the flow through points $(0, y)$ exist for all t. But for points (x, y), with $x \neq 0$, the flow exists only for $t \in]-1/x, +\infty[$ if $x > 0$ and for $t \in]-\infty, -1/x[$ if $x > 0$. The domain of the flow is then given by

$$D_t(Y) = \begin{cases} \{(x,y) \in \mathbb{R}^2 : x > -1/t\} & \text{if } t > 0, \\ \mathbb{R}, & \text{if } t = 0, \\ \{(x,y) \in \mathbb{R}^2 : x < -1/t\} & \text{if } t < 0. \end{cases}$$

Let $\Phi : M \to N$ be a smooth map. In general, given a vector field X in M, it is not possible to use Φ to map X to obtain a vector field Y in N. However, given *a priori* two vector fields, one in M and one in N, it makes sense to ask if they are related by a map.

Definition 11.3. Let $\Phi : M \to N$ be a smooth map. A vector field $X \in \mathfrak{X}(M)$ is said to be **Φ-related** to a vector field $Y \in \mathfrak{X}(N)$ if

$$Y_{\Phi(p)} = d\Phi(X_p), \quad \forall p \in M.$$

If X and Y are Φ-related vector fields then, as derivations of $C^\infty(M)$,

$$Y(f) \circ \Phi = X(f \circ \Phi), \quad \forall f \in C^\infty(N).$$

When Y is determined from X via Φ we write $Y = \Phi_*(X)$, and call $\Phi_*(X)$ the **push forward** of X by Φ. This is the case, for example, when Φ is a diffeomorphism, in which case

$$\Phi_*(X)(f) = X(f \circ \Phi) \circ \Phi^{-1}, \quad \forall f \in C^\infty(N).$$

The integral curves of vector fields which are Φ-related are also Φ-related. The proof is a simple exercise applying the chain rule and is left as an exercise.

Proposition 11.3. *Let $\Phi : M \to N$ be a smooth map and let $X \in \mathfrak{X}(M)$ and $Y \in \mathfrak{X}(N)$ be Φ-related vector fields. If $\gamma : I \to M$ is an integral curve of X, then $\Phi \circ \gamma : I \to N$ is an integral curve of Y. In particular, $\Phi(D_t(X)) \subset D_t(Y)$ and the flows of X and Y yield a commutative diagram*

$$
\begin{array}{ccc}
D_t(X) & \xrightarrow{\;\Phi\;} & D_t(Y) \\
{\scriptstyle \phi_X^t}\big\downarrow & & \big\downarrow{\scriptstyle \phi_Y^t} \\
D_{-t}(X) & \xrightarrow{\;\Phi\;} & D_{-t}(Y)
\end{array}
$$

If $X \in \mathfrak{X}(M)$ is a vector field and $f \in C^\infty(M)$ we have $X(f) \in C^\infty(M)$. The expression for $X(f)$ in local coordinates shows that X is a first-order differential operator. If we iterate, we obtain the powers X^k, which are the kth-order differential operators defined by

$$ X^{k+1}(f) := X(X^k(f)). $$

Proposition 11.4 (Taylor Formula). *Let $X \in \mathfrak{X}(M)$ be a vector field and let $f \in C^\infty(M)$. For each positive integer k, one has the expansion*

$$ f \circ \phi_X^t = f + tX(f) + \frac{t^2}{2!}X^2(f) + \cdots + \frac{t^k}{k!}X^k(f) + O(t^{k+1}), $$

where for each $p \in M$, $t \mapsto O(t^{k+1})(p)$ denotes a real smooth function defined in a neighborhood of $t = 0$ whose derivatives of order $\leq k$ all vanish at $t = 0$.

Proof. By the usual Taylor formula for real functions applied to $t \mapsto f(\phi_X^t(p))$, it is enough to show that

$$ \left. \frac{\mathrm{d}^k}{\mathrm{d}t^k} f(\phi_X^t(p)) \right|_{t=0} = X^k(f)(p). $$

To prove this, we show by induction that

$$ \frac{\mathrm{d}^k}{\mathrm{d}t^k} f(\phi_X^t(p)) = X^k(f)(\phi_X^t(p)). $$

When $k = 1$, this follows because

$$\frac{d}{dt} f(\phi_X^t(p)) = d_p f \cdot X_{\phi_X^t(p)} = X_{\phi_X^t(p)}(f) = X(f)(\phi_X^t(p)).$$

On the other hand, if we assume that the formula is valid for $k - 1$, we obtain

$$\frac{d^k}{dt^k} f(\phi_X^t(p)) = \frac{d}{dt}\left(\frac{d^{k-1}}{dt^{k-1}} f(\phi_X^t(p)) \right)$$

$$= \frac{d}{dt} X^{k-1}(f)(\phi_X^t(p))$$

$$= X(X^{k-1}(f))(\phi_X^t(p)) = X^k(f)(\phi_X^t(p)). \qquad \square$$

Another common notation for the flow of a vector field, which is justified by the previous result, is the exponential notation

$$\exp(tX) := \phi_X^t.$$

In this notation, the properties of the flow are written as

$$\exp(tX)^{-1} = \exp(-tX), \quad \exp((t+s)X) = \exp(tX) \circ \exp(sX),$$

while the Taylor expansion takes the suggestive form

$$f(\exp(tX)) = f + tX(f) + \frac{t^2}{2!} X^2(f) + \cdots + \frac{t^k}{k!} X^k(f) + O(t^{k+1}).$$

We will not use this notation in these Lectures.

If $X \in \mathfrak{X}(M)$ is a vector field, a point $p \in M$ is called a **singular point** or a **fixed point** of X if $X_p = 0$. It should be obvious that the integral curve through a singular point of X is the constant path $\phi_X^t(p) := p$, for all $t \in \mathbb{R}$. On the other hand, for non-singular points we have a unique local canonical form X.

Theorem 11.1 (Flow Box Theorem). *Let $X \in \mathfrak{X}(M)$ be a vector field and $p \in M$ a non-singular point: $X_p \neq 0$. There are local coordinates (U, x^i) centered at p, such that*

$$X|_U = \frac{\partial}{\partial x^1}.$$

Proof. First we choose a chart $(V, \psi) = (V, y^i)$, centered at p, such that

$$X|_p = \left.\frac{\partial}{\partial y^1}\right|_p.$$

The map $\sigma : \mathbb{R}^d \to M$ given by

$$\sigma(t_1, \ldots, t_d) = \phi_X^{t_1}(\psi^{-1}(0, t_2, \ldots, t_d)),$$

is well defined and C^∞ in a neighborhood of the origin. Its differential at the origin is given by

$$\mathrm{d}_0\sigma \cdot \left.\frac{\partial}{\partial t_1}\right|_0 = \left.\frac{\mathrm{d}}{\mathrm{d}t_1}\phi_X^{t_1}(\psi^{-1}(0,0,\ldots,0))\right|_{t_1=0} = X_p = \left.\frac{\partial}{\partial y^1}\right|_p,$$

$$\mathrm{d}_0\sigma \cdot \left.\frac{\partial}{\partial t_i}\right|_0 = \left.\frac{\partial}{\partial t_i}\psi^{-1}(0, t_2, \ldots, t_d))\right|_0 = \left.\frac{\partial}{\partial y^i}\right|_p.$$

We conclude that σ is a local diffeomorphism in a neighborhood of the origin. Hence, there exists an open set U containing p such that $\phi = \sigma^{-1} : U \to \mathbb{R}^d$ is a chart. If we write $(U, \phi) = (U, (x^i))$, we have

$$\left.\frac{\partial}{\partial x^1}\right|_{\sigma(t_1,\ldots,t_d)} = \left.\mathrm{d}\sigma \cdot \frac{\partial}{\partial t_1}\right|_{(t_1,\ldots,t_d)} = \left.\frac{\mathrm{d}}{\mathrm{d}t}\phi_X^t(\psi^{-1}(0, t_2, \ldots, t_d))\right|_{t=t_1}$$

$$= X(\phi_X^{t_1}(\psi^{-1}(0, t_2, \ldots, t_d))) = X_{\sigma(t_1,\ldots,t_d)}. \qquad \square$$

Exercises

Exercise 11.1

Let M be a connected manifold. Show that for any pair of points $p, q \in M$, with $p \neq q$, there exists a smooth path $\gamma : [0, 1] \to M$ such that

(a) $\gamma(0) = p$ and $\gamma(1) = q$;
(b) $\frac{\mathrm{d}\gamma}{\mathrm{d}t}(t) \neq 0$, for every $t \in [0, 1]$;
(c) γ is simple (i.e., γ is injective).

Use this to prove that any connected manifold of dimension 1 is diffeomorphic to either $\mathbb{R}$ or $\mathbb{S}^1$.

Exercise 11.2

Let $X \in \mathfrak{X}(M)$ be a vector field and $f \in C^\infty(M)$ a nowhere vanishing function. Find the relationship between the integral curves of X and of fX.

Exercise 11.3

Verify the properties of the flow of a vector field given by Proposition 11.2.

Exercise 11.4

Determine the flow of the vector field $X = y\frac{\partial}{\partial x} - x\frac{\partial}{\partial y} + \frac{\partial}{\partial z}$ in $\mathbb{R}^3$.

Exercise 11.5

Give an example of an embedded submanifold $N \subset \mathbb{R}^2$ and a vector field $X \in \mathfrak{X}(N)$ which is not the restriction of a vector field $\tilde{X} \in \mathfrak{X}(\mathbb{R}^2)$.

Exercise 11.6

Give an example of a manifold M and two vector fields X_1 and X_2 which are complete but for which their sum $X_1 + X_2$ is not complete. On the other hand, show that if M is compact then every vector field $X \in \mathfrak{X}(M)$ is complete.

Hint: Show that if $K \subset M$ is a compact set then there exists $a > 0$ such that for every $x \in K$ the maximal integral curve through x exists for $t \in [-a, a]$.

Exercise 11.7

Let $A \subset M$. Call a map $X : A \to TM$ a *vector field along A* if $X_p \in T_pM$ for all $p \in A$. Show that if $A \subset O \subset M$, with A closed and O open, then every smooth vector field X along A can be extended to a smooth vector field in M such that $X_p = 0$ for $p \notin O$.

Exercise 11.8

Let $X \in \mathfrak{X}(M)$ be a vector field without singular points. Show that the integral curves of X form a foliation $\mathcal{F}$ of M of dimension 1. Conversely, show that locally the leaves of a foliation of dimension 1 are the orbits of a vector field. What about globally?

Exercise 11.9

A **Riemannian structure** on a manifold M is a smooth choice of an inner product $\langle \, , \, \rangle_p$ in each tangent space T_pM. Here by smooth we mean that for any vector fields $X, Y \in \mathfrak{X}(M)$, the function $p \mapsto \langle X(p), Y(p) \rangle_p$ is C^∞. Show that every smooth manifold admits a Riemannian structure M.

Exercise 11.10

Let $\langle\,,\,\rangle$ be a Riemannian structure on a manifold M. Given a function $f \in C^\infty(M)$ show that there exists a unique vector field $\mathrm{grad}(f) \in \mathfrak{X}(M)$ such that

$$X(f) = \langle X, \mathrm{grad}(f) \rangle, \quad \forall X \in \mathfrak{X}(M).$$

One calls $\mathrm{grad}(f)$ the **gradient** of f. Verify that

(i) If $M = \mathbb{R}^d$ and $\langle\,,\,\rangle$ is the usual Riemannian structure defined by

$$\left\langle \sum_{i=1}^d X^i \frac{\partial}{\partial x^i}, \sum_{i=1}^d Y^i \frac{\partial}{\partial x^i} \right\rangle := \sum_{i=1}^d X^i Y^i,$$

show that this yields the usual definition where $\mathrm{grad}(f) = \sum_{i=1}^d \frac{\partial f}{\partial x^i} \frac{\partial}{\partial x^i}$;

(ii) $p \in M$ is a singular point of $\mathrm{grad}(f)$ if and only if it is a singular point of f;

(iii) If $x \in \mathbb{R}$ is a regular value of f the integral curves of $\mathrm{grad}(f)$ are orthogonal to the level set $f^{-1}(x)$, i.e., any integral curve γ of $\mathrm{grad}(f)$ with $\gamma(t_0) \in f^{-1}(x)$, satisfies

$$\langle \dot{\gamma}(t_0), v \rangle = 0, \quad \forall v \in T_{\gamma(0)} f^{-1}(x).$$

Lecture 12

Lie Bracket and Lie Derivative

Definition 12.1. Let $X, Y \in \mathfrak{X}(M)$ be smooth vector fields. The **Lie bracket** of X and Y is the vector field $[X, Y] \in \mathfrak{X}(M)$ given by

$$[X, Y](f) := X(Y(f)) - Y(X(f)), \quad \forall f \in C^{\infty}(M).$$

Note that the formula for the Lie bracket $[X, Y]$ shows that it is a differential operator of order ≤ 2. A simple computation shows that $[X, Y]$ is a linear derivation of $C^{\infty}(M)$, i.e., that

$$[X, Y](fg) = [X, Y](f)g + f[X, Y](g), \quad \forall f, g \in C^{\infty}(M).$$

In order words, the terms of the second order cancel each other and we have in fact that $[X, Y] \in \mathfrak{X}(M)$.

In a local chart one can compute the Lie bracket in a straightforward way if we think of vector fields as differential operators.

Example 12.1. Let $M = \mathbb{R}^3$ with coordinates (x, y, z), and consider the vector fields

$$X = z\frac{\partial}{\partial y} - y\frac{\partial}{\partial z}, \quad Y = x\frac{\partial}{\partial z} - z\frac{\partial}{\partial x}, \quad Z = y\frac{\partial}{\partial x} - x\frac{\partial}{\partial y}.$$

Then we compute

$$
\begin{aligned}
[X, Y] &= \left(z\frac{\partial}{\partial y} - y\frac{\partial}{\partial z} \right) \left(x\frac{\partial}{\partial z} - z\frac{\partial}{\partial x} \right) \\
&\quad - \left(x\frac{\partial}{\partial z} - z\frac{\partial}{\partial x} \right) \left(z\frac{\partial}{\partial y} - y\frac{\partial}{\partial z} \right) \\
&= y\frac{\partial}{\partial x} - x\frac{\partial}{\partial y} = Z.
\end{aligned}
$$

We leave it as an exercise the computation of the other Lie brackets. The result is:

$$[Y, Z] = X, \quad [Z, X] = Y.$$

Our next result shows that the Lie bracket $[X, Y]$ measures the failure in the commutativity of the flows of X and Y.

Proposition 12.1. *Let* $X, Y \in \mathfrak{X}(M)$ *be vector fields. For each* $p \in M$, *the commutator*

$$\gamma_p(\varepsilon) := \phi_Y^{-\sqrt{\varepsilon}} \circ \phi_X^{-\sqrt{\varepsilon}} \circ \phi_Y^{\sqrt{\varepsilon}} \circ \phi_X^{\sqrt{\varepsilon}}(p)$$

is well defined for a small enough $\varepsilon \geq 0$, *and we have*

$$[X, Y]_p = \frac{\mathrm{d}}{\mathrm{d}\varepsilon} \gamma_p(\varepsilon) \bigg|_{\varepsilon=0+}.$$

Proof. One can work on a local chart $(U, x^1, \ldots, x^d)$, centered at p. Writing

$$X = \sum_{i=1}^{d} X^i \frac{\partial}{\partial x^i}, \quad Y = \sum_{i=1}^{d} Y^i \frac{\partial}{\partial x^i},$$

the Lie bracket of X and Y is

$$[X, Y](x^i) = X(Y^i) - Y(X^i).$$

Consider the points p_1, p_2, and p_3 defined by (see Figure 12.1):

$$p_1 = \phi_X^{\sqrt{\varepsilon}}(p), \quad p_2 = \phi_Y^{\sqrt{\varepsilon}}(p_1), \quad p_3 = \phi_X^{-\sqrt{\varepsilon}}(p_2),$$

Then $\gamma_p(\varepsilon) = \phi_Y^{-\sqrt{\varepsilon}}(p_3)$, and Taylor's formula (Proposition 11.4) applied to each coordinate x^i yields

$$x^i(p_1) = x^i(p) + \sqrt{\varepsilon} X^i(p) + \frac{1}{2} \varepsilon X^2(x^i)(p) + O(\varepsilon^{\frac{3}{2}}).$$

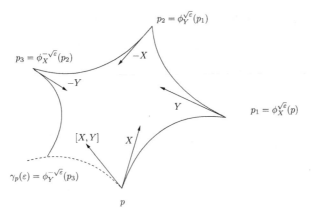

Fig. 12.1.   Flows and Lie bracket.

Similarly, one finds

$$x^i(p_2) = x^i(p_1) + \sqrt{\varepsilon}Y^i(p_1) + \frac{1}{2}\varepsilon Y^2(x^i)(p_1) + O(\varepsilon^{\frac{3}{2}})$$

$$= x^i(p) + \sqrt{\varepsilon}X^i(p) + \frac{1}{2}\varepsilon X^2(x^i)(p)$$

$$+ \sqrt{\varepsilon}Y^i(p_1) + \frac{1}{2}\varepsilon Y^2(x^i)(p_1) + O(\varepsilon^{\frac{3}{2}}).$$

The last two terms can also be estimated using again Taylor's formula as follows:

$$Y^i(p_1) = Y^i(\phi_X^{\sqrt{\varepsilon}}(p)) = Y^i(p) + \sqrt{\varepsilon}X(Y^i)(p) + O(\varepsilon)$$
$$Y^2(x^i)(p_1) = Y^2(x^i)(\phi_X^{\sqrt{\varepsilon}}(p)) = Y^2(x^i)(p) + \sqrt{\varepsilon}X(Y^2(x^i))(p) + O(\varepsilon).$$

Hence, we obtain

$$x^i(p_2) = x^i(p) + \sqrt{\varepsilon}(Y^i(p) + X^i(p))$$
$$+ \varepsilon\left(\frac{1}{2}Y^2(x^i)(p) + X(Y^i)(p) + \frac{1}{2}X^2(x^i)(p)\right) + O(\varepsilon^{\frac{3}{2}}).$$

Proceeding in a similar fashion, we can estimate $x^i(p_3)$ and $x^i(\gamma_p(\varepsilon))$, obtaining

$$x^i(p_3) = x^i(p_2) - \sqrt{\varepsilon}X^i(p_2) + \frac{1}{2}\varepsilon X^2(x^i)(p_2) + O(\varepsilon^{\frac{3}{2}})$$

$$= x^i(p) + \sqrt{\varepsilon}Y^i(p)$$

$$+ \varepsilon\left(X(Y^i)(p) - Y(X^i)(p) + \frac{1}{2}Y^2(x^i)(p)\right) + O(\varepsilon^{\frac{3}{2}}),$$

$$x^i(\gamma_p(\varepsilon)) = x^i(p_3) - \sqrt{\varepsilon}Y^i(p_3) + \frac{1}{2}\varepsilon Y^2(x^i)(p_3) + O(\varepsilon^{\frac{3}{2}})$$

$$= x^i(p) + \varepsilon\left(X(Y^i)(p) - Y(X^i)(p)\right) + O(\varepsilon^{\frac{3}{2}}).$$

Therefore, we can compute the limit in the statement

$$\lim_{\varepsilon \to 0^+} \frac{x^i(\gamma_p(\varepsilon)) - x^i(p)}{\varepsilon} = X(Y^i)(p) - Y(X^i)(p) = [X, Y]_p(x^i).$$

$\square$

The following proposition gives the most basic properties of the Lie bracket of vector fields. The proof is elementary and is left as an exercise.

Proposition 12.2. *The Lie bracket satisfies the following properties:*

(i) *Skew-symmetry:* $[X, Y] = -[Y, X]$.
(ii) *Bi-linearity:* $[aX + bY, Z] = a[X, Z] + b[Y, Z]$, $\forall a, b \in \mathbb{R}$.
(iii) *Jacobi identity:* $[X, [Y, Z]] + [Y, [Z, X]] + [Z, [X, Y]] = 0$.
(iv) *Leibniz identity:* $[X, fY] = X(f)Y + f[X, Y]$, $\forall f \in C^\infty(M)$.

Moreover, if $\Phi : M \to N$ *is a smooth map,* $X, Y \in \mathfrak{X}(M)$ *are* Φ-*related with, respectively,* $Z, W \in \mathfrak{X}(N)$, *then* $[X, Y]$ *is* Φ-*related with* $[Z, W]$.

The geometric interpretation of the Lie bracket given by Proposition 12.1 shows that the Lie bracket and the flow of vector fields are intimately related. There is another form of this relationship which we now explain. For that, we need the following definition:

Definition 12.2. Let $X, Y \in \mathfrak{X}(M)$ be vector fields and $f \in C^\infty(M)$.

(i) The **Lie derivative of f along** X is the function $\mathcal{L}_X f \in C^\infty(M)$ given by

$$(\mathcal{L}_X f)(p) := \lim_{t \to 0} \frac{1}{t} \left(f(\phi_X^t(p)) - f(p) \right).$$

(ii) The **Lie derivative of** Y **along** X is the vector field $\mathcal{L}_X Y \in \mathfrak{X}(M)$ given by

$$(\mathcal{L}_X Y)_p := \lim_{t \to 0} \frac{1}{t} \left(d\phi_X^{-t} \cdot Y_{\phi_X^t(p)} - Y_p \right).$$

One can merge these two definitions by observing that a diffeomorphism $\Phi : M \to M$ acts on functions $C^\infty(M)$ by

$$(\Phi^* f)(p) = f(\Phi(p)),$$

and acts on vector fields $Y \in \mathfrak{X}(M)$ by

$$(\Phi^* Y)_p = d\Phi^{-1} \cdot Y_{\Phi(p)}.$$

Note that $\Phi^* Y = (\Phi^{-1})_* Y$, so the two operations are related by

$$\Phi^* Y(f) = Y((\Phi^{-1})^* f).$$

It follows that the Lie derivative of an object P (a function or a vector field) is given by

$$\mathcal{L}_X P = \frac{d}{dt}(\phi_X^t)^* P \bigg|_{t=0} = \lim_{t \to 0} \frac{1}{t} \left((\phi_X^t)^* P - P \right). \tag{12.1}$$

We will see later that one can take Lie derivatives of other objects using precisely this last formula as the definition. We can now give another geometric interpretation of the Lie bracket.

Theorem 12.1. *Let* $X \in \mathfrak{X}(M)$ *be a vector field. Then,*

(i) *For any function* $f \in C^\infty(M)$: $\mathcal{L}_X f = X(f) d.$
(ii) *For any vector field* $Y \in \mathfrak{X}(M)$: $\mathcal{L}_X Y = [X, Y].$

Proof. To prove (i), we simply observe that

$$\mathcal{L}_X f = \frac{d}{dt} f \circ \phi_X^t \bigg|_{t=0} = df \cdot X = X(f).$$

To prove (ii), we note first that

$$(\mathcal{L}_X Y)(f)(p) = \lim_{t \to 0} \frac{1}{t} \left(d\phi_X^{-t} \cdot Y_{\phi_X^t(p)} - Y_p \right)(f)$$

$$= \lim_{t \to 0} \frac{1}{t} \left(Y_{\phi_X^t(p)}(f \circ \phi_X^{-t}) - Y_p(f) \right).$$

On the other hand, Taylor's formula gives

$$f \circ \phi_X^{-t} = f - tX(f) + O(t^2).$$

Hence, using also (i), we find

$$(\mathcal{L}_X Y)(f)(p) = \lim_{t \to 0} \frac{1}{t} \left(Y_{\phi_X^t(p)}(f) - tY_{\phi_X^t(p)}(X(f)) - Y_p(f) \right)$$

$$= \lim_{t \to 0} \frac{1}{t} \left(Y_{\phi_X^t(p)}(f) - Y_p(f) \right) - Y_p(X(f))$$

$$= X_p(Y(f)) - Y_p(X(f)) = [X, Y](f)(p). \qquad \square$$

Exercises

Exercise 12.1

Check the properties of the Lie bracket given in Proposition 12.2.

Exercise 12.2

Complete the computation of the Lie brackets in Example 12.1 and show that all three vector fields X, Y, and Z are tangent to the sphere $\mathbb{S}^2 \subset \mathbb{R}^3$. Show that there are unique vector fields $\tilde{X}$, $\tilde{Y}$ and $\tilde{Z}$ on $\mathbb{RP}^2$ such that $\pi_* X = \tilde{X}$, $\pi_* Y = \tilde{Y}$, and $\pi_* Z = \tilde{Z}$ where $\pi : \mathbb{S}^2 \to \mathbb{RP}^2$ is the projection. What are the Lie brackets between $\tilde{X}$, $\tilde{Y}$, and $\tilde{Z}$?

Exercise 12.3

Find 3 everywhere linearly independent vector fields X, Y, and Z on the sphere $\mathbb{S}^3$ such that $[X, Y] = Z$, $[Y, Z] = X$, and $[Z, X] = Y$.

Hint: Recall that $\mathbb{S}^3$ can be identified with the unit quaternions.

Exercise 12.4

In $\mathbb{R}^2$, consider the vector fields $X = \frac{\partial}{\partial x}$ and $Y = x\frac{\partial}{\partial y}$. Compute the Lie bracket $[X, Y]$ in three distinct ways: (i) using the definition, (ii) using Proposition 12.1 and (iii) using Theorem 12.1.

Exercise 12.5

Let $\Phi : M \to N$ be a surjective submersion. We say that $X \in \mathfrak{X}(M)$ is **vertical** if $X_p \in \ker(d_p\Phi)$ for all $p \in M$, and that $Y \in \mathfrak{X}(M)$ is **projectable** if it is Φ-related to some $\tilde{Y} \in \mathfrak{X}(N)$. Show the following:

(i) The Lie bracket of vertical (respectively, projectable) vector fields is vertical (respectively, projectable).
(ii) If X is vertical and Y is projectable then $[X, Y]$ is vertical.
(iii) If Φ has connected fibers and $[X, Y]$ is vertical for any vertical vector field $X \in \mathfrak{X}(M)$, then Y is projectable.

Hint: Vertical vector fields are projectable!

Exercise 12.6

Let $X, Y \in \mathfrak{X}(M)$ be vector fields with flows ϕ_X^t and ϕ_Y^s. Show the following:

(i) $[X, Y] = 0$ if and only if $\phi_X^t \circ \phi_Y^s(p) = \phi_Y^s \circ \phi_X^t(p)$ for all $p \in M$ and all s and t sufficiently small (which may depend on p);
(ii) Given an example where $[X, Y] = 0$ and $\phi_X^t \circ \phi_Y^s \neq \phi_Y^s \circ \phi_X^t$ for some s and t.

Exercise 12.7

Let $X_1, \ldots, X_k \in \mathfrak{X}(M)$ be vector fields such that

(a) $\{X_1|_p, \ldots, X_k|_p\}$ are linearly independent, for all $p \in M$;
(b) $[X_i, X_j] = 0$, for all $i, j = 1, \ldots, k$.

Show that for each $p \in M$ there exists a neighborhood U of p and a unique k-dimensional foliation $\mathcal{F}$ of U such that

$$T_q L = \langle X_1|_q, \ldots, X_k|_q \rangle, \quad \forall q \in U,$$

where $L \in \mathcal{F}$ is the leaf containing q.

Hint: Use the previous exercise to show that the leaf L is obtained by flowing from q along the flows of the vector fields $X_1, \ldots, X_k$.

Lecture 13

Distributions and the Frobenius Theorem

A vector field $X \in \mathfrak{X}(M)$ which is *nowhere vanishing* determines a partition of M into 1-dimensional submanifolds

$$\mathcal{F} = \{\gamma(I) : \gamma : I \to M \text{ a maximal integral curve of } X\}.$$

By the Flow Box Theorem, this is a 1-dimensional foliation of M. Note that if $Y \in \mathfrak{X}(M)$ is another vector field such that $Y = fX$, for some *nowhere vanishing* smooth function $f \in C^\infty(M)$, then Y determines the same foliation of M. So, this foliation only depends on the family of 1-dimensional subspaces

$$M \ni p \mapsto \langle X_p \rangle \subset T_pM.$$

We will now generalize all this to higher dimensions.

Definition 13.1. Let M be a smooth manifold of dimension d and let $1 \le k \le d$ be an integer. A k-dimensional **distribution** D in M is a map

$$M \ni p \mapsto D_p \subset T_pM,$$

which associates to each $p \in M$ a subspace $D_p \subset T_pM$ of dimension k. We say that a distribution D is of class C^∞ if for each $p \in M$ there exists a neighborhood U of p and smooth vector fields $X_1, \ldots, X_k \in \mathfrak{X}(U)$, such that

$$D_q = \langle X_1|_q, \ldots, X_k|_q \rangle, \quad \forall q \in U.$$

If D is a distribution in M we denote the set of **vector fields tangent to** D by

$$\mathfrak{X}(D) := \{X \in \mathfrak{X}(M) : X_p \in D_p, \forall p \in M\}.$$

Note that $\mathfrak{X}(D)$ is a module over the ring $C^\infty(M)$: if $f \in C^\infty(M)$ and $X \in \mathfrak{X}(D)$ then $fX \in \mathfrak{X}(D)$.

Example 13.1. Every nowhere vanishing smooth vector field X defines a 1-dimensional smooth distribution by

$$D_p := \langle X_p \rangle = \{\lambda X_p : \lambda \in \mathbb{R}\}.$$

Note that $Y \in \mathfrak{X}(D)$ if and only $Y = fX$ for some uniquely defined smooth function $f \in C^\infty(M)$.

Example 13.2. A set of smooth vector fields $X_1, \ldots, X_k$ which at each $p \in M$ are linearly independent define a k-dimensional smooth distribution D by

$$D_p := \langle X_1|_p, \ldots, X_k|_p \rangle.$$

A vector field $X \in \mathfrak{X}(D)$ if and only if

$$X = f_1 X_1 + \cdots + f_k X_k,$$

for uniquely defined functions $f_i \in C^\infty(M)$.

For example, in $M = \mathbb{R}^3$, we have the 2-dimensional smooth distribution $D = \langle X_1, X_2 \rangle$ generated by the vector fields

$$X_1 = \frac{\partial}{\partial x} + z^2 \frac{\partial}{\partial y}, \quad X_2 = \frac{\partial}{\partial y} + z^2 \frac{\partial}{\partial z}.$$

and every vector field $X \in \mathfrak{X}(D)$ is a linear combination $aX_1 + bX_2$, where the smooth functions $a = a(x, y)$ and $b = b(x, y)$ are uniquely determined.

Example 13.3. More generally, a set of smooth vector fields $X_1, \ldots, X_s$ which at each $p \in M$ span a k-dimensional subspace define a k-dimensional smooth distribution D by setting

$$D_p := \langle X_1|_p, \ldots, X_s|_p \rangle.$$

We have that $X \in \mathfrak{X}(D)$ if and only if

$$X = f_1 X_1 + \cdots + f_s X_s,$$

for some smooth functions $f_i \in C^\infty(M)$. The difference from the previous example is that the functions f_i are not uniquely defined. Moreover, we may not be able to find k vector fields tangent to D which globally generate D.

For example, in $M = \mathbb{R}^3 \setminus \{0\}$ consider the vector fields X, Y, and Z defined in Example 12.1. The matrix whose columns are the components of the vector fields X, Y, and Z relative to the usual coordinates (x, y, z) of $\mathbb{R}^3$ is

$$\begin{pmatrix} 0 & -z & y \\ z & 0 & -x \\ -y & x & 0 \end{pmatrix}$$

and has rank 2 everywhere. Hence, we have the 2-dimensional distribution $D = \langle X, Y, Z \rangle$. We leave it as an exercise to check that this distribution is not globally generated by only 2 vector fields.

We can think of a distribution as a generalization of the notion of a vector field. In this sense, the concept of an integral curve of a vector field is replaced by the following:

Definition 13.2. Let D be a distribution in M. A connected submanifold (N, Φ) of M is called an **integral manifold** of D if

$$d_p\Phi(T_pN) = D_{\Phi(p)}, \quad \forall p \in N.$$

Note that if D is a k-dimensional distribution, its integral manifolds, if they exist, are k-dimensional manifolds.

Example 13.4. Consider the 2-distribution of $\mathbb{R}^3$ given in Example 13.2. The plane $N = \{z = 0\}$ is an integral manifold of this distribution, since it is a connected submanifold and

$$D_{(x,y,0)} = \left\langle \left.\frac{\partial}{\partial x}\right|_{(x,y,0)}, \left.\frac{\partial}{\partial y}\right|_{(x,y,0)} \right\rangle = T_{(x,y,0)}N.$$

Example 13.5. Consider the 2-distribution D of $\mathbb{R}^3 \setminus \{0\}$ defined by the vector fields X, Y, and Z in Example 13.3. The spheres

$$S_c = \{(x, y, z) \in \mathbb{R}^3 \setminus 0 : x^2 + y^2 + z^2 = c\},$$

are integral manifolds of D: each sphere is a connected submanifold, has dimension 2 and:

$$X_p, Y_p.Z_p \in T_p S_c, \quad \forall p \in S_c.$$

Since D has dimension 2, we have $T_p S_c = D_p$, for all $p \in S_c$.

As suggested by the last example, given a smooth k-dimensional foliation $\mathcal{F}$ of a manifold M, we associate to it a k-dimensional distribution d defined by

$$D_p := T_p L,$$

where $L \in \mathcal{F}$ denotes the leaf containing the point $p \in M$. Henceforth, we will denote this distribution by $T\mathcal{F}$ and we will write $T_p \mathcal{F}$ instead of $(T\mathcal{F})_p$. The existence of foliated charts shows that $T\mathcal{F}$ is a smooth distribution. A vector field is tangent to $T\mathcal{F}$ if and only if it is tangent to each leaf of the foliation.

Definition 13.3. A smooth distribution D in M is called **integrable** if there exists a foliation $\mathcal{F}$ in M such that $D = T\mathcal{F}$.

A distribution D in M may fail to be integrable. In fact, there may not even exist integral manifolds through each point of M. The following proposition gives a necessary condition for this to happen:

Proposition 13.1. *Let D be a smooth distribution in M. If there exists an integral manifold of D through $p \in M$, then for any $X, Y \in \mathfrak{X}(D)$ we must have that $[X, Y]_p \in D_p$.*

Proof. Let $X, Y \in \mathfrak{X}(D)$ and fix $p \in M$. Assume there exists an integral manifold (N, Φ) of D through p and choose $q \in N$, such that $\Phi(q) = p$. For any $q' \in N$, the map $\mathrm{d}_{q'}\Phi : T_{q'}N \to T_{\Phi(q')}M$ is injective and its image is $D_{\Phi(q')}$. By the local normal form for submanifolds, there exist smooth vector fields $\tilde{X}, \tilde{Y} \in \mathfrak{X}(N)$ which are Φ-related with X and Y, respectively. It follows from Proposition 12.2 that $[\tilde{X}, \tilde{Y}]$ is also Φ-related with $[X, Y]$ and we must have

$$[X, Y]_p = \mathrm{d}_{q_0}\Phi([\tilde{X}, \tilde{Y}]_q) \in \mathrm{d}_q\Phi(T_q N) = D_p.$$

$\square$

Example 13.6. For the smooth distribution $D = \langle X_1, X_2 \rangle$ of $\mathbb{R}^3$ given in Example 13.2, we saw that the plane $z = 0$ is an integral manifold. On the other hand, we find that

$$[X_1, X_2] = -2z^3 \frac{\partial}{\partial y}.$$

If $z \neq 0$ this vector field is not tangent to the distribution. Hence, the only points through which there exist integral manifolds are the points in the plane $z = 0$.

For an integrable distribution $D = T\mathcal{F}$ we have an integral manifold through every point. Hence, for any pair of vector fields $X, Y \in \mathfrak{X}(T\mathcal{F})$, we must have $[X, Y] \in \mathfrak{X}(T\mathcal{F})$.

Definition 13.4. A smooth distribution D in M is called **involutive** if for any $X, Y \in \mathfrak{X}(D)$ one has $[X, Y] \in \mathfrak{X}(D)$.

The following important result says that the lack of involutivity is the only obstruction to the integrability of a distribution.

Theorem 13.1 (Frobenius). *A smooth distribution D is integrable if and only if it is involutive. In this case, the integral foliation tangent to D is unique.*

Proof. Proposition 13.1 show that one of the implications hold. To check the other implication we assume that D is an involutive distribution.

We claim that, for each $p \in M$, there exist vector fields $X_1, \ldots, X_k \in \mathfrak{X}(U)$, defined in an open neighborhood U of p, such that

(a) $D|_U = \langle X_1, \ldots, X_k \rangle$;
(b) $[X_i, X_j] = 0$, for every $i, j = 1, \ldots, k$.

Then, by Exercise 12.7, we obtain an open cover $\{U_i\}_{i \in I}$ of M, such that for each $i \in I$ there exists a unique foliation $\mathcal{F}_i$ in U_i which satisfies $T\mathcal{F}_i = D|_{U_i}$. By uniqueness, whenever $U_i \cap U_j \neq 0$, we obtain $\mathcal{F}_i|_{U_i \cap U_j} = \mathcal{F}_j|_{U_i \cap U_j}$. Hence, there exists a unique foliation $\mathcal{F}$ of M such that $\mathcal{F}|_{U_i} = \mathcal{F}_i$.

To prove the claim, fix $p \in M$. Since D is smooth, there exist vector fields $Y_1, \ldots, Y_k$ defined in some neighborhood V of p, such that $D|_V = \langle Y_1, \ldots, Y_k \rangle$. We can also assume that V is the domain of some coordinate system $(x^1, \ldots, x^d)$ of M, so that

$$Y_i = \sum_{l=1}^{d} a_{il} \frac{\partial}{\partial x^l}, \quad (i = 1, \ldots, k),$$

where $a_{il} \in C^\infty(V)$. The matrix $A(q) = [a_{il}(q)]_{i,l=1}^{k,d}$ has rank k at p and we can assume, eventually after some relabeling of the coordinates, that the $k \times k$ minor formed by the first k rows and k columns of A has non-zero determinant in a smaller open neighborhood U of p. Let B be the $k \times k$ inverse matrix of this minor, and define vector fields $X_1, \ldots, X_k \in \mathfrak{X}(U)$ by

$$X_i := \sum_{j,l=1}^{k,d} b_{ij} a_{jl} \frac{\partial}{\partial x^l} = \frac{\partial}{\partial x^i} + \sum_{l=k+1}^{d} c_{il} \frac{\partial}{\partial x^l}, \quad (i = 1, \ldots, k),$$

where $c_{il} \in C^\infty(U)$. On the one hand, we have that

$$D|_U = \langle Y_1, \ldots, Y_k \rangle = \langle X_1, \ldots, X_k \rangle,$$

so (a) is satisfied. On the other hand, a simple computation shows that

$$[X_i, X_j] = \sum_{l=k+1}^{d} d_{ij}^l \frac{\partial}{\partial x^l}, \quad (i, j = 1, \ldots, k),$$

for certain functions $d_{ij}^l \in C^\infty(U)$. Since D is involutive, this commutator must be a $C^\infty(M)$-linear combination of $X_1, \ldots, X_k$. Therefore, the functions d_{ij}^l must be identically zero, so (b) also holds. $\square$

The Frobenius Theorem establishes a one-to-one correspondence:

$$\left\{ \text{involutive distributions on } M \right\} \longleftrightarrow \left\{ \text{foliations on } M \right\}$$

This is an example of an *integrability theorem*: a distribution D is an *infinitesimal object* on M while a foliation $\mathcal{F}$ is a *global object* on M and the integrability condition is the involutivity of D. We will see other integrability theorems later.

Exercises

Exercise 13.1
Give an example of a smooth distribution D of dimension 1 on the cylinder $\mathbb{S}^1 \times \mathbb{R}$ which is not globally generated by a vector field.

Exercise 13.2
Show that the 2-dimensional distribution D in Example 13.3 is not globally generated by only 2 vector fields.

Exercise 13.3
Show that the 2-dimensional distribution in $\mathbb{R}^3$ defined by the vector fields

$$X_1 = \frac{\partial}{\partial x}, \quad X_2 = e^{-x}\frac{\partial}{\partial y} + \frac{\partial}{\partial z},$$

has no integral manifolds.

Exercise 13.4
Consider the distribution D in $\mathbb{R}^3$ generated by the vector fields:

$$\frac{\partial}{\partial x} + \cos x \cos y \frac{\partial}{\partial z}, \quad \frac{\partial}{\partial y} - \sin x \sin y \frac{\partial}{\partial z}.$$

Check that D is involutive and determine the foliation $\mathcal{F}$ that integrates it.

Exercise 13.5
Consider the 3-sphere:

$$\mathbb{S}^3 = \{(x, y, z, w) \in \mathbb{R}^4 : x^2 + y^2 + z^2 + w^2 = 1\}.$$

Check that the vector field in $\mathbb{R}^3$ given by

$$X = -y\frac{\partial}{\partial x} + x\frac{\partial}{\partial y} - w\frac{\partial}{\partial z} + z\frac{\partial}{\partial w},$$

restricts to a nowhere vanishing vector field on $\mathbb{S}^3$, so determines a 1-dimensional distribution D. Find the foliation $\mathcal{F}$ integrating this distribution.

Exercise 13.6

Let $X_1, \ldots, X_k \in \mathfrak{X}(M)$ be vector fields which pairwise commute, i.e., such that

$$[X_i, X_j] = 0, \quad (i, j = 1, \ldots, k).$$

Assume that there exists $p \in M$ such that the tangent vectors $\{X_1|_p, \ldots, X_k|_p\}$ form a linearly independent set. Using the Flow Box Theorem, show that there exist a chart (U, x^i) centered at p such that

$$X_i = \frac{\partial}{\partial x^i}, \quad (i = 1, \ldots, k).$$

Lecture 14

Lie Groups and Lie Algebras

The next definition axiomatizes some of the properties of the Lie bracket of vector fields (see Proposition 12.2).

Definition 14.1. A **Lie algebra** is a real vector space $\mathfrak{g}$ together with a binary operation $[\ ,\ ] : \mathfrak{g} \times \mathfrak{g} \to \mathfrak{g}$, called the **Lie bracket**, which satisfies:

(i) Skew-symmetry: $[X, Y] = -[Y, X]$.
(ii) Bilinearity: $[aX + bY, Z] = a[X, Z] + b[Y, Z]$, $\forall a, b \in \mathbb{R}$.
(iii) Jacobi identity: $[X, [Y, Z]] + [Y, [Z, X]] + [Z, [X, Y]] = 0$.

We can also define Lie algebras over the complex numbers or over other fields.

Example 14.1.

(1) $\mathbb{R}^d$ with the zero Lie bracket $[\ ,\ ] \equiv 0$ is a Lie algebra, called the **abelian Lie algebra** of dimension d.
(2) In $\mathbb{R}^3$ one can define a Lie algebra structure with the Lie bracket, the vector product

$$[\vec{v}, \vec{w}] := \vec{v} \times \vec{w}.$$

(3) If V is any vector space, the vector space of all linear transformations $T : V \to V$ is a Lie algebra with Lie bracket the commutator

$$[T, S] := T \circ S - S \circ T.$$

This Lie algebra is called the **general linear Lie algebra** and denoted $\mathfrak{gl}(V)$. When $V = \mathbb{R}^n$, we denote it by $\mathfrak{gl}(\mathbb{R}, n)$ or simply $\mathfrak{gl}(n)$. After fixing a basis, we can identify $\mathfrak{gl}(n)$ with the space of all $n \times n$ real matrices. Under this identification, the Lie bracket becomes the commutator of matrices.

(4) If $\mathfrak{g}_1, \ldots, \mathfrak{g}_k$ are Lie algebras, their cartesian product $\mathfrak{g}_1 \times \cdots \times \mathfrak{g}_k$ is a Lie algebra with Lie bracket

$$[(X_1, \ldots, X_k), (Y_1, \ldots, Y_k)] := ([X_1, Y_1]_{\mathfrak{g}_1}, \ldots, [X_k, Y_k]_{\mathfrak{g}_k}).$$

(5) The space of all smooth vector fields $\mathfrak{X}(M)$ with the usual Lie bracket is a Lie algebra, which is infinite dimensional if $\dim M > 0$. We will be mainly interested in finite-dimensional Lie algebras.

We shall see shortly that Lie algebras are the "infinitesimal versions" of groups with a smooth structure, as in the following definition.

Definition 14.2. A **Lie group** is a group G with a smooth structure such that its structure maps are smooth:

$$\mu : G \times G \to G, \ (g, h) \mapsto gh \quad \text{(multiplication)},$$

$$\iota : G \to G, \ g \mapsto g^{-1} \quad \text{(inverse)}.$$

One can also define topological groups, analytic groups, etc.

Example 14.2.

(1) Any countable group with the discrete topology is a Lie group of dimension 0 (we need it to be countable so that the discrete topology is second countable).

(2) $\mathbb{R}^d$ with the usual addition of vectors is an abelian Lie group. The groups of all non-zero real numbers $\mathbb{R}^*$ and all non-zero complex numbers $\mathbb{C}^*$, with the usual multiplication operations, are also abelian Lie groups. Note that $\mathbb{C}^*$ is also a complex Lie group, thinking of $\mathbb{C}^*$ as a complex manifold, but we will mostly restrict ourselves to real Lie groups.

(3) The circle $\mathbb{S}^1 = \{z \in \mathbb{C} : \|z\| = 1\} \subset \mathbb{C}^*$ with the usual complex multiplication is also an abelian Lie group. The unit quaternions $\mathbb{S}^3$, with quaternionic multiplication, is a non-abelian Lie group. It can be shown that the only spheres $\mathbb{S}^d$ that admit Lie group structures are $d = 0, 1, 3$.

(4) If V is a finite-dimensional real vector space, the set of all *invertible* linear transformations $T : V \to V$ is a Lie group

with multiplication composition of transformations. It is called the **general linear group** and is denoted by $GL(V)$. Fixing a basis, we can identify $V = \mathbb{R}^n$, and under this identification $GL(V)$ becomes the group of all *invertible* $n \times n$ matrices with matrix multiplication and we denote it by $GL(\mathbb{R}, n)$ or simply $GL(n)$.

(5) If $G_1, \ldots, G_k$ are Lie groups their cartesian product $G \times \cdots \times G_k$ is also a Lie group. For example, the torus $\mathbb{T}^d = \mathbb{S}^1 \times \cdots \times \mathbb{S}^1$ is an abelian Lie group.

(6) If G is a Lie group, its **connected component of the identity** is a Lie group, denoted by G^0. For example, the connected component of the identity of the multiplicative group of non-zero real numbers $(\mathbb{R}^*, \cdot)$ is the group of positive real numbers $(\mathbb{R}_+, \cdot)$.

In a Lie group G, a **left-invariant vector field** is a vector field X such that

$$(L_g)_* X = X, \quad \forall g \in G,$$

where $L_g : G \to G$, $h \mapsto gh$ denotes the **left translation** by g. One defines analogously a **right invariant vector field** a vector field X such that

$$(R_g)_* X = X, \quad \forall g \in G,$$

where $R_g : G \to G$, $h \mapsto hg$ is the **right translation** by g. We choose to use left-invariant vector fields and we denote the set of all such smooth vector fields by

$$\mathfrak{X}_{\text{L-inv}}(G) := \{X \in \mathfrak{X}(G) : (L_g)_* X = X, \ \forall g \in G\}.$$

Proposition 14.1. *Let G be a Lie group.*

(i) *Every left-invariant vector field is smooth.*

(ii) *If $X, Y \in \mathfrak{X}_{\text{L-inv}}(G)$ then $[X, Y] \in \mathfrak{X}_{\text{L-inv}}(G)$.*

(iii) *$\mathfrak{X}_{\text{L-inv}}(G) \subset \mathfrak{X}(G)$ is a finite-dimensional subspace of dimension* $\dim G$.

Proof. We leave the proof of (i) as an exercise. To check (ii), it is enough to observe that if $X, Y \in \mathfrak{X}_{\text{L-inv}}(G)$ then

$$(L_g)_* [X, Y] = [(L_g)_* X, (L_g)_* Y] = [X, Y], \quad \forall g \in G.$$

Hence, $[X, Y] \in \mathfrak{X}_{\text{L-inv}}(G)$.

Now to see that (iii) holds, it is clear from the definition of a left invariant vector field that $\mathfrak{X}_{\text{L-inv}}(G) \subset \mathfrak{X}(G)$ is a linear subspace. On the other hand, the restriction map

$$\mathfrak{X}_{\text{L-inv}}(G) \to T_e G, \ X \mapsto X_e,$$

is a linear isomorphism. Its inverse associates to $\mathbf{v} \in T_e G$ the left invariant vector field X defined by

$$X_g := dL_g \cdot \mathbf{v}.$$

We conclude that $\dim \mathfrak{X}_{\text{L-inv}}(G) = \dim T_e G = \dim G.$ $\qquad\square$

This proposition shows that for a Lie group G the set $\mathfrak{X}_{\text{L-inv}}(G)$ forms a Lie algebra. We call it the **Lie algebra of the Lie group** G and denote it by $\mathfrak{g}$. The proof also shows that $\mathfrak{g}$ can be identified with $T_e G$.

Example 14.3.

(1) The Lie algebra of any discrete Lie group is the 0-dimensional vector space $\mathfrak{g} = \mathbb{R}^0 = \{0\}$. The Lie algebra of any 1-dimensional Lie group, such as $\mathbb{S}^1$ or $\mathbb{R}^*$, is the 1-dimensional (abelian) Lie algebra $\mathbb{R}$.

(2) Let $G = (\mathbb{R}^d, +)$. A vector field X in $\mathbb{R}^d$ is left invariant if and only if it is constant, i.e., $X = \sum_{i=1}^d a_i \frac{\partial}{\partial x^i}$, with $a_i \in \mathbb{R}$. The Lie bracket of any two constant vector fields is zero, hence the Lie algebra of $(\mathbb{R}^d, +)$ is the abelian Lie algebra $\mathbb{R}^d$.

(3) The Lie algebra of the cartesian product of two Lie groups $G \times H$, is the cartesian product of their Lie algebras $\mathfrak{g} \times \mathfrak{h}$. For example, the Lie algebra of the torus $\mathbb{T}^d$ is the abelian Lie algebra $\mathbb{R}^d$.

(4) The tangent space at the identity to the general linear group $G = GL(n)$ can be identified with $\mathfrak{gl}(n)$. The restriction map $\mathfrak{g} \to \mathfrak{gl}(n)$, maps the commutator of left-invariant vector fields to the commutator of matrices (exercise). Hence, we can identify the Lie algebra of $GL(n)$ with $\mathfrak{gl}(n)$.

(5) One may wonder if the Lie algebra $\mathfrak{X}(M)$ is associated with some Lie group. Since this Lie algebra is infinite dimensional (if $\dim M > 0$), this Lie group must be infinite dimensional too. One can show that it is the group $\text{Diff}(M)$ of all diffeomorphisms of M under composition. The study of such infinite-dimensional Lie groups is an important topic which is beyond the scope of this course.

Definition 14.3.

(i) A **homomorphism of Lie algebras** is a linear map $\phi : \mathfrak{g} \to \mathfrak{h}$ between two Lie algebras which preserves the Lie brackets:

$$\phi([X, Y]_{\mathfrak{g}}) = [\phi(X), \phi(Y)]_{\mathfrak{h}}, \quad \forall X, Y \in \mathfrak{g}.$$

(ii) A **homomorphism of Lie groups** is a smooth map $\Phi : G \to H$ between two Lie groups which is also a group homomorphism:

$$\Phi(gh^{-1}) = \Phi(g)\Phi(h)^{-1}, \quad \forall g, h \in G.$$

We have seen that to each Lie group there is associated a Lie algebra. Similarly, to each homomorphism of Lie groups there is associated a homomorphism of their Lie algebras. To see this, note that if $\Phi : G \to H$ is a Lie group homomorphism we have an induced map $\Phi_* : \mathfrak{g} \to \mathfrak{h}$ which to $X \in \mathfrak{g}$ associates $\Phi_*(X) \in \mathfrak{h}$, the unique left-invariant vector field such that $\Phi_*(X)|_e = d_e\Phi \cdot X_e$.

Proposition 14.2. *If $\Phi : G \to H$ is a Lie group homomorphism, then,*

(i) *For all $X \in \mathfrak{g}$, Φ_*X is Φ-related with X;*
(ii) *$\Phi_* : \mathfrak{g} \to \mathfrak{h}$ is a Lie algebra homomorphism.*

Proof. Part (ii) follows from (i), since the Lie bracket of Φ-related vector fields is preserved — see Proposition 12.2. In order to show that (i) holds, we observe that since Φ is a group homomorphism, $\Phi \circ L_g = L_{\Phi(g)} \circ \Phi$. Hence, we find

$$\begin{aligned}
\Phi_*(X)_{\Phi(g)} &= d_e L_{\Phi(g)} \cdot d_e\Phi \cdot X_e \\
&= d_e(L_{\Phi(g)} \circ \Phi) \cdot X_e \\
&= d_e(\Phi \circ L_g) \cdot X_e \\
&= d_g\Phi \cdot d_e L_g \cdot X_e = d_g\Phi \cdot X_g. \qquad \square
\end{aligned}$$

Example 14.4.

(1) Let $T^2 = \mathbb{S}^1 \times \mathbb{S}^1$. For each $a \in \mathbb{R}$ one has the Lie group homomorphism

$$\Phi_a : \mathbb{R} \to T^2, \quad t \mapsto (e^{it}, e^{iat}).$$

If a is rational, the image Φ_a is a closed curve, while if a is irrational the image is a dense curve in the torus. The induced Lie algebra homomorphism is

$$(\Phi_a)_* : \mathbb{R} \to \mathbb{R}^2, \quad X \mapsto (X, aX).$$

(2) The determinant defines a Lie group homomorphism det : $GL(n) \to \mathbb{R}^*$. The induced Lie algebra homomorphism $(\det)_*$: $\mathfrak{gl}(n) \to \mathbb{R}$ coincides with the trace $(\det)_*(X) = \operatorname{tr} X$ (exercise).

(3) Conjugation by a fixed matrix $A \in GL(n)$ yields a Lie group automorphism

$$\Phi_A : GL(n) \to GL(n), \quad B \mapsto ABA^{-1}.$$

Since this map is linear the associated Lie algebra automorphism is also given by conjugation

$$(\Phi_A)_* : \mathfrak{gl}(n) \to \mathfrak{gl}(n), \quad X \mapsto AXA^{-1}.$$

(4) More generally, for any Lie group G we can consider conjugation by a fix $g \in G$:

$$i_g : G \to G, \quad h \mapsto ghg^{-1}.$$

This is a Lie group automorphism and the induced Lie algebra automorphism is denoted by

$$\operatorname{Ad}_g : \mathfrak{g} \to \mathfrak{g}, \quad \operatorname{Ad}_g(X) := (i_g)_* X.$$

As another instance of the Lie group/algebra correspondence, we will show now that to each subgroup of a Lie group G corresponds a Lie subalgebra of its Lie algebra $\mathfrak{g}$.

Definition 14.4. A subspace $\mathfrak{h} \subset \mathfrak{g}$ is called a **Lie subalgebra** if, for all $X, Y \in \mathfrak{h}$, we have $[X, Y] \in \mathfrak{h}$.

Example 14.5.

(1) Any subspace of the abelian Lie algebra $\mathbb{R}^d$ is a Lie subalgebra.

(2) In the Lie algebra $\mathfrak{gl}(n)$, we have the Lie subalgebra formed by all matrices of zero trace

$$\mathfrak{sl}(n) := \{X \in \mathfrak{gl}(n) : \operatorname{tr} X = 0\},$$

and also the Lie subalgebra formed by all skew-symmetric matrices

$$\mathfrak{o}(n) := \{X \in \mathfrak{gl}(n) : X + X^T = 0\}.$$

(3) The complex $n \times n$ matrices, denoted by $\mathfrak{gl}(n, \mathbb{C})$, can be seen as a real Lie algebra. It has the Lie subalgebra of all skew-Hermitian matrices

$$\mathfrak{u}(n) := \{X \in \mathfrak{gl}(n, \mathbb{C}) : X + \bar{X}^T = 0\},$$

and the Lie subalgebra of all skew-Hermitian matrices of trace zero

$$\mathfrak{su}(n) := \{X \in \mathfrak{gl}(n, \mathbb{C}) : X + \bar{X}^T = 0, \operatorname{tr} X = 0\}.$$

(4) If $\phi : \mathfrak{g} \to \mathfrak{h}$ is a homomorphism of Lie algebras, then its kernel is a Lie subalgebra of $\mathfrak{g}$ and its image is a Lie subalgebra of $\mathfrak{h}$.

A notion of a Lie subgroup is defined similarly.

Definition 14.5. A **Lie subgroup** of G is a submanifold (H, Φ) of G such that

(i) H is Lie group;
(ii) $\Phi : H \to G$ is a Lie group homomorphism.

As we discussed in Lecture 7, we can always replace the submanifold (H, Φ) by the subset $\Phi(H) \subset G$, and the immersion Φ by the inclusion i. Since $\Phi(H)$ is a subgroup of G, in the definition of a Lie subgroup we can assume that $H \subset G$ is a subgroup and that Φ is the inclusion. Note, however, that the topology on H may be different from the subspace topology. On the other hand, since the induced map $\Phi_* : \mathfrak{h} \to \mathfrak{g}$ is injective, we see that the Lie algebra of a Lie subgroup $H \subset G$ corresponds to the Lie subalgebra $\mathfrak{h} \subset \mathfrak{g}$.

Example 14.6.

(1) In Example 14.1 (1), for each $a \in \mathbb{R}$ we have a Lie subgroup $\Phi_a(\mathbb{R})$ of $\mathbb{T}^2$. If a is rational, this Lie subgroup is embedded, while if a is irrational this Lie subgroup is only immersed.

(2) The general linear group $GL(n)$ has the following (embedded) subgroups:

(i) The **special linear group** of all matrices of determinant 1

$$SL(n) := \{A \in GL(n) : \det A = 1\}.$$

To this subgroup corresponds the Lie subalgebra $\mathfrak{sl}(n)$.

(ii) The **orthogonal group** of all orthogonal matrices

$$O(n) := \{A \in GL(n) : AA^T = I\}.$$

To this subgroup corresponds the Lie subalgebra $\mathfrak{o}(n)$.

(iii) The **special orthogonal group** of all orthogonal matrices of positive determinant

$$SO(n) := \{A \in O(n) : \det A = 1\}.$$

To this subgroup corresponds the Lie subalgebra $\mathfrak{so}(n) = \mathfrak{o}(n)$.

(3) The (real) Lie group $GL(n, \mathbb{C})$ has the following (embedded) subgroups:

(i) The **unitary group** of all unitary matrices

$$U(n) := \{A \in GL(n, \mathbb{C}) : A\bar{A}^T = I\}.$$

To this subgroup corresponds the Lie subalgebra $\mathfrak{u}(n)$.

(ii) The **special unitary group** of all unitary matrices of determinant 1

$$SU(n) := \{A \in U(n) : \det A = 1\}.$$

To this subgroup corresponds the Lie subalgebra $\mathfrak{su}(n)$.

(4) Let $\Phi : G \to H$ is a Lie group homomorphism and let $(\Phi)_* : \mathfrak{g} \to \mathfrak{h}$ be the induced Lie algebra homomorphism. Then $\operatorname{Ker} \Phi \subset G$ and $\operatorname{Im} \Phi \subset H$ are Lie subgroups whose Lie algebras coincide with $\operatorname{Ker}(\Phi)_* \subset \mathfrak{g}$ and $\operatorname{Im}(\Phi)_* \subset \mathfrak{h}$, respectively.

Exercises

Exercise 14.1
Show that in the definition of a Lie group, it is enough to assume that

(a) The inverse map $G \to G$, $g \mapsto g^{-1}$ is smooth, or that
(b) The map $G \times G \to G$, $(g, h) \mapsto gh^{-1}$, is smooth.

Exercise 14.2
Show that every left-invariant vector field in a Lie group G is smooth and complete.

Exercise 14.3
Show that the tangent space at the identity of $GL(n)$ can be identified with $\mathfrak{gl}(n)$. Show also that, under this identification, the linear isomorphism $\mathfrak{g} \to \mathfrak{gl}(n)$ takes the Lie bracket of left-invariant vector fields to the commutator of matrices. What happens if one defines the Lie algebra of G using right invariant vector fields, instead of left invariant vector fields?

Exercise 14.4
Show that the tangent bundle TG of a Lie group G is trivial, i.e., there exist vector fields $X_1, \ldots, X_d \in \mathfrak{X}(G)$ which at each $g \in G$ give a basis for $T_g G$. Conclude that an even dimension sphere $\mathbb{S}^{2n}$ does not admit the structure of a Lie group.

Exercise 14.5
Show that the Lie algebra homomorphism induced by the determinant $\det : GL(n) \to \mathbb{R}^*$ coincides with the trace: $(\det)_* = \operatorname{tr}$.

Exercise 14.6
Consider $\mathbb{S}^3 \subset \mathbb{H}$ as the set of quaternions of norm 1. Show that $\mathbb{S}^3$, with the product of quaternions, is a Lie group and determine its Lie algebra.

Exercise 14.7
Show that $\mathbb{S}^3$ and $SU(2)$ are isomorphic Lie groups.

Hint: For any pair of complex numbers $z, w \in \mathbb{C}$ with $|z|^2 + |w|^2 = 1$, the matrix:

$$\begin{pmatrix} z & w \\ -\bar{w} & \bar{z} \end{pmatrix}$$

is an element in $SU(2)$.

Exercise 14.8

Identify the vectors $v \in \mathbb{R}^3$ with the purely imaginary quaternions. For each quaternion $q \in \mathbb{S}^3$ of norm 1 define a linear map $T_q : \mathbb{R}^3 \to \mathbb{R}^3$ by $v \mapsto qvq^{-1}$. Show that T_q is a special orthogonal transformation and that the map $\mathbb{S}^3 \to SO(3)$, $q \mapsto T_q$, is a Lie group homomorphism. Is this map surjective? Injective?

Exercise 14.9

Let G be a Lie group. Show that the connected component of the identity is a Lie group G^0 whose Lie algebra is isomorphic to the Lie algebra of G.

Exercise 14.10

Let G be a connected Lie group with Lie algebra $\mathfrak{g}$. Show that G is abelian if and only if $\mathfrak{g}$ is abelian. What can you say if G is not connected?

Exercise 14.11

Show that a compact connected abelian Lie group G is isomorphic to a torus $\mathbb{T}^d$.

Exercise 14.12

Let (H, Φ) be a Lie subgroup of G. Show that Φ is an embedding if and only if $\Phi(H)$ is closed in G.

Exercise 14.13

Let $A \subset G$ be a subgroup of a Lie group G. Show that if (A, i) has a smooth structure making it into a submanifold of G, then this smooth structure is unique and that for that smooth structure A is a Lie group and (A, i) a Lie subgroup.

Hint: Show that (A, i) is a regularly immersed submanifold.

Lecture 15

Integrations of Lie Algebras

We saw in the previous lecture that

- To each Lie group corresponds a Lie algebra;
- To each Lie group homomorphism corresponds a Lie algebra homomorphism;
- To each Lie subgroup corresponds to a Lie subalgebra.

It is natural to wonder about the inverse of each of these correspondences. We have seen that two distinct Lie groups can have isomorphic Lie algebras (e.g., $\mathbb{R}^n$ and $\mathbb{T}^n$, $O(n)$ and $SO(n)$, or $SU(2)$ and $SO(3)$). There are indeed topological issues that one must take care of to make the inverse correspondences work. For that, we start with the following result that shows that a connected Lie group is determined by a neighborhood of the identity.

Proposition 15.1. *Let G be a connected Lie group and U a neighborhood of the identity $e \in G$. Then,*

$$G = \bigcup_{n=1}^{\infty} U^n,$$

where $U^n := \{g_1 \cdots g_n : g_i \in U, i = 1, \ldots, n\}$.

Proof. If $U^{-1} := \{g^{-1} : g \in U\}$ then $V := U \cap U^{-1}$ is a neighborhood of the origin such that $V = V^{-1}$. Let

$$H := \bigcup_{n=1}^{\infty} V^n \subset \bigcup_{n=1}^{\infty} U^n.$$

To complete the proof we show that $H = G$. For that, we note

(i) H is a subgroup: if $g = g_1 \ldots g_n, h = h_1 \ldots h_m \in H$, where $g_i, h_j \in V$, then $gh^{-1} = g_1 \ldots g_n h_m^{-1} \ldots h_1^{-1} \in V^{n+m} \subset H$.

(ii) H is open: if $g \in H$ then $gV \subset gH = H$ is an open set containing g.

(iii) H is closed: for each $g \in G$, gH is an open set and we have $H^c = \bigcup_{g \notin H} gH$.

Since G is connected and $H \neq \emptyset$ is open and closed, we must have $H = G$. □

Theorem 15.1. *Let G be a Lie group with Lie algebra $\mathfrak{g}$. Given a Lie subalgebra $\mathfrak{h} \subset \mathfrak{g}$, there exists a unique connected Lie subgroup $H \subset G$ with Lie algebra $\mathfrak{h}$.*

Proof. A Lie subalgebra $\mathfrak{h}$ defines a distribution D in G by setting

$$D_g := \{X_g : X \in \mathfrak{h}\}.$$

This distribution is smooth and involutive. In fact, if $X_1, \ldots, X_k$ is a basis for $\mathfrak{h}$, then these vector fields are smooth and generate D everywhere, hence D is smooth. On the other hand, if $Y, Z \in \mathfrak{X}(D)$, then,

$$Y = \sum_{i=1}^{k} a_i X_i, \quad Z = \sum_{j=1}^{k} b_j X_j.$$

Using that $\mathfrak{h}$ is a Lie subalgebra it follows that

$$[Y, Z] = \sum_{i,j=1}^{k} a_i b_j [X_i, X_j] + a_i X_i(b_j) X_j - b_j X_j(a_i) X_i \in \mathfrak{X}(D),$$

proving that D is involutive.

Applying Frobenius Theorem, let (H, i) be the leaf of D that contains the identity $e \in G$, where $i : H \hookrightarrow G$ denotes the inclusion. We claim that (H, i) is the desired Lie subgroup. To prove this claim, note that if $g \in H$ then $(H, L_{g^{-1}} \circ i)$ is also an integral manifold of D which contains e, since

$$d_h(L_{g^{-1}} \circ i)(T_h H) = d_h L_{g^{-1}}(D_h) = D_{g^{-1}h}.$$

Hence, $L_{g^{-1}} \circ i(H) \subset i(H)$, and we conclude that for all $g, h \in H$, we have $g^{-1}h \in H$, proving that H is a subgroup of G. To verify that (H, i) is a Lie subgroup, it remains to prove that the map $\hat{\nu}$: $H \times H \to H$, $(g, h) \mapsto g^{-1}h$, is smooth. For this, we observe that the map $\nu : H \times H \to G$, $(g, h) \mapsto i(g)^{-1}i(h)$ is smooth, being the composition of smooth maps, so that we have a commutative diagram

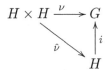

Since the leaves of any foliation are regularly immersed submanifolds (Proposition 9.1), we conclude that $\hat{\nu} : H \times H \to H$ is smooth.

Uniqueness follows from Proposition 15.1 (exercise!). □

The question of deciding if every finite-dimensional Lie algebra $\mathfrak{g}$ is associated with some Lie group G is a much harder question which is beyond the scope of these lecture notes. There are several ways to prove that this is indeed true. For example, one can construct an integration G explicitly (see, e.g., Duistermaat and Kolk, 2000) or one can develop the structure theory of Lie algebras and show that any finite-dimensional Lie algebra is isomorphic to a matrix Lie algebra (see, e.g., Varadarajan, 1984), obtaining the following result.

Theorem 15.2 (Ado). *Let $\mathfrak{g}$ be a finite-dimensional Lie algebra. There exists an integer n and an injective Lie algebra homomorphism $\phi : \mathfrak{g} \to \mathfrak{gl}(n)$.*

Since $\mathfrak{gl}(n)$ is the Lie algebra of $GL(n)$, as a corollary of Ado's Theorem and Theorem 15.1 it follows that.

Theorem 15.3. *For any finite-dimensional Lie algebra $\mathfrak{g}$, there exists a Lie group G with Lie algebra isomorphic to $\mathfrak{g}$.*

The previous theorem gives a matrix group integrating any finite-dimensional Lie algebra. However, we will see later an example of a connected Lie group which *is not* isomorphic to a matrix group.

We already saw that there can be several non-isomorphic, connected, Lie groups integrating the same Lie algebra. In order to clarify the issue of multiple (connected) Lie groups integrating the same

Lie algebra, recall that if $\pi : N \to M$ is a covering of a manifold M, then there is a unique differentiable structure on N for which the covering map is a local diffeomorphism. In particular, if M is connected then the **universal covering space** of M, which is characterized as the 1-connected (i.e., connected and simply connected) covering of M, has a natural smooth structure. For Lie groups, this leads to the following result.

Proposition 15.2. *Given a connected Lie group G its universal covering space $\widetilde{G}$ has a unique Lie group structure for which the covering map $\pi : \widetilde{G} \to G$ is a Lie group homomorphism. Moreover, the Lie algebras of G and $\widetilde{G}$ are isomorphic and $\ker \pi \subset \widetilde{G}$ is a discrete, normal, subgroup of the center of $\widetilde{G}$. In particular, $\pi_1(G) \simeq \ker \pi$ is abelian.*

Proof. Recall that one can identify the universal covering space of G explicitly as

$$\widetilde{G} := \{[\gamma] \mid \gamma : [0,1] \to G, \gamma(0) = e\}, \quad \pi : \widetilde{G} \to G, \quad [\gamma] \mapsto \gamma(1),$$

where $[\gamma]$ denotes the homotopy class of the path γ relative to end points. One can define a group structure in $\widetilde{G}$ as follows:

(i) The product $[\gamma][\eta]$ in $\widetilde{G}$ is the homotopy class of the path $t \mapsto \gamma(t)\eta(t)$.
(ii) The identity $\tilde{e} \in \widetilde{G}$ is the homotopy class of the constant path based at the identity $\gamma(t) = e$.
(iii) The inverse map $i : \widetilde{G} \to \widetilde{G}$ associated to an element $[\gamma]$ is the homotopy class of the path $t \mapsto \gamma(t)^{-1}$.

With these choices, the covering map $\pi : \widetilde{G} \to G$ is a group homomorphism.

Now, consider on $\widetilde{G}$ the unique smooth structure for which the covering map is a local diffeomorphism. To check that $\widetilde{G}$ is a Lie group, observe that the map $\tilde{\nu} : \widetilde{G} \times \widetilde{G} \to \widetilde{G}$, $(g,h) \to g^{-1}h$, is smooth since it fits into the commutative diagram

$$\begin{array}{ccc} \widetilde{G} \times \widetilde{G} & \xrightarrow{\tilde{\nu}} & \widetilde{G} \\ \scriptstyle{\pi \times \pi} \downarrow & & \downarrow \scriptstyle{\pi} \\ G \times G & \xrightarrow{\nu} & G \end{array}$$

where the vertical arrows are local diffeomorphisms and ν is differentiable. Since $\pi : \widetilde{G} \to G$ is a local diffeomorphism it induces an isomorphism between the Lie algebras of $\widetilde{G}$ and G.

Uniqueness follows, because the condition that $\pi : \widetilde{G} \to G$ induces an isomorphism between the Lie algebras of $\widetilde{G}$ and G implies that π is a local diffeomorphism, so both the smooth structure and the group structure are uniquely determined.

We leave as an exercise the remaining statements in the theorem.

□

Example 15.1. The special unitary group $SU(2)$ is formed by the matrices:

$$SU(2) = \left\{ \begin{pmatrix} a & b \\ -\bar{b} & \bar{a} \end{pmatrix} : a, b \in \mathbb{C}, |a|^2 + |b|^2 = 1 \right\}.$$

Therefore $SU(2)$ is isomorphic as a manifold to $\mathbb{S}^3$, hence it is 1-connected. In fact, by an exercise in the previous lecture, $SU(2)$ is isomorphic, as a Lie group, to the group $\mathbb{S}^3$ consisting of the quaternions of length 1.

The Lie algebra of $SU(2)$ consists of the skew-Hermitian matrices of trace zero:

$$\mathfrak{su}(2) = \left\{ \begin{pmatrix} i\alpha & \beta \\ -\bar{\beta} & -i\alpha \end{pmatrix} : \alpha \in \mathbb{R}, \beta \in \mathbb{C} \right\}.$$

Setting $x = \frac{\alpha}{\sqrt{2}}, y = \frac{\mathrm{Re}\,\beta}{\sqrt{2}}, z = \frac{\mathrm{Im}\,\beta}{\sqrt{2}}$, we obtain identifications

$$\begin{pmatrix} i\alpha & \beta \\ -\bar{\beta} & -i\alpha \end{pmatrix} \quad \longleftrightarrow \quad (x, y, z),$$

giving a Lie algebra isomorphism $\mathfrak{su}(2) \simeq \mathbb{R}^3$, where on $\mathbb{R}^3$ the Lie bracket is given by the vector product. We make use of this identification and consider the standard Euclidean inner product on $\mathbb{R}^3$. Then, by Example 14.1 (4), for each $g \in SU(2)$ we have a linear transformation $\mathrm{Ad}_g : \mathbb{R}^3 \to \mathbb{R}^3$. This satisfies (exercise):

(a) For each $g \in SU(2)$, Ad_g preserves the inner product and the usual orientation, hence determines an element in $SO(3)$.
(b) $\mathrm{Ad} : SU(2) \to SO(3)$, $g \mapsto \mathrm{Ad}_g$, is a surjective group homomorphism with kernel the group $\mathbb{Z}_2 = \{\pm I\}$.

It follows that $\mathrm{Ad} : SU(2) \to SO(3)$ is a covering map (see Exercise 15.1). Since $SU(2) \simeq \mathbb{S}^3$ is 1-connected, we conclude that $SU(2)$ is the universal covering space of $SO(3)$. The covering map identifies the antipodal points in the sphere, so we can identify $SO(3)$ with the real projective space $\mathbb{RP}^3$ and $\pi_1(SO(3)) = \mathbb{Z}_2$.

Considering now the question of integrating homomorphisms of Lie algebras to homomorphisms of Lie groups, one notices again that there are topological obstructions. For example, the identity $\mathrm{Id} : \mathbb{R} \to \mathbb{R}$ is a Lie algebra isomorphism between the Lie algebras of $\mathbb{S}^1$ and $(\mathbb{R}, +)$. However, the only Lie group homomorphism $\Phi : \mathbb{S}^1 \to \mathbb{R}$ is the trivial one because the image $\Phi(\mathbb{S}^1)$ is a compact subgroup of $(\mathbb{R}, +)$, and $\{0\}$ is the only such subgroup. Therefore, there is no Lie group homomorphism $\Phi : \mathbb{S}^1 \to \mathbb{R}$ with $\Phi_* = \mathrm{Id}$. The problem in this example is that $\mathbb{S}^1$ is not simply connected.

Theorem 15.4. *Let G and H be Lie groups with Lie algebras $\mathfrak{g}$ and $\mathfrak{h}$. If G is 1-connected then for every Lie algebra homomorphism $\phi : \mathfrak{g} \to \mathfrak{h}$, there exists a unique Lie group homomorphism $\Phi : G \to H$ such that $\Phi_* = \phi$.*

Proof. Let $\mathfrak{k} = \{(X, \phi(X)) : X \in \mathfrak{g}\} \subset \mathfrak{g} \times \mathfrak{h}$ be the graph of ϕ. Since ϕ is a Lie algebra homomorphism, $\mathfrak{k}$ is a Lie subalgebra of $\mathfrak{g} \times \mathfrak{h}$. Hence, there exists a unique connected Lie subgroup $K \subset G \times H$ with Lie algebra $\mathfrak{k}$. Let us consider the restriction to K of the projections on each factor as in the diagram

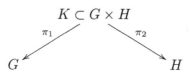

The restriction of the first projection $\pi_1|_K : K \to G$ gives a Lie group homomorphism such that

$$(\pi_1)_*(X, \phi(X)) = X.$$

Hence, the map $(\pi_1|_K)_* : \mathfrak{k} \to \mathfrak{g}$ is a Lie algebra isomorphism and it follows that $\pi_1|_K : K \to G$ is a covering map (see Exercise 15.1).

Since G is 1-connected, we conclude that $\pi_1|_K$ is a Lie isomorphism. Then, the composition

$$\Phi = \pi_2 \circ (\pi_1|_K)^{-1} : G \to H$$

is a Lie group homomorphism and we have that

$$(\Phi)_*(X) = (\pi_2)_* \circ (\pi_1|_K)_*^{-1}(X) = (\pi_2)_*(X, \phi(X)) = \phi(X).$$

We leave the proof of uniqueness as an exercise. $\qquad\square$

We summarize the previous results in the following statements, sometimes known as Lie's theorems.

Theorem 15.5 (Lie I). *If G is a connected Lie group with Lie algebra $\mathfrak{g}$, there is a unique (up to isomorphism) 1-connected Lie group $\widetilde{G}$ with Lie algebra $\mathfrak{g}$ and a surjective Lie group morphism $\Phi : \widetilde{G} \to G$.*

Theorem 15.6 (Lie II). *Let G and H be two Lie groups, with Lie algebras denoted $\mathfrak{g}$ and $\mathfrak{h}$, respectively. If G is 1-connected, then a Lie algebra morphism $\phi : \mathfrak{g} \to \mathfrak{h}$ integrates to a unique Lie group morphism $\Phi : G \to H$.*

Theorem 15.7 (Lie III). *Any finite-dimensional Lie algebra $\mathfrak{g}$ is integrable to a Lie group.*

Note also that, given a finite-dimensional Lie algebra $\mathfrak{g}$, one can obtain any connected Lie group G integrating it (up to isomorphism) as follows:

(i) Construct the 1-connected Lie group $\widetilde{G}$ integrating $\mathfrak{g}$.
(ii) Find a discrete normal subgroup N of the center of $\widetilde{G}$.
(iii) $G = \widetilde{G}/N$ is a connected Lie group integrating $\mathfrak{g}$.

If one drops the condition of G being connected this problem is not solvable since it would include as a special case the classification of all finite groups, a problem which is well-known not to have any reasonable solution.

As we observed before, Ado's Theorem does not prevent the existence of Lie groups which are not isomorphic to any group of matrices. As an application of the integration of morphisms, we give one such example.

Example 15.2. Consider the special linear group

$$SL(2) = \left\{ \begin{pmatrix} a & b \\ c & d \end{pmatrix} : ad - bc = 1 \right\}.$$

To exhibit its topological structure, it is convenient to perform the change of variables $(a, b, c, d) \mapsto (p, q, r, s)$ defined by

$$a = p + q, \quad d = p - q, \quad b = r + s, \quad c = r - s.$$

Then

$$ad - bc = 1 \quad \Longleftrightarrow \quad p^2 + s^2 = q^2 + r^2 + 1.$$

Therefore, we can also describe $SL(2)$ as

$$SL(2) = \{(p, q, r, s) \in \mathbb{R}^4 : p^2 + s^2 = q^2 + r^2 + 1\}.$$

This description shows that $SL(2)$ is diffeomorphic to $\mathbb{R}^2 \times \mathbb{S}^1$. In particular,

$$\pi_1(SL(2)) = \pi_1(\mathbb{S}^1) = \mathbb{Z}.$$

Let $\widetilde{SL(2)}$ be the universal covering group of $SL(2)$. We claim that $\widetilde{SL(2)}$ is not isomorphic to any group of matrices. Although $SL(2)$ is not 1-connected, by Exercise 15.5 one has:

- Given any Lie algebra morphism $\phi : \mathfrak{sl}(2) \to \mathfrak{gl}(n)$, there exists a unique Lie group morphism $\Phi : SL(2) \to GL(n)$ such that $\Phi_* = \phi$.

We prove the claim by contradiction. Assume that for some n there exists an injective Lie group homomorphism

$$\widetilde{\Phi} : \widetilde{SL(2)} \to GL(n).$$

Then $\widetilde{\Phi}$ induces a morphism $\phi := (\widetilde{\Phi})_* : \mathfrak{sl}(2) \to \mathfrak{gl}(n)$, so there exists a unique Lie group homomorphism $\Phi : SL(2) \to GL(n)$ such that

$\Phi_* = \phi$. By uniqueness in Lie II, we obtain a commutative diagram

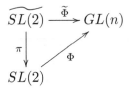

In this diagram, the morphism π is not injective, while the morphism $\widetilde{\Phi}$ is injective, which is a contradiction.

Exercises

Exercise 15.1
Let $\Phi : G \to H$ be a Lie group homomorphism between connected Lie groups G and H such that $(\Phi)_* : \mathfrak{g} \to \mathfrak{h}$ is an isomorphism. Show that Φ is a covering map.

Exercise 15.2
Complete the proof of Theorem 15.1 by showing that the integrating Lie subgroup is unique.

Exercise 15.3
Let G be a Lie group and let $\pi : H \to G$ be a covering map. Show that H is a Lie group.

Exercise 15.4
Let $SL(2, \mathbb{C})$ be the group of complex 2×2 matrices with determinant 1. Show that $SL(2, \mathbb{C})$ is 1-connected.

Hint: Show that a matrix in $SL(2, \mathbb{C})$ can be written uniquely as a product AB, where $A \in SU(2)$ and B is upper triangular with determinant 1.

Exercise 15.5
Show that any homomorphism of Lie algebras $\phi : \mathfrak{sl}(2) \to \mathfrak{gl}(n)$ integrates to a unique homomorphism of Lie groups $\Phi : SL(2) \to GL(n)$.

Hint: Consider the complexification $\phi^c : \mathfrak{sl}(2, \mathbb{C}) \to \mathfrak{gl}(n, \mathbb{C})$ of ϕ and use the previous exercise.

Exercise 15.6

Let G be a connected Lie group and let $D \subset G$ be a discrete normal subgroup. Show that D is contained in the center of G, so in particular it must be abelian. Conclude that any connected Lie group has abelian fundamental group.

Exercise 15.7

Find all (up to isomorphism) connected Lie groups integrating the abelian Lie algebra $\mathfrak{g} = \mathbb{R}^d$.

Exercise 15.8

Find all (up to isomorphism) connected Lie groups integrating the Lie algebra $\mathfrak{so}(3)$.

Lecture 16

The Exponential Map

We will now construct the *exponential map* for Lie groups/algebras, which generalizes the exponential of matrices. So let G be a Lie group with Lie algebra $\mathfrak{g}$. Given a left-invariant vector field $X \in \mathfrak{g}$, the map

$$\phi_X : \mathbb{R} \to \mathfrak{g}, \quad t \mapsto tX,$$

is a Lie algebra homomorphism. Since $\mathbb{R}$ is 1-connected it follows from Lie II that there exists a unique Lie group homomorphism $\Phi_X : \mathbb{R} \to G$ such that $(\Phi_X)_* = \phi_X$. Noting that $\Phi_X(0) = e$ and

$$\Phi_X(t+s) = \Phi_X(t)\Phi_X(s) = L_{\Phi_X(t)}\Phi_X(s),$$

we find

$$
\begin{aligned}
\frac{\mathrm{d}}{\mathrm{d}t}\Phi_X(t) &= \left.\frac{\mathrm{d}}{\mathrm{d}s}\Phi_X(t+s)\right|_{s=0} \\
&= \mathrm{d}_e L_{\Phi_X(t)} \cdot \left.\frac{\mathrm{d}}{\mathrm{d}s}\Phi_X(s)\right|_{s=0} \\
&= \mathrm{d}_e L_{\Phi_X(t)} \cdot X_e = X_{\Phi_X(t)}.
\end{aligned}
$$

This means that $t \mapsto \Phi_X(t)$ is actually the integral curve of X through $e \in G$.

Definition 16.1. The **exponential map** $\exp : \mathfrak{g} \to G$ is the map

$$\exp(X) := \Phi_X(1) = \phi_X^1(e).$$

The proof of the following proposition listing basic properties of the exponential map is left as an exercise.

Proposition 16.1. *The exponential map* $\exp : \mathfrak{g} \to G$ *satisfies:*

(i) $\exp((t+s)X) = \exp(sX)\exp(tX)$;
(ii) $\exp(-tX) = [\exp(tX)]^{-1}$;
(iii) $\exp$ *is a smooth map and* $d_0 \exp = \mathrm{Id}$;
(iv) *For any Lie group homomorphism* $\Phi : G \to H$, *the following diagram commutes*

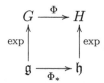

Property (iii) implies that the exponential is a diffeomorphism from a neighborhood of $0 \in \mathfrak{g}$ to a neighborhood of $e \in G$. In general, the exponential $\exp : \mathfrak{g} \to G$ is neither surjective, nor injective. Also, it may fail to be a local diffeomorphism at other points of G. There are however examples of Lie groups/algebras in which some of these properties do hold (see the exercises in this lecture).

Example 16.1. Recall that the Lie algebra of $G = GL(n)$ is $\mathfrak{g} = \mathfrak{gl}(n)$. Given a matrix $A = (a_{ij}) \in \mathfrak{gl}(n)$ the corresponding left-invariant vector field in $GL(n)$ is

$$X_A = \sum_{ijk} x_{ik} a_{kj} \frac{\partial}{\partial x_{ij}}.$$

Hence, the integral curves of this vector field are the solutions of the system of ode's

$$\dot{x}_{ij} = \sum_{k=1}^{n} x_{ik} a_{kj}, \quad (i, j = 1, \dots, n).$$

The solutions are

$$x_{ij}(t) = \sum_{k=1}^{n} x_{ik}(0)(e^{tA})_{kj},$$

where the matrix exponential is defined as usual by the series

$$e^A := \sum_{k=0}^{+\infty} \frac{A^n}{n!}.$$

We conclude that the exponential map $\exp : \mathfrak{gl}(n) \to GL(n)$ coincides with the usual matrix exponential.

By item (iv) in Proposition 16.1, it follows from the previous example that if $\mathfrak{h} \subset \mathfrak{gl}(n)$ is a Lie subalgebra and $H \subset GL(n)$ is the associated connected Lie subgroup, then the exponential map $\exp : \mathfrak{h} \to H$ also coincides with the matrix exponential.

The exponential map is very useful in the study of Lie groups and Lie algebras since it provides a direct link between the Lie algebra (the infinitesimal object) and the Lie group (the global object). For example, we have the following result whose proof is left as an exercise:

Proposition 16.2. *Let H be a subgroup of a Lie group G and let $\mathfrak{h} \subset \mathfrak{g}$ be a subspace of the Lie algebra of G. If $U \subset \mathfrak{g}$ is a neighborhood of 0 which is diffeomorphic via the exponential map to a neighborhood $V \subset G$ of e, and*

$$\exp(\mathfrak{h} \cap U) = H \cap V,$$

then, for the relative topology, H is a Lie subgroup of G with Lie algebra $\mathfrak{h}$.

Using this proposition, one can then prove the following important result.

Theorem 16.1. *Let G be a Lie group and $H \subset G$ a closed subgroup. Then H, with the relative topology, is a Lie subgroup.*

Sketch of the proof. The idea of the proof is the consider the set

$$\mathfrak{h} := \{X \in \mathfrak{g} : \exp(tX) \in H, \forall t \in \mathbb{R}\}$$

and apply the previous proposition.

Clearly, the set $\mathfrak{h}$ is closed under multiplication by scalars. On the other hand, if $X, Y \in \mathfrak{g}$ one shows that

$$\lim_{n \to +\infty} \left(\exp\left(\frac{t}{n}X\right) \exp\left(\frac{t}{n}Y\right) \right)^n = \exp(t(X + Y)),$$

and then it follows that $\mathfrak{h}$ is also closed under addition since H is a closed subset. Hence, $\mathfrak{h}$ is a linear subspace.

Finally, arguing by contradiction using again that H is closed in G, one shows that there exists neighborhoods $U \subset \mathfrak{g}$ of 0 and $V \subset G$ of e, such that $\exp : U \to V$ is a diffeomorphism and:

$$\exp(\mathfrak{h} \cap U) = H \cap V.$$

The details of this proof can be found, e.g., in Warner (1983). □

The previous results allow one to check quickly if subgroups of $GL(n)$ are Lie subgroups and to determine their Lie algebras.

Example 16.2. Consider the subgroup $SL(n) \subset GL(n)$. It is a closed subgroup, so by Theorem 16.1 it is a Lie subgroup. To find its Lie algebra, one observes first that the set $\mathfrak{sl}(n)$ of matrices of trace zero is a subspace of $\mathfrak{gl}(n)$ and second that we have the well-known formula

$$\det(e^X) = e^{\operatorname{tr} X}.$$

Hence, we see that $\exp(X) \in SL(n)$ if and only if $\operatorname{tr} X = 0$. By Proposition 16.2, we conclude that the Lie algebra of $SL(n)$ is $\mathfrak{sl}(n)$.

Using the same method, it should be now easy to check that the Lie algebras of the matrix groups $SO(n)$, $SU(n)$ and $U(n)$ are $\mathfrak{so}(n)$, $\mathfrak{su}(n)$ and $\mathfrak{u}(n)$, respectively.

We will not get into a deeper study of the theory of Lie groups and Lie algebras. We refer the reader to standard textbooks in the subject such as Helgason (2001), Humphreys (1978), Samelson (1990), Serre (2006), and Varadarajan (1984).

Exercises

Exercise 16.1
Verify the properties of the exponential map given in Proposition 16.1.

Exercise 16.2
Show that the exponential map $\exp : \mathfrak{gl}(2) \to GL(2)$ is not surjective.

Exercise 16.3

Let $N \subset GL(n)$ be the subgroup formed by all upper triangular matrices with diagonal elements all equal to 1. Show that N is a Lie subgroup, find its Lie algebra $\mathfrak{n}$ and prove that the exponential map $\exp : \mathfrak{n} \to N$ is a bijection.

Exercise 16.4

Let G be a compact Lie group. Show that $\exp : \mathfrak{g} \to G$ is surjective.

Hint: Use the fact, to be proved later, that any compact Lie group has a bi-invariant metric, i.e., a metric invariant under both right and left translations.

Exercise 16.5

Let G and H be Lie groups. Show that

(a) Every continuous homomorphism $\Phi : \mathbb{R} \to G$ is smooth.
(b) Every continuous homomorphism $\Phi : G \to H$ is smooth.
(c) If G and H are isomorphic as topological groups, then G and H are isomorphic as Lie groups.

Exercise 16.6

Let G be a Lie group with Lie algebra $\mathfrak{g}$ and let $H \subset G$ be a Lie subgroup with Lie algebra $\mathfrak{h} \subset \mathfrak{g}$. Show that $X \in \mathfrak{g}$ belongs to $\mathfrak{h}$ if and only if $\exp(tX) \in H$ for all $t \in \mathbb{R}$.

Exercise 16.7

Prove Proposition 16.2.

Hint: Show that H has a smooth structure compatible with the relative topology making (H, i) a submanifold of G, by considering the charts:

$$\{(H \cap hV, \exp^{-1} \circ L_h) : h \in H\}.$$

Then check that multiplication in H is smooth and use the previous exercise to complete the proof.

Lecture 17

Groups of Transformations

Let G be a group. Recall from Lecture 10 that we denote an action of G on a set M by a map $\Psi : G \times M \to M$, which we write as $(g, p) \mapsto g \cdot p$ and satisfies:

(i) $e \cdot p = p$, for all $p \in M$;

(ii) $g \cdot (h \cdot p) = (gh) \cdot p$, for all $g, h \in G$ and $p \in M$.

An action can also be viewed as a group homomorphism from G to the group of bijections of M. For each $g \in G$, we denote by Ψ_g the bijection

$$\Psi_g : M \to M, \quad p \mapsto g \cdot p$$

When G is a Lie group, M is a smooth manifold and the map $\Psi : G \times M \to M$ is smooth, we say that we have a **smooth action**. In this case, each $\Psi_g : M \to M$ is a diffeomorphism of M, so one also says that G is a **group of transformations** of M. For such a smooth action, the **isotropy subgroup** at $p \in M$ is the closed subgroup

$$G_p := \{ g \in G : g \cdot p = p \} \subset G,$$

hence it is an embedded Lie subgroup of G – see Theorem 16.1.

The results in Lecture 10 concerning smooth structures on orbits spaces of discrete group actions extend to arbitrary smooth actions of Lie groups. First, we call a smooth action $\Psi : G \times M \to M$ a

proper action if the map

$$G \times M \to M \times M, \quad (g, p) \mapsto (p, g \cdot p),$$

is proper. Here are a few examples.

Example 17.1.

(1) The action by translations of a Lie group G on itself, $G \times G \to G$, $(g, h) \mapsto gh$, is always proper.
(2) Smooth actions of *compact* Lie groups are always proper. Also, for a proper action $G \times M \to M$ the isotropy groups G_p are all compact (why?). So, for example, the action of $O(n)$ on $\mathbb{R}^n$ by matrix multiplication is proper (since $O(n)$ is compact), while the action of $SL(n)$ on $\mathbb{R}^n$ by matrix multiplication is not proper (since the isotropy group of 0 is $SL(n)$ which is not compact).
(3) Given a smooth proper action $G \times M \to M$ and a closed subgroup $H \subset G$, the restricted action $H \times M \to M$ is still a smooth proper action. For example, restricting the action by translations of $(\mathbb{R}^n, +)$ on itself, we obtain the smooth proper action of $(\mathbb{R}, +)$ on $\mathbb{R}^n$ given by

$$t \cdot (x^1, \ldots, x^n) := (x^1 + t, x^2, \ldots, x^n).$$

Next, recall that an action is **free** if the isotropy groups G_p are all trivial. We leave as an exercise to check that for free actions the orbits are copies of G.

Lemma 17.1. *Given a smooth free action* $G \times M \to M$ *and* $p \in M$ *the map*

$$\Psi_p : G \to M, \quad g \mapsto g \cdot p$$

is an injective immersion. In particular, its orbits are submanifolds of M diffeomorphic to G.

In general, the orbits are not embedded submanifolds: for example, the irrational lines on the torus $\mathbb{T}^2$ are the orbits of a free, smooth, action of $(\mathbb{R}, +)$. We will see later that the orbits of *any* action are immersed submanifolds.

For proper actions, the geometry of the orbits is much nicer. In particular, proper actions which are free have smooth orbit spaces.

Theorem 17.1. *Let* $\Psi : G \times M \to M$ *be a smooth action of a Lie group* G *on a manifold* M. *If the action is free and proper, then* $G \backslash M$ *has a unique smooth structure, compatible with the quotient topology, such that* $\pi : M \to G \backslash M$ *is a submersion, and*

$$\dim G \backslash M = \dim M - \dim G.$$

In particular, the orbits of a smooth, proper, and free action of G *are embedded submanifolds diffeomorphic to* G.

Proof. We apply Theorem 10.1 to the orbit equivalence relation defined by the action. This means that we need to verify that its graph:

$$R = \{(p, g \cdot p) : p \in M, g \in G\} \subset M \times M,$$

is a proper submanifold and that the restriction of the projection $p_1|_R : R \to M$ is a submersion.

Consider the map

$$\Phi : G \times M \to M \times M, \quad (g, p) \mapsto (p, g \cdot p),$$

whose image is precisely R. Since the action is assumed to be free, Φ is injective. On the other hand, its differential $d_{(g,p)}\Phi : T_g G \times T_p M \to T_p M \times T_{g \cdot p} M$ is the map

$$(\mathbf{v}, \mathbf{w}) \mapsto (\mathbf{w}, d\Psi_p \cdot \mathbf{v} + d\Psi_g \cdot \mathbf{w}).$$

Since this map is injective, we conclude that Φ is an injective immersion with image R. By assumption, Φ is proper so R is a proper submanifold of $M \times M$.

To verify that $p_1|_R : R \to M$ is a submersion, it is enough to show that the composition $p_1 \circ \Phi : G \times M \to M$ is a submersion. But this composition is just the projection $(g, p) \mapsto p$, which is obviously a submersion. $\square$

Example 17.2. Consider the action of $\mathbb{S}^1 = \{w \in \mathbb{C} : |w| = 1\}$ on the 3-sphere $\mathbb{S}^3 = \{(z_1, z_2) \in \mathbb{C}^2 : |z_1|^2 + |z_2|^2 = 1\}$, defined by

$$w \cdot (z_1, z_2) = (wz_1, wz_2).$$

This action is free and proper. Hence, the orbits of this action are embedded submanifolds of $\mathbb{S}^3$ diffeomorphic to $\mathbb{S}^1$. The orbit space

$\mathbb{S}^1\backslash\mathbb{S}^3$ is a smooth manifold. We will see later that this manifold is diffeomorphic to $\mathbb{S}^2$.

Let G be a Lie group and consider the action of G on itself by left translations

$$G \times G \to G, \quad (g,h) \mapsto gh.$$

This action is free and proper. If $H \subset G$ is a closed subgroup, then H is a Lie subgroup and the action of H on G, by left translation is also free and proper. The orbit space for this action consist of the right cosets

$$H\backslash G = \{Hg : g \in G\}.$$

From Theorem 17.1, we conclude that

Corollary 17.1. *Let G be a Lie group and let $H \subset G$ be a closed subgroup. Then $H\backslash G$ has a unique smooth structure, compatible with the quotient topology, such that $\pi : G \to H\backslash G$ is a submersion. In particular,*

$$\dim H\backslash G = \dim G - \dim H.$$

Remark 17.1. So far we have considered **left actions**. One can also consider **right actions** $M \times G \to M$, $(m,g) \to m \cdot g$, with axioms (i) and (ii) replaced by

(i) $p \cdot e = p$, for all $p \in M$;
(ii) $(p \cdot h) \cdot g = p \cdot (hg)$, for all $g, h \in G$ and $p \in M$.

Given a left action $(g,m) \mapsto g\cdot m$ one obtains a right action by setting $m\cdot g := g^{-1}\cdot m$, and conversely. Hence, every result about left actions yields a result about right actions, and conversely. For example, if G is a Lie group and $H \subset G$ is a closed subgroup, the right action of H on G by right translations is free and proper. Hence, the set of left cosets

$$G/H = \{gH : g \in G\},$$

also has a natural smooth structure.

Given two G-actions, $G \times M \to M$ and $G \times N \to N$, a **G-equivariant map** is a map $\Phi : M \to N$ such that

$$\Phi(g \cdot p) = g \cdot \Phi(p), \quad \forall g \in G, p \in M.$$

We say that two G-actions are **equivalent** if there exists a G-equivariant bijection between them. For example, given any action $\Psi : G \times M \to M$ and fixing $p \in M$, the map

$$\Psi_p : G \to M, \quad g \mapsto g \cdot p,$$

induces a bijection $\bar{\Psi}_p : G/G_p \to O_p$, where $O_p \subset M$ is the orbit through p. Note that G acts on the set of right cosets by left translations

$$G \times G/G_p \to G/G_p, \quad (h, gG_p) \mapsto (hg)G_p.$$

The map $\bar{\Psi}_p : G/G_p \to O_p$ is a G-equivariant bijection.

For a smooth action $\Psi : G \times M \to M$, the results above with $H = G_p$ show that G/G_p has a smooth structure and that the map

$$\bar{\Psi}_p : G/G_p \to M, \quad gG_p \mapsto g \cdot p,$$

is an injective immersion with image the orbit through p.

Theorem 17.2. *Let $\Psi : G \times M \to M$ be a smooth action of a Lie group G on a manifold M. The orbits of the action are regularly immersed submanifolds of M. Moreover, for every $p \in M$, the map*

$$\bar{\Psi}_p : G/G_p \to M, \quad gG_p \mapsto g \cdot p,$$

is a G-equivariant diffeomorphism between G/G_p and the orbit O_p.

Proof. Since G_p is a closed subgroup, by Corollary 17.1, the space G/G_p has a smooth structure. The map

$$\bar{\Psi}_p : G/G_p \to M, \quad gG_p \mapsto g \cdot p,$$

is an injective immersion whose image is the orbit through p. This makes the orbit an immersed submanifold and we leave it as an exercise to show that it is regularly immersed.

This smooth structure on the orbit does not depend on the choice of M since two points $p, q \in M$ which belong to the same orbit have conjugate isotropy groups:

$$q = g \cdot p \quad \Longrightarrow \quad G_q = gG_pg^{-1}.$$

It follows that $\Phi : G/G_p \to G/G_q$, $hG_p \mapsto ghg^{-1}G_q$, is a G-equivariant diffeomorphism which makes the following diagram commute

$$
\begin{array}{ccc}
G/G_p & \xrightarrow{\bar{\Psi}_p} & M \\
{\scriptstyle \Phi} \downarrow & & \downarrow {\scriptstyle \Psi_g} \\
G/G_q & \xrightarrow{\bar{\Psi}_q} & M
\end{array}
$$

Since $\Psi_g : M \to M$, $m \mapsto g \cdot m$, is a diffeomorphism, the two immersions give equivalent smooth structures on the orbit. $\qquad\square$

A **transitive action** $\Psi : G \times M \to M$ is an action with only one orbit. This means that for any pair of points $p, q \in M$, there exists $g \in G$ such that $q = g \cdot p$. In this case, fixing any point $p \in M$, we obtain an equivariant bijection $G/G_p \to M$. When the action is smooth, this gives an equivariant diffeomorphism between M and the quotient G/G_p. In this case, one also calls M a **homogeneous space**.

The homogeneous G-spaces are just the manifolds of the form G/H where $H \subset G$ is a closed subgroup. For a homogenous space G/H we have the natural G-action, induced from the action of G on itself by left translations. Homogenous spaces are particularly nice examples of manifolds. A manifold can be a homogeneous G-space for different choices of Lie groups, as illustrated in the following examples.

Example 17.3. Let $\mathbb{S}^3$ be the unit quarternions. Identifying $\mathbb{R}^3$ with the purely imaginary quaternions, we obtain an action of $\mathbb{S}^3$ on $\mathbb{R}^3$:

$$q \cdot v = qvq^{-1}.$$

It is easy to see that the orbits of this action are the spheres of radius r and the origin. Let us restrict the action to $\mathbb{S}^2$, the sphere of radius 1. An easy computation shows that the isotropy group of $p = (1, 0, 0)$ is the subgroup $\mathbb{S}^1 = (\mathbb{S}^3)_p \subset \mathbb{S}^3$ formed by quaternions of the form $q_0 + iq_1 + 0j + 0k$. It follows that the sphere is a homogeneous space

$$\mathbb{S}^2 \simeq \mathbb{S}^3/\mathbb{S}^1.$$

The surjective submersion $\pi : \mathbb{S}^3 \to \mathbb{S}^2$, $q \mapsto q \cdot (1, 0, 0)$, whose fibers are diffeomorphic to $\mathbb{S}^1$, is known as the **Hopf fibration**.

Example 17.4. Let $O(3) \times \mathbb{R}^3 \to \mathbb{R}^3$ be the standard action by matrix multiplication:

$$(A, \vec{v}) \mapsto A\vec{v}.$$

The orbits of this action are also the spheres the spheres of radius r and the origin. Again, we consider the sphere $\mathbb{S}^2$ of radius 1 and we let $p_N = (0, 0, 1) \in \mathbb{S}^2$ be the north pole. The isotropy group at p_N is

$$O(3)_{p_N} = \left\{ \left(\begin{array}{c|c} A & 0 \\ \hline 0 & 1 \end{array} \right) : A \in O(2) \right\}.$$

It follows that the map

$$O(3)/O(2) \to \mathbb{S}^2, \quad A\,O(3) \mapsto A \cdot p_N,$$

is a diffeomorphism.

An entirely similar reasoning shows that $\mathbb{S}^2$ is also diffeomorphic to the homogeneous space $SO(3)/SO(2)$. We leave as an exercise to check that this generalizes to higher dimensional spheres, so that

$$\mathbb{S}^d \simeq O(d+1)/O(d) \simeq SO(d+1)/SO(d).$$

Example 17.5. Let $G_k(\mathbb{R}^d)$ denote that set of all linear subspaces of $\mathbb{R}^d$ of dimension k. The usual action of the orthogonal group $O(d)$ on $\mathbb{R}^d$ by matrix multiplication induces an action $O(d) \times G_k(\mathbb{R}^d) \to G_k(\mathbb{R}^d)$, where an invertible linear transformation takes a linear subspace of dimension k to another one. It is easy to check that given any two k-dimensional linear subspaces $S_1, S_2 \subset \mathbb{R}^d$ there exists $A \in O(d)$ mapping S_1 onto S_2. In other words, the action $O(d) \times G_k(\mathbb{R}^d) \to G_k(\mathbb{R}^d)$ is transitive.

Then let $S_0 \in G_k(\mathbb{R}^d)$ be the subspace $\mathbb{R}^k \times \{0\} \subset \mathbb{R}^d$. The corresponding isotropy group is

$$O(d)_{S_0} = \left\{ \left(\begin{array}{c|c} A & 0 \\ \hline 0 & B \end{array} \right) \in O(d) : A \in O(k), B \in O(d-k) \right\},$$

so we have a bijection

$$O(d)/(O(k) \times O(d-k)) \to G_k(\mathbb{R}^d).$$

On $G_k(\mathbb{R}^d)$ we can consider the unique smooth structure for which this bijection becomes a diffeomorphism. This gives $G_k(\mathbb{R}^d)$ the structure of a manifold of dimension $k(d - k) = \dim O(d) - (\dim O(k) + \dim O(d - k))$. One can show that this smooth structure is independent of the choice of base point S_0. The manifold $G_k(\mathbb{R}^d)$ is called the **Grassmannian manifold** of k-planes in $\mathbb{R}^d$.

Since Lie groups have infinitesimal counterparts, it should come as no surprise that Lie group actions also have an infinitesimal counterpart. Let $\Psi : G \times M \to M$ be a smooth action, which we can view as a homomorphism

$$\widehat{\Psi} : G \to \operatorname{Diff}(M).$$

We think of $\operatorname{Diff}(M)$ as a Lie group with Lie algebra $\mathfrak{X}(M)$, then there should exist a homomorphism of Lie algebras

$$\psi = (\widehat{\Psi})_* : \mathfrak{g} \to \mathfrak{X}(M).$$

In fact, if $X \in \mathfrak{g}$ and $p \in M$ the curve

$$t \mapsto \exp(-tX) \cdot p,$$

goes through p at $t = 0$, and it is defined and smooth for $t \in \mathbb{R}$. We can then define a vector field $\psi(X)$ in M by

$$\psi(X)_p := \frac{\mathrm{d}}{\mathrm{d}t} \exp(-tX) \cdot p \Big|_{t=0}. \tag{17.1}$$

The proof of the following lemma is left as an exercise.

Lemma 17.2. *For each $X \in \mathfrak{g}$, the vector field $\psi(X)$ is smooth. The resulting map $\psi : \mathfrak{g} \to \mathfrak{X}(M)$ is linear and satisfies*

$$\psi([X, Y]_{\mathfrak{g}}) = [\psi(X), \psi(Y)], \quad \forall X, Y \in \mathfrak{g}.$$

Remark 17.2. The appearance of a minus sign in formula (17.1) is easy to explain. With our conventions, where the Lie algebra of a Lie group is formed by the left-invariant vector fields, the Lie algebra of the group of diffeomorphisms $\operatorname{Diff}(M)$ is formed by the vector fields $\mathfrak{X}(M)$ with a Lie bracket which is the *symmetric* of the usual Lie bracket of vector fields. One can see this, for example, by determining the 1-parameter subgroups of the group of diffeomorphims.

We could have defined the Lie bracket of vector fields with the opposite sign, but this would lead to the presence of negative signs in other formulas.

Also, given a right Lie group action $M \times G \to M$, $(p, g) \mapsto p \cdot g$, we obtain a Lie algebra homomorphism $\psi : \mathfrak{g} \to \mathfrak{X}(M)$ by setting now

$$\psi(X)_p := \frac{d}{dt} \exp(tX) \cdot p \bigg|_{t=0}. \tag{17.2}$$

The reason for the absence of a minus sign is discussed in the exercises.

The lemma above suggests the following definition.

Definition 17.1. An **infinitesimal action** of a Lie algebra $\mathfrak{g}$ on a manifold M is a homomorphism of Lie algebras $\psi : \mathfrak{g} \to \mathfrak{X}(M)$.

Example 17.6. The Lie algebra $\mathfrak{so}(3)$ has a basis consisting of the skew-symmetric matrices:

$$X = \begin{bmatrix} 0 & 0 & 0 \\ 0 & 0 & -1 \\ 0 & 1 & 0 \end{bmatrix}, \quad Y = \begin{bmatrix} 0 & 0 & 1 \\ 0 & 0 & 0 \\ -1 & 0 & 0 \end{bmatrix}, \quad Z = \begin{bmatrix} 0 & -1 & 0 \\ 1 & 0 & 0 \\ 0 & 0 & 0 \end{bmatrix}.$$

In this basis, we have the following Lie bracket relations:

$$[X, Y] = Z, \quad [Y, Z] = X, \quad [Z, X] = Y.$$

For the usual action of $SO(3)$ on $\mathbb{R}^3$ by rotations, we can compute the infinitesimal action as follows. First, we compute the exponential

$$\exp(tX) = \begin{bmatrix} 1 & 0 & 0 \\ 0 & \cos t & -\sin t \\ 0 & \sin t & \cos t \end{bmatrix}.$$

Then,

$$\psi(X)_{(x,y,z)} = \frac{d}{dt} \exp(-tX) \cdot (x, y, z) \bigg|_{t=0} = z \frac{\partial}{\partial y} - y \frac{\partial}{\partial z}.$$

Similarly, we compute:

$$\psi(Y) = x\frac{\partial}{\partial z} - z\frac{\partial}{\partial x}, \quad \psi(Z) = y\frac{\partial}{\partial x} - x\frac{\partial}{\partial y}.$$

The vector fields $\{\psi(X), \psi(Y), \psi(Z)\}$ are called the **infinitesimal generators** of the action. Using that ψ is a homomorphism of Lie algebras, one recovers the Lie brackets of Example 12.1.

A smooth action $\Psi : G \times M \to M$ induces an infinitesimal action $\psi : \mathfrak{g} \to \mathfrak{X}(M)$. The converse does not necessarily hold, as shown in the next example.

Example 17.7. Any non-zero vector field X on a manifold M determines an infinitesimal action of the Lie algebra $\mathfrak{g} = \mathbb{R}$ on M by setting

$$\psi : \mathbb{R} \to \mathfrak{X}(M), \quad \lambda \mapsto \lambda X.$$

This infinitesimal action integrates to a Lie group action of $(\mathbb{R}, +)$ on M if and only if the vector field X is complete. Namely, the action is given by the flow of the vector field, i.e.,

$$\Psi : \mathbb{R} \times M \to M, \quad (t, x) \mapsto \phi_X^t(x).$$

The Lie group $\mathbb{S}^1$ also has Lie algebra $\mathbb{R}$ but, even if the vector field X is complete, there may not exist an action $\Psi : \mathbb{S}^1 \times M \to M$ with $\Psi_* = \psi$, since the flow may not be periodic.

In order to integrate an infinitesimal Lie algebra action to a Lie group action there are two issues one needs to take care of. First, if an action $\psi : \mathfrak{g} \to \mathfrak{X}(M)$ is induced from a Lie group action $G \times M \to M$ then the infinitesimal generators $\psi(X) \in \mathfrak{X}(M)$ are all complete vector fields. Second, as suggested by Lie's second theorem, one should require G to be 1-connected. A proof that these two conditions are sufficient can be found in Lee (2013, Theorem 20.16).

Theorem 17.3. *Let $\psi : \mathfrak{g} \to \mathfrak{X}(M)$ be a Lie algebra action such that $\psi(X)$ is complete, for all $X \in \mathfrak{g}$. Then there exists a smooth action $\Psi : G \to \mathrm{Diff}(M)$ with $\Psi_* = \psi$, where G is the 1-connected Lie group with Lie algebra $\mathfrak{g}$.*

Corollary 17.2. *If M is a compact manifold and G is a 1-connected Lie group with Lie algebra $\mathfrak{g}$, then every infinitesimal Lie algebra* *action* $\Psi : \mathfrak{g} \to \mathfrak{X}(M)$ *integrates to a smooth Lie group action* $\Psi : G \times M \to M$.

Example 17.8. A **representation of a Lie group** G in a vector space V is a Lie group homomorphism $\widehat{\Psi} : G \to GL(V)$. A **representation of a Lie algebra** $\mathfrak{g}$ is a Lie algebra homomorphism $\rho : \mathfrak{g} \to \mathfrak{gl}(V)$.

Note that $GL(V) \subset \mathrm{Diff}(V)$, so representation of a Lie group G is the same as a smooth **linear Lie group action**, i.e., an action $\Psi : G \times V \to V$ where each Ψ_g is linear.

On the other hand, one calls a vector field X on a vector space V a **linear vector field** if for any linear function $l \in V^*$, the function $X(l)$ is a also linear. A linear vector field $X \in \mathfrak{X}_{\mathrm{lin}}(V)$ determines a linear map $X : V^* \to V^*$, and its transpose is an element $-X^* \in \mathfrak{gl}(V)$. The converse also holds, so we have a 1:1 correspondence associating to a linear map $T : V \to V$ a linear vector field $X_T \in \mathfrak{X}(V)$ such that

$$X_T(l) = -l \circ T,$$

for any linear function $l : V \to \mathbb{R}$. The minus sign guarantees that

$$[X_{T_1}, X_{T_2}] = X_{[T_1, T_2]},$$

so one has an injective Lie algebra morphism $\mathfrak{gl}(V) \hookrightarrow \mathfrak{X}(V)$, $T \mapsto X_T$, whose image is the subalgebra $\mathfrak{X}_{\mathrm{lin}}(V)$ of linear vector fields. Hence, a representation of a Lie algebra $\rho : \mathfrak{g} \to \mathfrak{gl}(V)$ is the same as a **linear Lie algebra action** $\psi : \mathfrak{g} \to \mathfrak{X}_{\mathrm{lin}}(V)$.

Since linear vector fields are always complete, we conclude from Theorem 17.3 that for a 1-connected Lie group there is a 1:1 correspondence between representations of G and representations of its Lie algebra $\mathfrak{g}$.

Exercises

Exercise 17.1
Show that the orbits of a smooth action are regularly immersed submanifolds.

Exercise 17.2

Let $\Psi : G \times M \to M$ be a proper and free smooth action and denote by $B = G\backslash M$ its orbit space. Show that the projection $\pi : M \to B$ is locally trivial, i.e., for any $b \in B$ there exists a neighborhood $b \in U \subset B$ and diffeomorphism

$$\sigma : \pi^{-1}(U) \to G \times U, \quad q \mapsto (\chi(q), \pi(q)),$$

such that

$$\sigma(g \cdot q) = (g\chi(q), \pi(q)), \quad \forall q \in \pi^{-1}(U), g \in G.$$

Exercise 17.3

Let G be a connected Lie group and $H \subset G$ a closed connected subgroup with Lie algebra $\mathfrak{h} \subset \mathfrak{g}$. Show the following:

(a) H is a normal subgroup of G if and only if $\mathfrak{h}$ is an **ideal** of $\mathfrak{g}$, i.e.,

$$\forall X \in \mathfrak{g}, Y \in \mathfrak{h}, \quad [X, Y] \in \mathfrak{h}.$$

(b) If H is normal in G, then G/H is a Lie group and $\pi : G \to G/H$ is a Lie group homomorphism.

Exercise 17.4

Let G be a Lie group and let $H \subset G$ be a closed subgroup. Show that if G/H and H are both connected then G is connected. Conclude from this that the groups $SO(d)$, $SU(d)$, and $U(d)$ are all connected. Show that $O(d)$ and $GL(d)$ have two connected components.

Exercise 17.5

Let $\Psi : G \times M \to M$ be a smooth transitive action with M connected. Show the following:

(a) The connected component of the identity G^0 also acts transitively on M.

(b) For all $p \in M$, G/G^0 is diffeomorphic to $G_p/(G_p \cap G^0)$.

(c) If G_p is connected for some $p \in M$, then G is connected.

Exercise 17.6

For a Lie group G one defines the **adjoint representation**

$$\mathrm{Ad} : G \to GL(\mathfrak{g}), \quad g \mapsto \mathrm{Ad}_g := \mathrm{d}_e i_g,$$

where $i_g : G \to G$, $h \mapsto ghg^{-1}$ denotes conjugation by g. Show that the induced Lie algebra representation is

$$\mathrm{ad} : \mathfrak{g} \to \mathfrak{gl}(\mathfrak{g}), \quad \mathrm{ad}_X(Y) = [X, Y].$$

Exercise 17.7

Find the orbits and the isotropy groups for the adjoint representations of the 3-dimensional Lie groups $SL(2)$, $SO(3)$, and $SU(2)$.

Exercise 17.8

Show that the real and complex projective spaces can be exhibited as homogenous spaces

$$\mathbb{RP}^d \simeq SO(d+1)/O(d), \quad \mathbb{CP}^d \simeq SU(d+1)/U(d).$$

Exercise 17.9

For a vector space V of dimension d denote by $S_k(V)$ the set of all k-frames of V:

$$S_k(V) = \{(\mathbf{v}_1, \ldots, \mathbf{v}_k) \in V$$
$$\times \ldots \times V : \mathbf{v}_1, \ldots, \mathbf{v}_k \text{ are linearly independent}\}.$$

Show that $S_k(V)$ is a homogenous space of dimension dk. $S_k(V)$ is called the **Stiefel manifold** of k-frames of V.

Hint: Fix a basis of V and consider the action $GL(d)$ in V by matrix multiplication.

Exercise 17.10

Give a proof of Lemma 17.2 and explain the appearance/absence of signs in formulas (17.1) and (17.2).

Hint: If G is a Lie group with Lie algebra $\mathfrak{g}$, for each $X \in \mathfrak{g}$ denoted by $\overline{X} \in \mathfrak{X}(G)$ the *right invariant* vector field in G with $\overline{X}_e = X$. Show that

$$[\overline{X}, \overline{Y}] = -\overline{[X, Y]}, \quad \forall X, Y \in \mathfrak{g},$$

and express the infinitesimal action $\phi : \mathfrak{g} \to \mathfrak{X}(M)$ in terms of right invariant vector fields.

Exercise 17.11

Let $\Psi : G \times M \to M$ be a smooth action with associated infinitesimal action $\psi : \mathfrak{g} \to \mathfrak{X}(M)$. If G_p is the isotropy group at p, show that its

Lie algebra is the **isotropy subalgebra**:

$$\mathfrak{g}_p = \{X \in \mathfrak{g} : \psi(X)_p = 0\}.$$

Exercise 17.12

Let $\Psi : G \times M \to M$ be a smooth action with associated infinitesimal action $\psi : \mathfrak{g} \to \mathfrak{X}(M)$. We call $p_0 \in M$ a **fixed point of the action** if:

$$g \cdot p_0 = p_0, \forall g \in G.$$

Show that if p_0 is a fixed point of the action then

(a) Ψ induces a representation $\Psi_{p_0} : G \to GL(T_{p_0}M)$;
(b) ψ induces a representation $\psi_{p_0} : \mathfrak{g} \to \mathfrak{gl}(T_{p_0}M)$;
(c) The group representation Ψ_{p_0} integrates the Lie algebra representation ψ_{p_0}.

PART 3
Differential Forms

Lecture 18

Differential Forms and Tensor Fields

For a finite-dimensional real vector space V, we denote the dual vector space by

$$V^* := \{\alpha : \alpha : V \to \mathbb{R} \text{ is a linear map}\}.$$

Its **tensor algebra** is

$$\bigotimes V^* = \bigoplus_{k=0}^{+\infty} \otimes^k V^*,$$

where $\otimes^k V^*$ is identified with the space of k-multilinear maps $V \times \cdots \times V \to \mathbb{R}$. It is furnished with the **tensor product** $\otimes^k V^* \times \otimes^l V^* \to \otimes^{k+l} V^*$ which to multilinear maps $\alpha \in \otimes^k V^*$ and $\beta \in \otimes^l V^*$ associates the multilinear map $\alpha \otimes \beta \in \otimes^{k+l} V^*$ defined by

$$\alpha \otimes \beta(v_1, \ldots, v_{k+l}) := \alpha(v_1, \ldots, v_k)\beta(v_{k+1}, \ldots, v_{k+l}).$$

The **exterior algebra** of V is the quotient of the tensor algebra by the two-sided ideal generated by all elements $\alpha \otimes \alpha$, $\alpha \in V^*$. It is furnished with the **exterior product** $\wedge : \wedge^k V^* \times \wedge^l V^* \to \wedge^{k+l} V^*$. It can also be viewed as the subspace (not a subalgebra!) of the tensor algebra

$$\bigwedge V^* = \bigoplus_{k=0}^{d} \wedge^k V^* \subset \bigotimes V^*,$$

where $\wedge^k V^*$ consists of all alternating k-multilinear maps $V \times \cdots \times V \to \mathbb{R}$. For example, if $\alpha_i \in V^*$, then $\alpha_1 \wedge \cdots \wedge \alpha_k$ is the alternating

175

k-multilinear map

$$\alpha_1 \wedge \cdots \wedge \alpha_k(\mathbf{v}_1, \ldots, \mathbf{v}_k) = \det(\alpha_i(\mathbf{v}_j))_{i,j=1}^k.$$

If $T : V \to W$ is a linear transformation between two finite-dimensional vector spaces, its transpose is the linear transformation $T^* : W^* \to V^*$ defined by

$$(T^*\alpha)(\mathbf{v}) := \alpha(T\mathbf{v}).$$

More generally, there exists an induced linear map $T^* : \wedge^k W^* \to \wedge^k V^*$ defined by

$$(T^*\omega)(\mathbf{v}_1, \ldots, \mathbf{v}_k) := \omega(T\mathbf{v}_1, \ldots, T\mathbf{v}_k).$$

There is the restriction of a similarly defined map $T^* : \otimes^k W^* \to \otimes^k V^*$.

Let now M be a smooth manifold. If $(x^1, \ldots, x^d)$ are local coordinates around $p \in M$, we know that the tangent vectors

$$\left\{ \frac{\partial}{\partial x^1}\bigg|_p, \ldots, \frac{\partial}{\partial x^d}\bigg|_p \right\},$$

yield a basis for $T_p M$, while the differentials

$$\left\{ d_p x^1, \ldots, d_p x^d \right\},$$

yield a dual basis for $T_p^* M$. If we take tensor products and exterior products of elements of these bases, we obtain bases for $\otimes^k T_p M$, $\wedge^k T_p M$, $\otimes^k T_p^* M$, $\wedge^k T_p^* M$, etc. For example, the space $\wedge^k T_p^* M$ has the basis

$$d_p x^{i_1} \wedge \cdots \wedge d_p x^{i_k} \quad (i_1 < \cdots < i_k).$$

As in the case of the tangent and cotangent spaces, we are interested in the spaces $\otimes^k T_p M$, $\wedge^k T_p M$, $\otimes^k T_p^* M$, $\wedge^k T_p^* M$, etc., when p varies. For example, we define

$$\wedge^k T^* M := \bigcup_{p \in M} \wedge^k T_p^* M.$$

and we have a projection $\pi : \wedge^k T^* M \to M$. We call $\wedge^k T^* M$ the **k-exterior bundle** of M. We leave as an exercise to check that, just like the case of the tangent bundle, one has natural smooth structures on this bundle.

Proposition 18.1. *There exists a canonical smooth structure on $\wedge^k T^* M$ such that the canonical projection in M is a submersion.*

One has also smooth structures on the bundles $\wedge^k TM$, $\otimes^k T^* M$, $\otimes^k TM$, $\otimes^k TM \otimes^s T^* M$, etc. For any map $\pi : E \to M$ a **section** is a map $s : M \to E$ such that $\pi \circ s = \mathrm{Id}$.

Definition 18.1. Let M be a manifold.

(i) A **differential form of degree** k is a section of π : $\wedge^k T^* M \to M$.
(ii) A **multivector field of degree** k is a section of π : $\wedge^k TM \to M$.
(iii) A **tensor field of degree** (k, s) is a section of π : $\otimes^k TM \otimes^s T^* M \to M$.

We will consider only smooth differential forms, smooth multivector fields and smooth tensor fields, meaning that the corresponding sections are smooth maps. Note that $\wedge^k TM$ and $\wedge^k T^* M$ are submanifolds of $\otimes^k TM \otimes^s T^* M$, so a multivector field of degree k and a differential form of degree k are examples of tensor fields of degree $(k, 0$ and $(0, k)$, respectively. Of course, there are tensor fields of degree $(k, 0$ and $(0, k)$ which are not alternating: for example, a Riemannian structure (see Exercise 11.9) is a tensor field of degree $(0, 2)$ which is symmetric, rather than alternating.

If $(U, \phi) = (U, x^1, \dots, x^d)$ is a chart then a tensor field θ of degree (k, s) takes the local expression:

$$\theta|_U = \sum_{i_1,\dots,i_k,j_1,\dots,j_s} \theta^{i_1,\dots,i_k}_{j_1,\dots,j_s} \frac{\partial}{\partial x^{i_1}} \otimes \cdots \otimes \frac{\partial}{\partial x^{i_k}} \otimes \mathrm{d}x^{j_1} \otimes \cdots \otimes \mathrm{d}x^{j_k}.$$

It should be clear that θ is smooth if and only if for any open cover by charts the components $\theta^{i_1,\dots,i_k}_{j_1,\dots,j_s}$ are smooth function in $C^\infty(U)$.

On the other hand, a smooth k-differential form ω can be written in a local chart as

$$\omega|_U = \sum_{i_1 < \cdots < i_k} \omega_{i_1 \cdots i_k} \mathrm{d}x^{i_1} \wedge \cdots \wedge \mathrm{d}x^{i_k} = \sum_{i_1 \cdots i_k} \frac{1}{k!} \omega_{i_1 \cdots i_k} \mathrm{d}x^{i_1} \wedge \cdots \wedge \mathrm{d}x^{i_k},$$

where the components $\omega_{i_1\cdots i_k} \in C^\infty(U)$ are alternating, i.e., for every permutation $\sigma \in S_k$, one has

$$\omega_{\sigma(i_1)\cdots\sigma(i_k)} = (-1)^{\operatorname{sgn}\sigma}\omega_{i_1\cdots i_k}.$$

Similarly, a smooth k-multivector field π can be written in a local chart as

$$\pi|_U = \sum_{i_1<\cdots<i_k} \pi^{i_1\cdots i_k}\frac{\partial}{\partial x^{i_1}}\wedge\cdots\wedge\frac{\partial}{\partial x^{i_k}} = \sum_{i_1\cdots i_k}\frac{1}{k!}\pi^{i_1\cdots i_k}\frac{\partial}{\partial x^{i_1}}\wedge\cdots\wedge\frac{\partial}{\partial x^{i_k}},$$

where the components $\pi^{i_1\cdots i_k} \in C^\infty(U)$ are alternating.

If $(U,\phi) = (U,x^1,\ldots,x^d)$ and $(V,\psi) = (V,y^1,\ldots,y^d)$ are charts with $U\cap V \neq \emptyset$, then on the intersection we have two local coordinate expressions for a k-form ω:

$$\omega|_{U\cap V} = \sum_{i_1<\cdots<i_k}\omega_{i_1\cdots i_k}dx^{i_1}\wedge\cdots\wedge dx^{i_k} = \sum_{j_1<\cdots<j_k}\overline{\omega}_{j_1\cdots j_k}dy^{j_1}\wedge\cdots\wedge dy^{j_k}.$$

Recalling the transformation formulas

$$\frac{\partial}{\partial x^i} = \sum_{j=1}^d \frac{\partial y^j}{\partial x^i}\frac{\partial}{\partial y^i}, \quad dx^i = \sum_{j=1}^d \frac{\partial x^i}{\partial y^j}dy^j,$$

one sees that the components of the forms relative to the two charts are related by

$$\overline{\omega}_{j_1\cdots j_k}(y) = \sum_{i_1<\cdots<i_k}\omega_{i_1\cdots i_k}(\phi\circ\psi^{-1}(y))\frac{\partial(x^{i_1}\cdots x^{i_k})}{\partial(y^{j_1}\cdots y^{j_k})}(y).$$

The symbol on the right side of this expression is an abbreviation for the minor consisting of the rows $i_1,\ldots,i_k$ and the columns $j_1,\ldots,j_k$ of the Jacobian matrix of the change of coordinates $\phi\circ\psi^{-1}: \psi(U\cap V) \to \phi(U\cap V)$.

We leave it as an exercise to determine the formulas of transformation of variables for multivector fields and tensor fields.

Remark 18.1. One maybe intrigued with the relative positions of the indices, as subscripts and superscripts, in the different objects. The convention that we follow is such that an index is only summed if it appears in a formula repeated both as a subscript and as a superscript. With this convention, one can even omit the summation sign from the formula, with the agreement that one sums over an index whenever that index is repeated. This convention is called the **Einstein convention sum**.

From now on we will concentrate on the study of differential forms. Although other objects, such as multivector fields and tensor fields, are also interesting, differential forms play a more fundamental role because, as we will see later, they are the objects one can integrate over a manifold.

We will denote the vector space of smooth differential forms of degree k on a manifold M by $\Omega^k(M)$. Given a differential form $\omega \in \Omega^k(M)$ its value at a point $\omega_p \in \wedge^k T_p^* M$ can be seen as an alternating, multilinear map

$$\omega_p : T_p M \times \cdots \times T_p M \to \mathbb{R}.$$

Hence, if $X_1, \ldots, X_k \in \mathfrak{X}(M)$ are smooth vector fields M we obtain a smooth function $\omega(X_1, \ldots, X_k) \in C^\infty(M)$:

$$p \mapsto \omega_p(X_1|_p, \ldots, X_k|_p).$$

Therefore, every differential form $\omega \in \Omega^k(M)$ can be seen as a map

$$\omega : \mathfrak{X}(M) \times \cdots \times \mathfrak{X}(M) \to C^\infty(M).$$

This map is $C^\infty(M)$-multilinear and alternating. Conversely, every $C^\infty(M)$-multilinear, alternating, map $\mathfrak{X}(M) \times \cdots \times \mathfrak{X}(M) \to C^\infty(M)$ defines a smooth differential form. This is usually the simplest way to specify a smooth differential form.

We introduced now several basic algebraic operations with differential forms.

Exterior Product of Differential Forms

The exterior (or wedge) product $\wedge$ in the exterior algebra $\wedge T_p^* M$ induces an exterior (or wedge) product of differential forms

$$\wedge : \Omega^k(M) \times \Omega^s(M) \to \Omega^{k+s}(M), \ (\omega \wedge \eta)_p \equiv \omega_p \wedge \eta_p.$$

We consider the space of all differential forms:

$$\Omega(M) := \bigoplus_{k=0}^{d} \Omega^k(M).$$

where we convention that $\Omega^0(M) := C^\infty(M)$. If we let $f \wedge \omega := f\omega$, then the exterior product satisfies:

(i) $(f\omega + g\eta) \wedge \theta = f\omega \wedge \theta + g\eta \wedge \theta$;

(ii) $\omega \wedge \eta = (-1)^{\deg \omega \deg \eta} \eta \wedge \omega$;

(iii) $(\omega \wedge \eta) \wedge \theta = \omega \wedge (\eta \wedge \theta)$;

(iv) $\alpha_1 \wedge \cdots \wedge \alpha_k(X_1, \ldots, X_k) = \det [\alpha_i(X_j)]_{i,j=1}^{k}$.

The first 3 properties say that $\Omega(M)$ is a **Grassmann algebra** over the ring $C^\infty(M)$. These 4 properties is all that we need to know to compute exterior products in local coordinates.

Example 18.1. In $\mathbb{R}^4$, with coordinates (x, y, z, w), consider the differential forms of degree 2:

$$\omega = (x + w^2)\mathrm{d}x \wedge \mathrm{d}y + e^z \mathrm{d}x \wedge \mathrm{d}w + \cos x \mathrm{d}y \wedge \mathrm{d}z,$$

$$\eta = x\mathrm{d}y \wedge \mathrm{d}z - e^z \mathrm{d}z \wedge \mathrm{d}w.$$

Then using only the first 3 properties above we find

$$\omega \wedge \eta = -(x + w^2)e^z \mathrm{d}x \wedge \mathrm{d}y \wedge \mathrm{d}z \wedge \mathrm{d}w + x e^z \mathrm{d}x \wedge \mathrm{d}w \wedge \mathrm{d}y \wedge \mathrm{d}z$$

$$= -w^2 e^z \mathrm{d}x \wedge \mathrm{d}y \wedge \mathrm{d}z \wedge \mathrm{d}w.$$

Also, if we would like to compute, e.g., η on the vector fields $X = y\frac{\partial}{\partial z} - \frac{\partial}{\partial y}$ and $Y = e^z\frac{\partial}{\partial w}$ we can use property (iv) to obtain

$$\eta(X, Y) = x\mathrm{d}y \wedge \mathrm{d}z(X, Y) - e^z \mathrm{d}z \wedge \mathrm{d}w(X, Y)$$

$$= x \begin{vmatrix} \mathrm{d}y(X) & \mathrm{d}y(Y) \\ \mathrm{d}z(X) & \mathrm{d}z(Y) \end{vmatrix} - e^z \begin{vmatrix} \mathrm{d}z(X) & \mathrm{d}z(Y) \\ \mathrm{d}w(X) & \mathrm{d}w(Y) \end{vmatrix}$$

$$= x \begin{vmatrix} -1 & 0 \\ y & 0 \end{vmatrix} - e^z \begin{vmatrix} y & 0 \\ 0 & e^z \end{vmatrix} = -y e^{2z}$$

Pull-Back of Differential Forms

Let $\Phi : M \to N$ be a smooth map. For each $p \in M$, the transpose of the differential $(\mathrm{d}_p\Phi)^* : T^*_{\Phi(p)}N \to T^*_p M$, induces a linear map

$$(\mathrm{d}_p\Phi)^* : \wedge^k T^*_{\Phi(p)}N \to \wedge^k T^*_p M.$$

The **pull-back of differential forms** $\Phi^* : \Omega^k(N) \to \Omega^k(M)$ is defined by

$$(\Phi^*\omega)(X_1, \ldots, X_k)_p = ((\mathrm{d}_p\Phi)^*\omega)(X_1|_p, \ldots, X_k|_p)$$
$$= \omega_{\Phi(p)}(\mathrm{d}_p\Phi \cdot X_1|_p, \ldots, \mathrm{d}_p\Phi \cdot X_k|_p).$$

The last expression shows that $\Phi^*\omega$ is a $C^\infty(M)$-multilinear, alternating, map $\mathfrak{X}(M) \times \cdots \times \mathfrak{X}(M) \to C^\infty(M)$, hence it is a smooth differential form in M.

It is easy to check that for any smooth map $\Phi : M \to N$, the pull-back $\Phi^* : \Omega(N) \to \Omega(M)$ satisfies:

(i) $\Phi^*(a\omega + b\eta) = a\Phi^*\omega + b\Phi^*\eta$, $a, b \in \mathbb{R}$;
(ii) $\Phi^*(\omega \wedge \eta) = \Phi^*\omega \wedge \Phi^*\eta$;
(iii) $\Phi^*(f\omega) = (f \circ \Phi)\Phi^*\omega$, $f \in C^\infty(M)$;
(iv) $\Phi^*(\mathrm{d}f) = \mathrm{d}(f \circ \Phi)$.

In the last property, for a smooth function $f : N \to \mathbb{R}$ we view $\mathrm{d}f : TN \to \mathbb{R}$ as a differential form of degree 1. So (iv) is just the chain rule. On the other hand, since $f \circ \Phi = \Phi^* f$, property (iii) is a special case of (ii). The first 2 properties say that $\Phi^* : \Omega(N) \to \Omega(M)$ is a homomorphism of Grassmann algebras. These properties is all that it is needed to compute pull-backs in local coordinates.

Example 18.2. Let $\Phi : \mathbb{R}^2 \to \mathbb{R}^4$ be the smooth map

$$\Phi(u, v) = \left(u + v, u - v, v^2, \frac{1}{1 + u^2} \right).$$

In order to compute the pull-back under Φ of the form

$$\eta = x\,\mathrm{d}y \wedge \mathrm{d}z - e^z \mathrm{d}z \wedge \mathrm{d}w \in \Omega^2(\mathbb{R}^4),$$

we proceed as follows using the properties above:

$$\Phi^*\eta = (x \circ \Phi)\mathrm{d}(y \circ \Phi) \wedge \mathrm{d}(z \circ \Phi) - e^{(z \circ \Phi)}\mathrm{d}(z \circ \Phi) \wedge \mathrm{d}(w \circ \Phi)$$

$$= (u + v)\mathrm{d}(u - v) \wedge \mathrm{d}(v^2) - e^{v^2}\mathrm{d}(v^2) \wedge \mathrm{d}\left(\frac{1}{1 + u^2}\right)$$

$$= (u + v)\mathrm{d}u \wedge 2v\mathrm{d}v - 2ve^{v^2}\mathrm{d}v \wedge \frac{-2u\mathrm{d}u}{(1 + u^2)^2}$$

$$= \left(2v(u + v) - \frac{4uve^{v^2}}{(1 + u^2)^2}\right)\mathrm{d}u \wedge \mathrm{d}v.$$

In other words, to compute the pull-back $\Phi^*\eta$ one simply replaces in η the coordinates (x, y, z, w) by its expressions in terms of the coordinates (u, v).

When (N, i) is a submanifold of M the pull-back of a differential form $\omega \in \Omega^k(M)$ by the inclusion map $i : N \hookrightarrow M$ is called the **restriction of the differential form** ω to N. Often one denotes the restriction $\omega|_N$ instead of $i^*\omega$. Sometimes, one even drops the restriction sign.

Example 18.3. For the sphere

$$\mathbb{S}^2 = \{(x, y, z) \in \mathbb{R}^3 : x^2 + y^2 + z^2 = 1\},$$

one can write

$$\omega = (x\,\mathrm{d}y \wedge \mathrm{d}z + y\,\mathrm{d}z \wedge \mathrm{d}x + z\,\mathrm{d}x \wedge \mathrm{d}y)|_{\mathbb{S}^2},$$

meaning that ω is the pull-back by the inclusion $i : \mathbb{S}^2 \hookrightarrow \mathbb{R}^3$ of the differential form $x\mathrm{d}y \wedge \mathrm{d}z + y\mathrm{d}z \wedge \mathrm{d}x + z\mathrm{d}x \wedge \mathrm{d}y \in \Omega^2(\mathbb{R}^3)$. In spherical coordinates (r, θ, φ) (see Exercise 5.2) one finds

$$\omega = -r^3 \sin\varphi\,\mathrm{d}\theta \wedge \mathrm{d}\varphi,$$

so that

$$\omega|_{\mathbb{S}^2} = -\sin\varphi\,\mathrm{d}\theta \wedge \mathrm{d}\varphi.$$

One should also note that if $\Phi : M \to N$ and $\Psi : N \to Q$ are smooth maps, then $\Psi \circ \Phi : M \to Q$ is a smooth map and we have:

$$(\Psi \circ \Phi)^*\omega = \Phi^*(\Psi^*\omega).$$

In categorical language, we have a *contravariant functor* from the category of smooth manifolds to the category of Grassmann algebras, which to a smooth manifold M associates the algebra $\Omega(M)$ and to a smooth map $\Phi : M \to N$ associates a homomorphism $\Phi^* : \Omega(N) \to \Omega(M)$.

Interior Product

Given a vector field $X \in \mathfrak{X}(M)$ and a differential form $\omega \in \Omega^k(M)$, the **interior product** of ω by X, denoted $i_X\omega \in \Omega^{k-1}(M)$, is the differential form of degree $(k-1)$ defined by

$$i_X\omega(X_1, \ldots, X_{k-1}) = \omega(X, X_1, \ldots, X_{k-1}).$$

Since $i_X\omega : \mathfrak{X}(M) \times \cdots \times \mathfrak{X}(M) \to C^\infty(M)$ is a $C^\infty(M)$-multilinear, alternating, map, it is indeed a smooth differential form of degree $k - 1$.

It is easy to check that the following properties hold:

(i) $i_X(f\omega + g\theta) = f i_X\omega + g i_X\theta$;
(ii) $i_X(\omega \wedge \theta) = (i_X\omega) \wedge \theta + (-1)^{\deg \omega}\omega \wedge (i_X\theta)$;
(iii) $i_{(fX+gY)}\omega = f i_X\omega + g i_Y\omega$;
(iv) $i_X(\mathrm{d}f) = X(f)$.

Again, these properties is all that it is needed to compute interior products in local coordinates.

Example 18.4. In $\mathbb{R}^3$, let $\omega = e^x \mathrm{d}x \wedge \mathrm{d}y + e^z \mathrm{d}y \wedge \mathrm{d}z$, and $X = x\frac{\partial}{\partial y} - y\frac{\partial}{\partial x}$. Then,

$$i_{\frac{\partial}{\partial x}}(\mathrm{d}x \wedge \mathrm{d}y) = \left(i_{\frac{\partial}{\partial x}}\mathrm{d}x\right) \wedge \mathrm{d}y - \mathrm{d}x \wedge \left(i_{\frac{\partial}{\partial y}}\mathrm{d}y\right) = \mathrm{d}y,$$

$$i_{\frac{\partial}{\partial y}}(\mathrm{d}x \wedge \mathrm{d}y) = \left(i_{\frac{\partial}{\partial y}}\mathrm{d}x\right) \wedge \mathrm{d}y - \mathrm{d}x \wedge \left(i_{\frac{\partial}{\partial y}}\mathrm{d}y\right) = -\mathrm{d}x,$$

$$i_{\frac{\partial}{\partial x}}(\mathrm{d}y \wedge \mathrm{d}z) = \left(i_{\frac{\partial}{\partial x}}\mathrm{d}y\right) \wedge \mathrm{d}z - \mathrm{d}y \wedge \left(i_{\frac{\partial}{\partial x}}\mathrm{d}z\right) = 0,$$

$$i_{\frac{\partial}{\partial y}}(\mathrm{d}y \wedge \mathrm{d}z) = \left(i_{\frac{\partial}{\partial y}}\mathrm{d}y\right) \wedge \mathrm{d}z - \mathrm{d}y \wedge \left(i_{\frac{\partial}{\partial y}}\mathrm{d}z\right) = \mathrm{d}z.$$

Hence, we conclude that

$$i_X\omega = -xe^x\mathrm{d}x - ye^x\mathrm{d}y + xe^z\mathrm{d}z.$$

Remark 18.2. One can extend the interior product in a more or less obvious way to other objects (multivector fields, tensor fields, etc.). For these objects it is frequent to use the designation **contraction**, instead of interior product. For example, one can define the contraction of a differential form ω of degree k by a multivector field π of degree $l < k$, to be a differential form $i_\pi \omega$ of degree $k - l$. In a local chart $(U, x^1, \ldots, x^d)$, if

$$\omega|_U = \sum_{i_1 \cdots i_k} \omega_{i_1 \cdots i_k} \mathrm{d}x^{i_1} \wedge \cdots \wedge \mathrm{d}x^{i_k}, \quad \pi|_U = \sum_{j_1 \cdots j_l} \pi^{j_1 \cdots j_l} \frac{\partial}{\partial x^{j_1}} \wedge \cdots \wedge \frac{\partial}{\partial x^{j_l}},$$

then:

$$(i_\pi \omega)|_U = \sum_{i_1 \cdots i_k} \omega_{i_1 \cdots i_k} \pi^{i_1 \cdots i_l} \mathrm{d}x^{i_{l+1}} \wedge \cdots \wedge \mathrm{d}x^{i_k}.$$

Exercises

Exercise 18.1
Construct the natural differentiable structure on $\wedge^k T^* M$, for which the canonical projection $\pi : \wedge^k T^* M \to M$ is a submersion.

Exercise 18.2
Determine the formulas of transformation of variables for multivector fields and tensor fields.

Exercise 18.3
Prove the basic properties of the pull-back and interior product of differential forms.

Exercise 18.4
Let $\Phi : M \to N$ be a smooth map and let $X \in \mathfrak{X}(M)$ and $Y \in \mathfrak{X}(N)$ be Φ-related smooth vector fields. Show that

$$\Phi^*(i_Y \omega) = i_X \Phi^* \omega, \quad \forall \omega \in \Omega(N).$$

Exercise 18.5
Let G be a Lie group. A differential form $\omega \in \Omega^k(G)$ is called **left invariant** if

$$(L_g)^* \omega = \omega, \quad \forall g \in G.$$

We denote by $\Omega^k_{\text{L-inv}}(G)$ the space of left-invariant k-forms. Show that the operations studied in this lecture are compatible with the group operation, namely:

(a) For any left-invariant vector field $X \in \mathfrak{g}$, interior product restricts to a map $i_X : \Omega^k_{\text{L-inv}}(G) \to \Omega^{k-1}_{\text{L-inv}}(G)$.
(b) The wedge product restricts to a product $\wedge : \Omega^k_{\text{L-inv}}(G) \times \Omega^s_{\text{L-inv}}(G) \to \Omega^{k+s}_{\text{L-inv}}(G)$.
(c) For any Lie group homomorphism $\Phi : G \to H$, pullback by Φ restricts to a map $\Phi^* : \Omega^k_{\text{L-inv}}(H) \to \Omega^k_{\text{L-inv}}(G)$.

Moreover, show that evaluation at the identity $e \in G$

$$\Omega^k_{\text{L-inv}}(G) \to \wedge^k \mathfrak{g}, \quad \omega \mapsto \omega_e,$$

is an isomorphism.

Exercise 18.6
Consider the matrix Lie group

$$G = \left\{ \begin{pmatrix} x & y \\ 0 & 1 \end{pmatrix} : x \in \mathbb{R} \setminus \{0\}, \ y \in \mathbb{R} \right\}.$$

Show that $\left\{ \frac{dx}{x}, \frac{dy}{x} \right\}$ is a basis for $\Omega^1_{\text{L-inv}}(G)$ and find a basis for $\Omega^2_{\text{L-inv}}(G)$ (see previous exercise for notation).

Exercise 18.7
Let $\Psi : G \times M \to M$ be a Lie group action with associated infinitesimal action $\psi : \mathfrak{g} \to \mathfrak{X}(M)$. A differential form $\omega \in \Omega^k(M)$ is called

(a) G-**invariant** if $(\Psi_g)^*\omega = \omega$, for all $g \in G$;
(b) **basic** if it is G-invariant and $i_{\psi(X)}\omega = 0$, for all $X \in \mathfrak{g}$.

Assuming that the action is free and proper show that $\omega \in \Omega^k(M)$ is basic if and only if there exists $\alpha \in \Omega^k(G \backslash M)$ such that

$$\omega = \pi^* \alpha,$$

where $\pi : M \to G \backslash M$ is the projection to the orbit space.

Exercise 18.8
Show that a Riemannian structure on a manifold M (see Exercise 11.9) defines a symmetric tensor field of degree $(0, 2)$.

Hint: In a chart (U, x^i), a symmetric tensor field of degree $(0, 2)$ is written as

$$g|_U = \sum_{i,j} g_{ij} \mathrm{d}x^i \otimes \mathrm{d}x^j,$$

where the components $g_{ij} \in C^\infty(U)$ satisfy $g_{ij} = g_{ji}$.

Exercise 18.9

Convince yourself that all the discussion in this lecture extends to manifolds with boundary. In particular, show that if M is a manifold with boundary ∂M and $i : \partial M \to M$ is the inclusion map then:

(a) for each differential form $\omega \in \Omega^k(M)$ the pullback (or restriction) $i^*\omega$ is a smooth differential form on ∂M;

(b) every differential form $\eta \in \Omega^k(\partial M)$ has an extension to a differential form $\omega \in \Omega^k(M)$ such that $\eta = i^*\omega$.

Lecture 19

Volume Forms and Orientation

If V is a linear vector space of dimension d and $\mu \in \wedge^d(V^*)$ is a non-zero element, then for any basis $\{\mathbf{v}_1, \ldots, \mathbf{v}_d\}$ of V we have

$$\mu(\mathbf{v}_1, \ldots, \mathbf{v}_d) \neq 0.$$

This implies that μ splits the ordered basis of V into two classes. We will say that a basis $\{\mathbf{v}_1, \ldots, \mathbf{v}_d\}$ has positive (respectively, negative) μ-orientation if this number is positive (respectively, negative). We also say that μ determines an **orientation** for the vector space V.

Example 19.1. Let $V = \mathbb{R}^d$ then we have a canonical element $\mu_0 \in \wedge^d(\mathbb{R}^d)^*$, namely the determinant

$$\mu_0(\mathbf{v}_1, \ldots, \mathbf{v}_d) = \det[\mathbf{v}_i^j]_{i,j=1}^n.$$

The standard basis of $\mathbb{R}^d$ is positively oriented for this canonical choice. Note also that $|\mu_0(\mathbf{v}_1, \ldots, \mathbf{v}_d)|$ represents the usual volume of the parallelepiped span by the vectors $\mathbf{v}_1, \ldots, \mathbf{v}_d$. For an arbitrary vector space V, there is no such canonical choice of orientation and one needs to choose an element $\mu \in \wedge^d(V^*)$ to orient its bases.

Definition 19.1. A **volume form** on a smooth manifold M of dimension d, is a top degree form $\mu \in \Omega^d(M)$ which is nowhere vanishing, i.e., $\mu_p \neq 0$, for all $p \in M$. A manifold M is said to be **orientable** if it admits a volume form.

Example 19.2. The sphere $\mathbb{S}^2$ is orientable since it admits the volume form

$$\omega = (x\,dy \wedge dz + y\,dz \wedge dx + z\,dx \wedge dy)|_{\mathbb{S}^2}.$$

More generally, on the d-dimensional sphere $\mathbb{S}^d$ we have the volume form

$$\omega = \sum_{i=1}^{d+1}(-1)^i x^i dx^1 \wedge \cdots \wedge \widehat{dx^i} \wedge \cdots \wedge dx^{d+1}\bigg|_{\mathbb{S}^d}.$$

We leave it as an exercise to check that this form never vanishes.

Example 19.3. We claim that the real projective space $\mathbb{RP}^2$ does not admit a volume form, hence it is not orientable. To see this let $\mu \in \Omega^2(\mathbb{RP}^2)$ be any differential 2-form. If $\pi : \mathbb{S}^2 \to \mathbb{RP}^2$ is the quotient map, then the pull-back $\pi^*\mu$ is a differential 2-form in $\mathbb{S}^2$. It follows that there exists a smooth function $f \in C^\infty(\mathbb{S}^2)$

$$\pi^*\mu = f\omega,$$

where $\omega \in \Omega^2(\mathbb{S}^2)$ is the volume form in the previous example.

Let $\Phi : \mathbb{S}^2 \to \mathbb{S}^2$, $p \mapsto -p$, be the anti-podal map. Since $\pi \circ \Phi = \pi$, we have

$$\Phi^*(\pi^*\mu) = (\pi \circ \Phi)^*\mu = \pi^*\mu.$$

On the other, it is easy to check that $\Phi^*\omega = -\omega$. Hence,

$$f\omega = \pi^*\mu = \Phi^*(\pi^*\mu) = \Phi^*(f\omega) = (f \circ \Phi)\Phi^*(\omega) = -(f \circ \Phi)\omega.$$

We conclude that $f(-p) = -f(p)$, for all $p \in \mathbb{S}^2$. By continuity, we must have $f(p_0) = 0$ for some $p_0 \in \mathbb{S}^2$. Hence, $\pi^*\mu$ vanishes at some point. Since π is a local diffeomorphism, we conclude that $\mu \in \Omega^2(\mathbb{RP}^2)$ vanishes at some point, so $\mathbb{RP}^2$ has no volume forms, as claimed.

If $\mu \in \Omega^d(M)$ is a volume form then any other differential form of top degree in M takes the form $f\mu$, for some $f \in C^\infty(M)$. In particular, if $\mu_1, \mu_2 \in \Omega^d(M)$ are two volume forms then there exists a unique *non-vanishing* function $f \in C^\infty(M)$ such that $\mu_2 = f\mu_1$.

We say that μ_1 and μ_2 define the *same orientation* if for all $p \in M$ and any ordered basis $\{\mathbf{v}_1, \dots, \mathbf{v}_d\}$ of T_pM, one has

$$\mu_1(\mathbf{v}_1, \dots, \mathbf{v}_d)\mu_2(\mathbf{v}_1, \dots, \mathbf{v}_d) > 0.$$

Note that if μ_1 and μ_2 define the same orientation, then a basis is μ_1-positive if and only if it is μ_2-positive.

Lemma 19.1. *Let M be manifold of dimension d. Two volume forms $\mu_1, \mu_2 \in \Omega^d(M)$ define the same orientation if and only if $\mu_2 = f\mu_1$ for a smooth everywhere positive function $f \in C^\infty(M)$.*

We leave the proof as an exercise. It follows that the property "same orientation" is an equivalence relation on the set of volume forms in an orientable manifold M.

Definition 19.2. An **orientation** for an orientable manifold M is a choice of an equivalence class $[\mu]$. A pair $(M, [\mu])$ is called an **oriented manifold**.

Hence, an orientation $[\mu]$ for a manifold M (if it exists!) amounts to a choice of orientation for each tangent space T_pM varying smoothly with p. It follows from the previous lemma that a connected orientable manifold has two orientations. More generally, an orientable manifold with k connected components has 2^k orientations.

Example 19.4.

(1) The **canonical orientation of $\mathbb{R}^d$** is the one defined by the volume form $dx^1 \wedge \cdots \wedge dx^d$. The usual basis of $T_p\mathbb{R}^d \simeq \mathbb{R}^d$ is positive for this orientation.
(2) The volume form in Example 19.2 defines the **canonical orientation of $\mathbb{S}^d$**.
(3) A Lie group G is always orientable. A orientation of its Lie algebra $\mathfrak{g}$ determines an orientation of G. In fact, if $\{\alpha_1, \dots, \alpha_d\}$ is a basis of left invariant 1-forms then $\mu = \alpha_1 \wedge \cdots \wedge \alpha_d$ is a (left invariant) volume form.

Given oriented manifolds $(M, [\mu_M])$ and $(N, [\mu_N])$ we say that a diffeomorphism $\Phi : M \to N$ **preserves orientations**, or that it is **positive**, if $[\Phi^*\mu_N] = [\mu_M]$.

Example 19.5. Let $[dx^1 \wedge \cdots \wedge dx^d]$ be the canonical orientation for $\mathbb{R}^d$. Given a diffeomorphism $\phi : U \to V$, where U, V are open sets in $\mathbb{R}^d$, we have

$$\phi^*(dx^1 \wedge \cdots \wedge dx^d) = \det[\phi'(x)]dx^1 \wedge \cdots \wedge dx^d.$$

Hence ϕ preserves the canonical orientation iff $\det[\phi'(x)] > 0$, for all $x \in \mathbb{R}^d$.

One can also express the possibility of orienting a manifold in terms of an atlas, as shown by the following proposition.

Proposition 19.1. *Let M be a manifold of dimension d. The following statements are equivalents:*

(i) *M is orientable, i.e., M has a volume form.*
(ii) *There exists an atlas $\{(U_i, \phi_i)\}_{i \in I}$ for M such that for all $i, j \in I$ the transition functions preserve the canonical orientation of $\mathbb{R}^d$.*

In particular, if $[\mu_M]$ is an orientation for M, then there exists an atlas $\{(U_i, \phi_i)\}_{i \in I}$ for M such that each chart $\phi_i : U_i \to \mathbb{R}^d$ is positive, where in $\mathbb{R}^d$ we consider the canonical orientation.

Proof. (i) $\Rightarrow$ (ii) Assume M is orientable and fix the canonical orientation for $\mathbb{R}^d$. Given a chart $(U, \phi) = (U, x^1, \dots, x^d)$, then either the chart is already position or it is a negative and then $(U, \bar{\phi}) := (U, -x^1, x^2, \dots, x^d)$ is a positive chart. Therefore, given any atlas we can construct another atlas whose charts are all positive by keeping all the positive charts and changing all the negative charts in this way. The corresponding transition functions will all preserve the canonical orientation for $\mathbb{R}^d$.

(ii) $\Rightarrow$ (i) Let $\{(U_i, \phi_i)\}_{i \in I}$ be an atlas for M such that all its transition functions preserve the canonical orientation of $\mathbb{R}^d$. Choose a partition of unity $\{\rho_i\}_{i \in I}$ subordinated to the cover $\{U_i\}_{i \in I}$. If $\mu_0 = dx^1 \wedge \cdots \wedge dx^d$ is the canonical volume form on $\mathbb{R}^d$ then

$$\mu := \sum_{i \in I} \rho_i \phi_i^* \mu_0,$$

is a top-degree form which is nowhere vanishing (exercise). Hence, M is orientable. $\qquad \square$

We observed in the previous lecture that it makes sense to talk about tensor fields, differential forms, etc., on a manifold with boundary. In particular, it makes sense to talk about an orientable manifold with boundary.

Proposition 19.2. *Let M be an orientable manifold with boundary ∂M. Then ∂M is also orientable.*

Proof. Let M be a manifold with boundary and $p \in \partial M$. In a boundary chart (U, x^i) centered at p, a tangent vector $\mathbf{v} \in T_p M$ can be written as $\mathbf{v} = \sum_{i=1}^{d} v^i \left. \frac{\partial}{\partial x^i} \right|_p$, and the tangent vectors in $T_p(\partial M)$ are exactly the tangent vectors whose last component vanishes

$$T_p(\partial M) = \{\mathbf{v} \in T_p M : v^d = 0\}.$$

We will say that a tangent vector is **exterior** to ∂M if $v^d < 0$. It is easy to see that this condition is independent of the choice of boundary chart.

Using this remark, given an orientation $[\mu]$ on M, we construct an orientation on ∂M as follows. If $p \in \partial M$, the orientation of $T_p(\partial M)$ is, by definition, $[i_{\mathbf{v}} \mu_p]$ where $\mathbf{v} \in T_p M$ is any exterior tangent vector to ∂M. Is easy to see that this definition is independent of the choice of exterior tangent vectors so we have a well-defined orientation on each tangent space $T_p(\partial M)$. Since these vary smoothly, ∂M is orientable. □

The previous proof shows that a choice of orientation on M induces an orientation on ∂M, which we call the **induced orientation**. Whenever M is an oriented manifold with boundary, we will always consider the induced orientation on ∂M.

Exercises

Exercise 19.1
Verify that the top degree form μ constructed in the second part of the proof of Proposition 19.1 is a volume form.

Exercise 19.2
Show that for any orientable manifolds M and N, the product $M \times N$ is orientable. Conclude that the torus $\mathbb{T}^d$ is orientable. Give an example of a volume form in $\mathbb{T}^d$.

Exercise 19.3

Let G be a finite group which acts freely on an oriented manifold M and assume that the action of each $g \in G$ is orientation preserving. Show that M/G is an orientable manifold. Using this result prove that $\mathbb{RP}^d$ is orientable if and only if d is odd.

Exercise 19.4

Let M be any manifold and consider the set of all orientations of its tangent spaces

$$\widetilde{M}_{\text{ori}} := \{[\mu_p] : \mu_p \in \wedge^d T_p M, \ p \in M\}.$$

Also, let $\pi : \widetilde{M}_{\text{ori}} \to M$, be the map $[\mu_p] \mapsto p$. Show the following:

(a) $\widetilde{M}_{\text{ori}}$ has a unique smooth structure such that $\pi : \widetilde{M}_{\text{ori}} \to M$ is a 2:1 cover;
(b) $\widetilde{M}_{\text{ori}}$ is an orientable manifold;
(c) If M is connected, then M is orientable if and only if M is not connected.

One calls $\widetilde{M}_{\text{ori}}$ the **orientation cover** of M.

Exercise 19.5

Verify that the Klein bottle form Example 8.5 is a non-orientable manifold.

Exercise 19.6

Let (M, g) be a Riemannian manifold of dimension d. Show the following:

(a) For each $p \in M$, the map $T_p M \to T_p^* M$, $v \mapsto g(v, \cdot)$, is an isomorphism, so the inner product on the tangent space $T_p M$ induces an inner product on the cotangent space $T_p^* M$.
(b) For each $p \in M$, there exists a neighborhood U of p and orthonormal smooth vector fields $X_1, \ldots, X_d \in \mathfrak{X}(U)$:

$$\langle X_i, X_j \rangle = \delta_{ij} \text{ (Kronecker symbol)}.$$

The set $\{X_1, \ldots, X_d\}$ is called a **(local) orthonormal frame**.

(c) For each $p \in M$, there exists a neighborhood U of p and orthonormal differential forms $\alpha_1, \ldots, \alpha_d \in \Omega^1(U)$:

$$\langle \alpha_i, \alpha_j \rangle = \delta_{ij} \text{ (Kronecker symbol)}.$$

The set $\{\alpha_1, \ldots, \alpha_d\}$ is called a **(local) orthonormal coframe**.

(d) Assume further that $[\mu]$ is an orientation for M. Show that there exists a volume form $\mu_0 \in \Omega^d(M)$ such that

$$\mu_0|_U = \alpha_1 \wedge \cdots \wedge \alpha_d,$$

for every local orthonormal coframe $\alpha_1, \ldots, \alpha_d \in \Omega^1(U)$ which is positive (i.e., $\alpha_1 \wedge \cdots \wedge \alpha_d$ is positive). One calls μ_0 the **canonical volume form of the oriented Riemannian manifold** $(M, g, [\mu])$.

Exercise 19.7

Let $(M, g, [\mu])$ be an oriented Riemannian manifold of dimension d. Show that there exists a unique linear map $* : \Omega^k(M) \to \Omega^{d-k}(M)$ such that for every positive local orthonormal coframe $\alpha_1, \ldots, \alpha_d$ the following properties hold:

(a) $*1 = \alpha_1 \wedge \cdots \wedge \alpha_d$ and $*(\alpha_1 \wedge \cdots \wedge \alpha_d) = 1$;
(b) $*(\alpha_1 \wedge \cdots \wedge \alpha_k) = \alpha_{k+1} \wedge \cdots \wedge \alpha_d$.

Moreover, check that for any k-form ω

$$* * \omega = (-1)^{k(d-k)} \omega.$$

One calls $*$ the **Hodge star operator**.

Lecture 20

Cartan Calculus

We will introduce now two important differentiation operations on differential forms: the *differential* of forms, which is an intrinsic derivative, and the *Lie derivative* of differential forms, which is a derivative along vector fields. These differential operations together with the algebraic operations on differential forms that we studied in the previous lecture, are the basis of a calculus on differential forms which is usually called *Cartan Calculus*.

The **differential** of $\omega \in \Omega^k(M)$ is the $(k+1)$-form $d\omega \in \Omega^{k+1}(M)$ defined by

$$
d\omega(X_0, \ldots, X_k) := \sum_{i=0}^{k} (-1)^i X_i(\omega(X_0, \ldots, \widehat{X}_i, \ldots, X_k))
$$
$$
+ \sum_{0 \leq i < j \leq k} (-1)^{i+j}
$$
$$
\times \omega([X_i, X_j], X_0, \ldots, \widehat{X}_i, \ldots, \widehat{X}_j \ldots, X_k), \quad (20.1)
$$

for any smooth vector fields $X_0, \ldots, X_k \in \mathfrak{X}(M)$. This formula defines a $C^\infty(M)$-multilinear, alternating, map $\mathfrak{X}(M) \times \cdots \times \mathfrak{X}(M) \to C^\infty(M)$, so that $d\omega$ is indeed a smooth differential $(k+1)$-form.

A smooth function $f \in \mathcal{C}^\infty(M)$ is a degree 0 form. In this case, formula (20.1) gives

$$
df(X) = X(f).
$$

Therefore this definition matches our previous definition of the differential of a smooth function.

Our next result shows that the differential is the only operation on the forms which extends the differential of functions in a reasonable way.

Theorem 20.1. *The differential* $d : \Omega^\bullet(M) \to \Omega^{\bullet+1}(M)$ *is the only operation on forms satisfying the following properties*:

(i) d *is* $\mathbb{R}$-*linear*: $d(a\omega + b\theta) = a d\omega + b d\theta$;

(ii) d *is a derivation*: $d(\omega \wedge \theta) = (d\omega) \wedge \theta + (-1)^{\deg \omega} \omega \wedge (d\theta)$;

(iii) d *extends the usual differential*: $df(X) = X(f)$ *if* $f \in C^\infty(M)$, $X \in \mathfrak{X}(M)$;

(iv) $d^2 = 0$.

Proof. We leave it for the exercises to check that d, as defined by (20.1), satisfies properties (i) through (iv). To prove uniqueness, we need to check that given $\omega \in \Omega^k(M)$, then $d\omega$ is determined by properties (i)–(iv).

Since d is a derivation, it is local: if $\omega|_U = 0$ on an open set U then $(d\omega)|_U = 0$. In fact, let $p \in U$ and $f \in C^\infty(M)$ with $f(p) > 0$ and $\operatorname{supp} f \subset U$. Since $f\omega \equiv 0$, we find that

$$0 = d(f\omega) = df \wedge \omega + f d\omega.$$

If we evaluate both sides of this identity at p, we conclude that $f(p)(d\omega)_p = 0$. Hence $d\omega|_U = 0$, as claimed.

Therefore, to prove uniqueness, it is enough to consider $\omega \in \Omega^k(U)$ where U is the domain of some local chart $(x^1, \ldots, x^d)$. In this case, we have

$$\omega = \sum_{i_1 < \cdots < i_k} \omega_{i_1 \cdots i_k} dx^{i_1} \wedge \cdots \wedge dx^{i_k}.$$

Using only properties (i)–(iv) we find

$$d\omega = \sum_{i_1 < \cdots < i_k} d(\omega_{i_1 \cdots i_k} dx^{i_1} \wedge \cdots \wedge dx^{i_k}) \qquad \text{(by (i))}$$

$$= \sum_{i_1 < \cdots < i_k} d(\omega_{i_1 \cdots i_k}) \wedge dx^{i_1} \wedge \cdots \wedge dx^{i_k} \qquad \text{(by (ii) and (iv))}$$

$$= \sum_{i_1 < \cdots < i_k} \sum_i \frac{\partial \omega_{i_1 \cdots i_k}}{\partial x^i} dx^i \wedge dx^{i_1} \wedge \cdots \wedge dx^{i_k} \qquad \text{(by (iii))}.$$

The last expression defines a differential form of degree $k + 1$ in U. Hence, $d\omega$ is completely determined by properties (i)–(iv), as claimed. □

As this proof shows, one can compute the differential of a form in local coordinates using only properties (i)–(iv). This is often much more efficient than applying directly formula (20.1).

Example 20.1. Let $\omega = e^y dx \wedge dz + e^z dy \wedge dz \in \Omega^2(\mathbb{R}^3)$. Using properties (i)–(iv), we find

$$d\omega = d(e^y dx \wedge dz + e^z dy \wedge dz)$$
$$= (de^y) \wedge dx \wedge dz + d(e^z) \wedge dy \wedge dz$$
$$= e^y dy \wedge dx \wedge dz + e^z dz \wedge dy \wedge dz = -e^y dx \wedge dy \wedge dz.$$

The pullback operation preserves differentials.

Proposition 20.1. *If $\Phi : M \to N$ is a smooth map, then for every $\omega \in \Omega^k(N)$*

$$\Phi^* d\omega = d\Phi^* \omega.$$

Proof. It is enough to prove this holds on local charts. We leave the (easy) computation to the exercises. □

The operation $d : \Omega^\bullet(M) \to \Omega^{\bullet+1}(M)$ is also referred to as *exterior differentiation*, since it increases the degree of a form. There is another type of differentiation of a form which preserves the degree.

Definition 20.1. The **Lie derivative** of a differential form $\omega \in \Omega^k(M)$ along a vector $X \in \mathfrak{X}(M)$ is the differential form $\mathcal{L}_X \omega \in \Omega^k(M)$ defined by

$$\mathcal{L}_X \omega = \frac{d}{dt}(\phi_X^t)^* \omega \bigg|_{t=0} = \lim_{t \to 0} \frac{1}{t} \left((\phi_X^t)^* \omega - \omega \right).$$

Example 20.2. Let $\omega = e^y dx \wedge dz + e^z dy \wedge dz \in \Omega^2(\mathbb{R}^3)$ and $X = x \frac{\partial}{\partial y} \in \mathfrak{X}(\mathbb{R}^3)$. The flow of X is given by $\phi_X^t(x, y, z) = (x, y + tx, z)$. Hence, we find that

$$(\phi_X^t)^* \omega = e^{y+tx} dx \wedge dz + e^z d(y + tx) \wedge dz$$
$$= e^{y+tx} dx \wedge dz + e^z dy \wedge dz + te^z dx \wedge dz.$$

Then

$$\mathcal{L}_X \omega = \frac{\mathrm{d}}{\mathrm{d}t}(\phi_X^t)^* \omega \bigg|_{t=0}$$

$$= \frac{\mathrm{d}}{\mathrm{d}t}\bigg|_{t=0} \left(e^{y+tx}\mathrm{d}x \wedge \mathrm{d}z + e^z\mathrm{d}y \wedge \mathrm{d}z + te^z\mathrm{d}x \wedge \mathrm{d}z \right)$$

$$= xe^y\mathrm{d}x \wedge \mathrm{d}z + e^z\mathrm{d}x \wedge \mathrm{d}z.$$

In most examples it is impossible to find explicitly the flow of a vector field. Still, the basic properties of the Lie derivative listed in the next proposition allow one to find the Lie derivative without knowledge of the flow. The proof is left as an exercise.

Proposition 20.2. *Let $X \in \mathfrak{X}(M)$ and $\omega, \eta \in \Omega^\bullet(M)$. Then,*

(i) $\mathcal{L}_X(a\omega + b\eta) = a\mathcal{L}_X\omega + b\mathcal{L}_X\eta$ *for all $a, b \in \mathbb{R}$;*
(ii) $\mathcal{L}_X(\omega \wedge \eta) = \mathcal{L}_X\omega \wedge \eta + \omega \wedge \mathcal{L}_X\eta$;
(iii) $\mathcal{L}_X(f) = X(f)$, *if $f \in \Omega^0(M) = C^\infty(M)$;*
(iv) $\mathcal{L}_X\mathrm{d}\omega = \mathrm{d}\mathcal{L}_X\omega$.

Example 20.3. Let us redo Example 20.2 using only properties (i)–(iv) in the previous proposition

$$\mathcal{L}_X \omega = \mathcal{L}_X(e^y\mathrm{d}x \wedge \mathrm{d}z + e^z\mathrm{d}y \wedge \mathrm{d}z)$$

$$= \mathcal{L}_X(e^y)\mathrm{d}x \wedge \mathrm{d}z + e^y\mathcal{L}_X(\mathrm{d}x) \wedge \mathrm{d}z + e^y\mathrm{d}x \wedge \mathcal{L}_X(\mathrm{d}z)$$

$$+ \mathcal{L}_X(e^z)\mathrm{d}y \wedge \mathrm{d}z + e^z\mathcal{L}_X(\mathrm{d}y) \wedge \mathrm{d}z + e^z\mathrm{d}y \wedge \mathcal{L}_X(\mathrm{d}z)$$

$$= X(e^y)\mathrm{d}x \wedge \mathrm{d}z + e^z\mathrm{d}X(y) \wedge \mathrm{d}z$$

$$= xe^y\mathrm{d}x \wedge \mathrm{d}z + e^z\mathrm{d}x \wedge \mathrm{d}z.$$

There is still another way to compute the Lie derivative by applying a formula which relates all three basic operations on forms: Lie derivative, exterior differential, and interior product. This "magic" formula often plays an unexpected role.

Theorem 20.2 (Cartan's Magic Formula). *Let $X \in \mathfrak{X}(M)$ and $\omega \in \Omega(M)$. Then*

$$\mathcal{L}_X\omega = i_X\mathrm{d}\omega + \mathrm{d}i_X\omega. \tag{20.2}$$

Proof. By Proposition 20.2, $\mathcal{L}_X : \Omega(M) \to \Omega(M)$ is a derivation. The properties of d and i_X give that $i_X\mathrm{d} + \mathrm{d}i_X : \Omega(M) \to \Omega(M)$ is also a derivation. Hence, it is enough to check that both derivations take the same values on differential forms of the type $\omega = f$ and $\omega = \mathrm{d}g$, where $f, g \in C^\infty(M)$.

On the one hand, the properties in Proposition 20.2, give

$$\mathcal{L}_X(f) = X(f), \quad \mathcal{L}_X(\mathrm{d}g) = \mathrm{d}\mathcal{L}_X g = \mathrm{d}(X(g)).$$

On the other hand, the properties of d and i_X yield

$$(i_X \mathrm{d} + \mathrm{d}i_X)f = i_X \mathrm{d}f = X(f),$$
$$(i_X \mathrm{d} + \mathrm{d}i_X)\mathrm{d}g = \mathrm{d}(i_X \mathrm{d}g) = \mathrm{d}(X(g)).$$

$\square$

Example 20.4. Let us redo Example 20.2 using Cartan's Magic Formula

$$\begin{aligned}
\mathcal{L}_X \omega &= i_X \mathrm{d}\omega + \mathrm{d}i_X \omega \\
&= i_X(-e^y \mathrm{d}x \wedge \mathrm{d}y \wedge \mathrm{d}z) + \mathrm{d}(xe^z \mathrm{d}z) \\
&= xe^y \mathrm{d}x \wedge \mathrm{d}z + e^z \mathrm{d}x \wedge \mathrm{d}z.
\end{aligned}$$

Exercises

Exercise 20.1
Show that d defined by formula (20.1), satisfies properties (i)–(iv) in Theorem 20.1.

Exercise 20.2
Let $\Phi : M \to N$ be a smooth map. Show that for any form $\omega \in \Omega^k(M)$

$$\Phi^* \mathrm{d}\omega = \mathrm{d}\Phi^* \omega.$$

Exercise 20.3
Let $I \subset \Omega(M)$ be an ideal generated by k linearly independent differential forms $\alpha_1, \ldots, \alpha_k \in \Omega^1(M)$ (i.e., such that $\{\alpha_1|_p, \ldots, \alpha_k|_p\}$ is a linearly independent set for every $p \in M$). Show that the following statements are equivalent:

(a) I is a differential ideal, i.e., if $\alpha \in I$ then $\mathrm{d}\alpha \in I$;
(b) $\mathrm{d}\alpha_i = \sum_j \omega_{ij} \wedge \alpha_j$, for some 1-forms $\omega_{ij} \in \Omega^1(M)$;

(c) If $\omega = \alpha_1 \wedge \cdots \wedge \alpha_k$, then $d\omega = \alpha \wedge \omega$, for some 1-form $\alpha \in \Omega^1(M)$.
(d) The distribution $D = \bigcap_{i=1}^k \ker \alpha_i$ is involutive.

Exercise 20.4

Prove the properties of the Lie derivative given in Proposition 20.2.

Exercise 20.5

Let $X, Y \in \mathfrak{X}(M)$ be vector fields and $\omega \in \Omega(M)$ a differential form. Show that

$$\mathcal{L}_{[X,Y]}\omega = \mathcal{L}_X(\mathcal{L}_Y\omega) - \mathcal{L}_Y(\mathcal{L}_X\omega).$$

Exercise 20.6

Let $\Phi : M \to N$ be smooth. Show that if $X \in \mathfrak{X}(M)$ and $Y \in \mathfrak{X}(N)$ are Φ-related vector fields, then

$$\Phi^*(\mathcal{L}_Y\omega) = \mathcal{L}_X(\Phi^*\omega),$$

for every differential form $\omega \in \Omega(N)$.

Exercise 20.7

Let $X \in \mathfrak{X}(M)$ and $\omega \in \Omega^k(M)$. Show that

$$\mathcal{L}_X(\omega(X_1, \ldots, X_k)) = \mathcal{L}_X\omega(X_1, \ldots, X_k)$$

$$+ \sum_{i=1}^k \omega(X_1, \ldots, \mathcal{L}_X X_i, \ldots, X_k). \qquad (20.3)$$

Exercise 20.8

Let M be a manifold equipped with a volume form μ. Given a vector field X, the **divergence** of X is the unique function $\operatorname{div}_\mu(X) \in C^\infty(M)$ that satisfies

$$\mathcal{L}_X\mu = \operatorname{div}_\mu(X)\mu.$$

Show that

(a) a complete vector field $X \in \mathfrak{X}(M)$ is divergence free (i.e., $\operatorname{div}_\mu(X) = 0$) if and only the flow of X preserves the volume form μ, i.e., if and only if

$$(\phi_X^t)^*\mu = \mu, \quad \forall t \in \mathbb{R};$$

(b) if $\mu = dx^1 \wedge \cdots \wedge dx^d$ is the canonical volume form on $M = \mathbb{R}^d$ then for a vector field $X = \sum_{i=1}^d X^i \frac{\partial}{\partial x^i}$ one has

$$\text{div}_\mu(X) = \sum_{i=1}^d \frac{\partial X^i}{\partial x^i};$$

(c) if $(M, g, [\mu])$ is an oriented Riemannian manifold with associated volume form μ and Hodge-star operator $*$ (see Exercises 19.6 and 19.7) then

$$\text{div}_\mu(X) = *\,d * X.$$

Exercise 20.9

Let $(M, g, [\mu])$ be an oriented Riemannian manifold. One defines the **laplacian** of $f : M \to \mathbb{R}$ to be the function $\Delta f : M \to \mathbb{R}$ given by

$$\Delta f := -\text{div}(\text{grad } f),$$

where $\text{grad } f$ denotes the gradient of f (see Exercise 11.10). Let $M = \mathbb{R}^3$ with its canonical Riemannian structure and canonical orientation. Find the divergence and the Laplacian in cylindrical and in spherical coordinates.

Exercise 20.10

Let $\mathfrak{X}^k(M)$ denote the space of multivector fields of degree k on M. Show that there exists a unique $\mathbb{R}$-bilinear operation $[\, , \,] : \mathfrak{X}^{p+1}(M) \times \mathfrak{X}^{q+1}(M) \to \mathfrak{X}^{p+q+1}(M)$ which coincides with the usual Lie bracket of vector fields when $p = q = 0$ and satisfies:

(a) $[P, Q] = -(-1)^{pq}[Q, P]$;
(b) $[P, Q \wedge R] = [P, Q] \wedge R + (-1)^{p(q+1)} Q \wedge [P, R]$.

Verify that this bracket satisfies the Jacobi type identity

$$(-1)^{pr}[P, [Q, R]] + (-1)^{qp}[Q, [R, P]] + (-1)^{rq}[R, [P, Q]] = 0.$$

In all these identities, $P \in \mathfrak{X}^{p+1}(M)$, $Q \in \mathfrak{X}^{q+1}(M)$ and $R \in \mathfrak{X}^{r+1}(M)$.

Note: This operation is known as the **Schouten bracket** and is the counterpart for multivector fields of the exterior differential for forms. It is an example of a graded Lie bracket.

Lecture 21

Integration on Manifolds

Ultimately, our interest in differential forms of degree d lies in the fact that they can be integrated over *oriented* d-manifolds, as we now explain.

Let us start with the case where $M = \mathbb{R}^d$, with the usual orientation. If $U \subset \mathbb{R}^d$ is open, then every differential form $\omega \in \Omega^d(U)$ can be written as

$$\omega = f \, dx^1 \wedge \cdots \wedge dx^d, \quad (f \in C^\infty(U)).$$

We say that ω is **integrable** in U and we define its integral by

$$\int_U \omega := \int_U f(x^1, \ldots, x^d) dx^1 \cdots dx^d,$$

provided the integral on the right-hand side exists and is finite. The usual change of variable formula for the integral in $\mathbb{R}^d$ yields the following result.

Lemma 21.1. *Let* $\Phi : U \to \mathbb{R}^d$ *be a diffeomorphism defined in an open connected set* $U \subset \mathbb{R}^d$. *If* $\omega \in \Omega^d(\Phi(U))$ *is integrable, then* $\Phi^*\omega \in \Omega^d(U)$ *is integrable and*

$$\int_{\Phi(U)} \omega = \pm \int_U \Phi^*\omega,$$

where $\pm$ *is the sign of* $\det(\Phi'(p))$ *for any* $p \in U$.

Therefore the integral is invariant under *orientation preserving* diffeomorphisms. For this reason, we will only consider the integral of differential forms over *oriented* manifolds. It is possible to define the integral over non-oriented manifolds, but this requires introducing **densities**, a generalization of volume forms.

In order to avoid convergence issues, we consider the following class of forms.

Definition 21.1. The support of $\omega \in \Omega^k(M)$ is the closed set

$$\operatorname{supp}\omega := \overline{\{p \in M : \omega_p \neq 0\}}.$$

We denote by $\Omega_c^k(M)$ the space of compactly supported differential forms.

If $\Phi : M \to N$ is a proper map (e.g., a diffeomorphism), pullback gives a map $\Phi^* : \Omega_c^k(M) \to \Omega_c^k(M)$. Obviously, if M is compact one has $\Omega_c^k(M) = \Omega^k(M)$.

After these preparations, we can define the integral of a top degree compactly supported form over an oriented manifold as follows.

Definition 21.2. If M is an oriented d-manifold and $\omega \in \Omega_c^d(M)$, we define its **integral** over M as follows:

- If $\operatorname{supp}\omega \subset U$, where (U, ϕ) is a positive coordinate chart, then

$$\int_M \omega := \int_{\phi(U)} (\phi^{-1})^*\omega.$$

- More generally, we consider an open cover of M by positive charts (U_α, ϕ_α) and a partition of unity $\{\rho_\alpha\}$ subordinated to this cover, and we set

$$\int_M \omega := \sum_\alpha \int_M \rho_\alpha \omega.$$

We remark that the sum in this definition is finite since we assume that $\operatorname{supp}\omega$ is compact. It is easy to check that the definition is independent of the choices of covering by positive charts and of partition of unity. We leave it to the exercises to check all these details.

The integral shares the properties of the usual integral of functions in $\mathbb{R}^d$. First, one checks easily that

(a) *Linearity*: If $\omega, \eta \in \Omega_c^d(M)$ and $a, b \in \mathbb{R}$, then

$$\int_M (a\omega + b\eta) = a\int_M \omega + b\int_M \eta.$$

(b) *Additivity*: If $M = M_1 \cup M_2$, $\omega \in \Omega_c^d(M)$ and $M_1 \cap M_2$ has zero measure, then

$$\int_M \omega = \int_{M_1} \omega + \int_{M_2} \omega.$$

Moreover, we have the following important property that ultimately justifies the relevance of differential forms.

Theorem 21.1 (Change of Variables Formula). *Let M and N be oriented manifolds of dimension d and let $\Phi : M \to N$ be an orientation preserving diffeomorphism. Then, for every differential form $\omega \in \Omega_c^d(N)$, one has*

$$\int_N \omega = \int_M \Phi^*\omega.$$

Proof. Since Φ is a diffeomorphism and preserves orientations, we can find an open cover of M by positive charts (U_α, ϕ_α), such that the open sets $\Phi(U_\alpha)$ are domains of positive charts $\psi_\alpha : \Phi(U_\alpha) \to \mathbb{R}^d$ for N. Let $\{\rho_\alpha\}$ be a partition of unity for N subordinated to the cover $\{\Phi(U_\alpha)\}$, so that $\{\rho_\alpha \circ \Phi\}$ is a partition of unity for M subordinated to the cover $\{U_\alpha\}$. By Lemma 21.1, we find

$$\int_{\Phi(U_\alpha)} \rho_\alpha \omega = \int_{U_\alpha} \Phi^*(\rho_\alpha \omega) = \int_{U_\alpha} (\rho_\alpha \circ \Phi)\Phi^*\omega.$$

Hence, we obtain

$$\int_N \omega = \sum_\alpha \int_N \rho_\alpha \omega = \sum_\alpha \int_{\Phi(U_\alpha)} \rho_\alpha \omega$$

$$= \sum_\alpha \int_{U_\alpha} (\rho_\alpha \circ \Phi)\Phi^*\omega$$

$$= \sum_\alpha \int_M (\rho_\alpha \circ \Phi)\Phi^*\omega = \int_M \Phi^*\omega.$$

$\square$

The computation of the integral of differential forms using the definition is not practical since it involves a partition of unity. The following result can often be applied to avoid the use of a partition of unity.

Proposition 21.1. *Let M be an oriented manifold of dimension d and let $C \subset M$ be a closed subset of zero measure. For any differential form $\omega \in \Omega_c^d(N)$, we have*

$$\int_M \omega = \int_{M\setminus C} \omega.$$

Proof. Using a partition of unity we can reduce the result to the case where M is an open subset of $\mathbb{R}^d$. For an open set $U \subset \mathbb{R}^d$, the result reduces to the equality

$$\int_U f\,\mathrm{d}x^1 \ldots \mathrm{d}x^d = \int_{U\setminus C} f\,\mathrm{d}x^1 \ldots \mathrm{d}x^d,$$

where $f : U \to \mathbb{R}$ is smooth and bounded. This is obvious since C has zero measure. $\qquad\square$

Example 21.1. Given a volume form μ is a compact manifold M, we can define the **volume of** M relative to μ to be the positive number:

$$\mathrm{vol}_\mu(M) := \int_M \mu,$$

where the integral is relative to the orientation $[\mu]$.

For example, let us find the volume of the sphere $\mathbb{S}^2$ relative to the volume form

$$\mu = (x\mathrm{d}y \wedge \mathrm{d}z + y\mathrm{d}z \wedge \mathrm{d}x + z\mathrm{d}x \wedge \mathrm{d}y)|_{\mathbb{S}^2}.$$

By the previous proposition, we have

$$\mathrm{vol}_\mu(\mathbb{S}^2) = \int_{\mathbb{S}^2} \mu = \int_{\mathbb{S}^2\setminus p_N} \mu,$$

where $p_N \in \mathbb{S}^2$ is the north pole. Stereographic projection $\pi_N : \mathbb{S}^2 \setminus \{p_N\} \to \mathbb{R}^2$ defines a global chart for $\mathbb{S}^2 \setminus \{p_N\}$ whose inverse is the parameterization

$$\pi_N^{-1}(u, v) = \frac{1}{u^2 + v^2 + 1}(2u, 2v, u^2 + v^2 - 1).$$

Using this parameterization we find

$$(\pi_N^{-1})^*\mu = -\frac{4}{(u^2 + v^2 + 1)^2}\,\mathrm{d}u \wedge \mathrm{d}v,$$

which shows that π_N is a negative chart. Therefore,

$$\int_{\mathbb{S}^2} \mu = \int_{\mathbb{R}^2} \frac{4}{(u^2 + v^2 + 1)^2}\,\mathrm{d}u \wedge \mathrm{d}v.$$

The integral on the right can be computed using polar coordinates, and the final result is

$$\mathrm{vol}_\mu(\mathbb{S}^2) = \int_{\mathbb{S}^2} \mu = \int_0^{+\infty} \int_0^{2\pi} \frac{4r}{(r^2 + 1)^2}\,\mathrm{d}\theta \mathrm{d}r = 4\pi.$$

Our next aim is to generalize Stokes Theorem to differential forms.

First, note that the previous discussion about integration makes sense for manifolds with boundary. For an oriented manifold with boundary M, as explained in Lecture 19, we have an induced orientation on ∂M.

Theorem 21.2 (Stokes Formula). *Let M be an oriented manifold with boundary of dimension d. If $\omega \in \Omega_c^{d-1}(M)$ then*

$$\int_M \mathrm{d}\omega = \int_{\partial M} \omega.$$

Proof. We divide the proof into several cases.

• $M = \mathbb{R}^d$. In this case, we can write

$$\omega = \sum_{i=1}^d f_i \mathrm{d}x^1 \wedge \cdots \wedge \widehat{\mathrm{d}x^i} \wedge \cdots \wedge \mathrm{d}x^d,$$

where f_i are compactly supported functions. We find its differential to be

$$\mathrm{d}\omega = \sum_{i=1}^d (-1)^{i-1} \frac{\partial f_i}{\partial x^i} \mathrm{d}x^1 \wedge \cdots \wedge \mathrm{d}x^d.$$

By Fubini's Theorem, we conclude that

$$\int_{\mathbb{R}^d} d\omega = \sum_{i=1}^{d}(-1)^{i-1}\int_{\mathbb{R}^{d-1}}\left(\int_{-\infty}^{+\infty}\frac{\partial f_i}{\partial x^i}dx^i\right)dx^1\cdots\widehat{dx^i}\cdots dx^d = 0,$$

where we used that f_i has compact support. Since $\partial\mathbb{R}^d = \emptyset$, Stokes Formula for $\mathbb{R}^d$ follows.

- $M = \mathbb{H}^d$. We proceed as in the case of $\mathbb{R}^n$, but this time we obtain

$$\int_{\mathbb{H}^d} d\omega = \sum_{i=1}^{d}(-1)^{i-1}\int_{\mathbb{H}^d}\frac{\partial f_i}{\partial x^i}dx^1\wedge\cdots\wedge\widehat{dx^i}\wedge\cdots\wedge dx^d$$

$$= \sum_{i=1}^{d-1}(-1)^{i-1}\int_{\mathbb{H}^{d-1}}\left(\int_{-\infty}^{+\infty}\frac{\partial f_i}{\partial x^i}dx^i\right)dx^1\cdots\widehat{dx^i}\cdots dx^d$$

$$+ (-1)^{d-1}\int_{\mathbb{R}^{d-1}}\left(\int_{0}^{+\infty}\frac{\partial f_d}{\partial x^d}dx^d\right)dx^1\cdots dx^{d-1}$$

$$= (-1)^d\int_{\mathbb{R}^{d-1}}f_d(x^1,\ldots,x^{d-1},0)dx^1\cdots dx^{d-1}.$$

On the other hand, $\partial\mathbb{H}^d = \{(x^1,\ldots,x^d) : x^d = 0\}$, hence

$$\int_{\partial\mathbb{H}^d}\omega = \int_{\partial\mathbb{H}^d}f_d(x^1,\ldots,x^{d-1},0)dx^1\wedge\cdots\wedge dx^{d-1}.$$

In $\partial\mathbb{H}^d = \mathbb{R}^{d-1}$ we must take the induced orientation from the canonical orientation $[\mu] = [dx^1\wedge\cdots\wedge dx^d]$ in $\mathbb{H}^d$. The induced orientation is given by: $[(-1)^d dx^1\wedge\cdots\wedge dx^{d-1}]$ so we conclude that

$$\int_{\partial\mathbb{H}^d}\omega = (-1)^d\int_{\partial\mathbb{R}^{d-1}}f_d(x^1,\ldots,x^{d-1},0)dx^1\cdots dx^{d-1}.$$

Therefore, Stokes Formula also holds for the half-space $\mathbb{H}^d$.

- *Any M.* We fix an open cover of M by positive charts (U_α,ϕ_α) and we choose a partition of unity $\{\rho_\alpha\}$ subordinated to this cover. We can also assume that the charts have been chosen so that $\phi_\alpha(U_\alpha)$ is either $\mathbb{R}^d$ or $\mathbb{H}^d$. The forms $\rho_\alpha\omega$ have compact support

$$\text{supp}\,\rho_\alpha\omega \subset \text{supp}\,\rho_\alpha \cap \text{supp}\,\omega \subset U_\alpha.$$

Since each U_α is diffeomorphic to either $\mathbb{R}^d$ or to $\mathbb{H}^d$, by the change of variable formula, we already know that

$$\int_{U_\alpha} \mathrm{d}(\rho_\alpha \omega) = \int_{\partial U_\alpha} \rho_\alpha \omega.$$

Note that $\partial U_\alpha = U_\alpha \cap \partial M$, so by the linearity and the additivity of the integral, we obtain

$$\int_M \mathrm{d}\omega = \sum_\alpha \int_M \mathrm{d}(\rho_\alpha \omega) = \sum_\alpha \int_{U_\alpha} \mathrm{d}(\rho_\alpha \omega)$$

$$= \sum_\alpha \int_{U_\alpha \cap \partial M} \rho_\alpha \omega = \int_{\partial M} \sum_\alpha \rho_\alpha \omega = \int_{\partial M} \omega. \qquad \square$$

Corollary 21.1. *Let M be an oriented, d-dimensional, manifold without boundary. Then, for any $\omega \in \Omega_c^{d-1}(M)$, one has*

$$\int_M \mathrm{d}\omega = 0.$$

Exercises

Exercise 21.1
Show that the integral of differential forms is linear and additive relative to the region of integration.

Exercise 21.2
In $\mathbb{H}^d$ consider the standard orientation $[\mu] = [\mathrm{d}x^1 \wedge \cdots \wedge \mathrm{d}x^d]$. Show that the induced orientation in $\partial \mathbb{H}^d = \mathbb{R}^{d-1}$ is given by $[\partial \mu] = [(-1)^d \mathrm{d}x^1 \wedge \cdots \wedge \mathrm{d}x^{d-1}]$.

Exercise 21.3
Consider the n-torus $\mathbb{T}^n$ as an embedded submanifold of $\mathbb{R}^{2n}$:

$$\mathbb{T}^n = \{(x^1, \ldots, x^n, y^1, \ldots, y^n) \in \mathbb{R}^{2n} : (x^i)^2 + (y^i)^2 = 1, i = 1, \ldots, n\},$$

and let $\omega \in \Omega^n(\mathbb{T}^n)$ be the form

$$\omega = \left(\mathrm{d}x^1 \wedge \cdots \wedge \mathrm{d}x^n\right)|_{\mathbb{T}^n}.$$

Compute the integral $\int_{\mathbb{T}^n} \omega$ for an orientation of your choice, in the following ways:

(a) using the definition;
(b) using Stokes formula.

Exercise 21.4

Find the volume of $\mathbb{S}^d$ for the standard volume form

$$\mu = \sum_{i=1}^{d+1} (-1)^{i+1} x^i dx^1 \wedge \cdots \wedge \widehat{dx^i} \wedge \cdots \wedge dx^{d+1} \bigg|_{\mathbb{S}^d}.$$

Exercise 21.5

Let $(M, g, [\mu])$ is an oriented Riemannian manifold with boundary, with associated volume form μ and Hodge-star operator $*$ (see Exercises 19.6 and 19.7). If $f : M \to \mathbb{R}$ is a smooth, compactly supported function, define the integral of f over M by

$$\int_M f := \int_M *f.$$

If X is any vector field, prove the classical **Divergence Theorem**, i.e.,

$$\int_M \operatorname{div}_\mu X = \int_{\partial M} X \cdot n,$$

where $n : \partial M \to T_{\partial M} M$ is the unit exterior normal vector field along ∂M.

Exercise 21.6

Let M be an oriented Riemannian manifold with boundary. For any smooth function $f : M \to \mathbb{R}$ denote by $\frac{\partial f}{\partial n}$ the function $n(f) : \partial M \to \mathbb{R}$, where n is the unit exterior normal vector field along ∂M. Verify the following Green identities

$$\int_{\partial M} f \frac{\partial g}{\partial n} = \int_M \langle \operatorname{grad} f, \operatorname{grad} g \rangle - \int_M f \Delta g,$$

$$\int_{\partial M} \left(f \frac{\partial g}{\partial n} - g \frac{\partial f}{\partial n} \right) = \int_M (g \Delta f - f \Delta g),$$

where $f, g \in C^\infty(M)$.

Exercise 21.7

Let G be a Lie group of dimension d.

(a) Show that if $\omega, \omega' \in \Omega^d(G)$ are left invariant and $[\omega] = [\omega']$, then

$$\int_G f\omega = a \int_G f\omega', \quad \forall f \in C^\infty(G),$$

for some real number $a > 0$.

Fix an orientation $\mu = [\omega]$ for G defined by a left-invariant form $\omega \in \Omega^d(G)$. Define the integral of $f : G \to \mathbb{R}$ relative to this orientation by

$$\int_G f := \int_G f\omega.$$

(b) Show that the integral is left invariant, i.e., for every $g \in G$ is valid the identity

$$\int_G f \circ L_g = \int_G f.$$

(c) Give an example of a Lie group where the integral is not right invariant.

For each $g \in G$, the differential form $R_g^* \omega$ is left invariant, hence

$$R_g^* \omega = \tilde{\lambda}(g)\omega,$$

for some smooth function $\tilde{\lambda} : G \to \mathbb{R}$. The **modular function** $\lambda : G \to \mathbb{R}_+$ is defined to be $\lambda(g) = |\tilde{\lambda}(g)|$.

(d) Show that the integral is right invariant if and only if G is **unimodular**, i.e., if and only if $\lambda \equiv 1$.

(e) Show that a compact Lie group is unimodular.

Exercise 21.8

Let G be a compact Lie group and let $\Phi : G \to GL(V)$ be a representation of G. Show that there exists an inner product $\langle \, , \, \rangle$ in V such that this representation is by orthogonal transformations, i.e., such that

$$\langle \Phi(g) \cdot \mathbf{v}, \Phi(g) \cdot \mathbf{w} \rangle = \langle \mathbf{v}, \mathbf{w} \rangle, \quad \forall g \in G.$$

Hint: Use the fact that a compact Lie group is unimodular.

Exercise 21.9

Let G be a compact Lie group. Show that G has a bi-invariant Riemannian metric, i.e., a Riemannian metric which is both right and left invariant.

Hint: A left-invariant Riemannian metric in G is also right invariant if and only if the inner product $\langle \, , \, \rangle$ induced in $\mathfrak{g} \simeq T_e G$ satisfies:

$$\langle \mathrm{Ad}(g) \cdot X, \mathrm{Ad}(g) \cdot Y \rangle = \langle X, Y \rangle, \quad \forall g \in G, X, Y \in \mathfrak{g}.$$

Lecture 22

de Rham Cohomology

The equation $\mathrm{d}^2 = 0$, which so far we have made little use, has in fact some deep consequences, as we shall see in this and the next lectures.

Definition 22.1. A form $\omega \in \Omega^k(M)$ is called

(i) a **closed form** if $\mathrm{d}\omega = 0$.
(ii) an **exact form** if $\omega = \mathrm{d}\eta$, for some $\eta \in \Omega^{k-1}(M)$.

We will denote by $Z^k(M)$, respectively $B^k(M)$, the subspaces of closed, respectively exact, differential forms of degree k.

In other words, the closed forms form the kernel of d, while the exact forms form the image of d. The pair $(\Omega(M), \mathrm{d})$ is called the *de Rham complex* of M and we will often represent it as

$$\cdots \longrightarrow \Omega^{k-1}(M) \xrightarrow{\ \mathrm{d}\ } \Omega^k(M) \xrightarrow{\ \mathrm{d}\ } \Omega^{k+1}(M) \longrightarrow \cdots$$

The fact that $\mathrm{d}^2 = 0$ means that every exact form is closed, in other words

$$B^k(M) \subset Z^k(M).$$

One should think of $(\Omega(M), \mathrm{d})$ as a set of differential equations associated with the manifold M. Finding the closed forms, means to solve the differential equation

$$\mathrm{d}\omega = 0.$$

On the other hand, the exact forms can be thought of as the trivial solutions of this equation. The space of all solutions modulus the trivial solutions will give important *global* information about the manifold.

Definition 22.2. The **de Rham cohomology space of degree** k is the vector space

$$H^k(M) := Z^k(M)/B^k(M).$$

In general, the computation of the cohomology spaces $H^k(M)$ directly from the definition is very hard. In the next lectures, we will study several properties enjoyed by the de Rham cohomology which allows for its computation. For now, we list some easy consequences of the definition and we look at a very simple example.

Proposition 22.1. *Let M be a smooth manifold. Then,*

(i) $H^0(M) = \mathbb{R}^l$, *where l is the number of connected components of M;*
(ii) $H^k(M) = \{0\}$, *if $k < 0$ or $k > \dim M$.*

Proof. We have $\Omega^0(M) = C^\infty(M)$ and if $f \in C^\infty(M)$ satisfies $df = 0$, then f is locally constant. Hence

$$Z^0(M) = \mathbb{R}^l,$$

where l is the number of connected components of M. Since $B^0(M) = \{0\}$, we have that $H^0(M) = \mathbb{R}^l$. On the other hand, since $\Omega^k(M) = \{0\}$ if $k > \dim M$, the result follows. □

Example 22.1. Let $M = \mathbb{S}^1 = \{(x,y) \in \mathbb{R}^2 : x^2 + y^2 = 1\}$. Since $\mathbb{S}^1$ is connected, it follows that

$$H^0(\mathbb{S}^1) = \mathbb{R}.$$

In order to compute $H^1(\mathbb{S}^1)$, consider the 1-form

$$\omega := (-y\mathrm{d}x + x\mathrm{d}y)|_{\mathbb{S}^1}.$$

Since $\dim(\mathbb{S}^1) = 1$, ω is closed. On the other hand, consider the parameterization $\sigma :]0, 2\pi[\to \mathbb{S}^1 - \{(1,0)\}$, given by $\sigma(t) = (\cos t, \sin t)$. Then

$$\int_{\mathbb{S}^1} \omega = \int_{]0,2\pi[} \sigma^*\omega = \int_{]0,2\pi[} (-\sin t)\mathrm{d}\cos t + \cos t\mathrm{d}\sin t = \int_0^{2\pi} \mathrm{d}t = 2\pi.$$

By Corollary 21.1, we conclude that ω is not exact, so it represents a non-zero cohomology class $[\omega] \in H^1(\mathbb{S}^1)$.

The form ω has a simple geometric meaning. Since $\sigma^*\omega = \mathrm{d}t$, we have that $\omega = \mathrm{d}\theta$ in $\mathbb{S}^1 - \{(1,0)\}$, where $\theta : \mathbb{S}^1 - \{(1,0)\} \to \mathbb{R}$ is the angle coordinate (the inverse of the parameterization σ). So ω extends $\mathrm{d}\theta$ to the whole circle and often one denotes ω by $\mathrm{d}\theta$. But note that ω is *not* an exact form. We claim that $[\omega]$ is a basis for $H^1(\mathbb{S}^1)$.

Since ω is nowhere vanishing, given any $\alpha \in \Omega^1(\mathbb{S}^1)$ we must have $\alpha = f\omega$, for some function $f : \mathbb{S}^1 \to \mathbb{R}$. Set

$$c := \frac{1}{2\pi} \int_{\mathbb{S}^1} \alpha = \frac{1}{2\pi} \int_0^{2\pi} f(\theta) \mathrm{d}\theta,$$

and define $g : \mathbb{R} \to \mathbb{R}$ by

$$g(t) = \int_0^t (\alpha - c\omega) = \int_0^t (f(\theta) - c) \mathrm{d}\theta.$$

We find

$$\begin{aligned}
g(t + 2\pi) &= \int_0^{t+2\pi} (f(\theta) - c) \mathrm{d}\theta \\
&= \int_0^t (f(\theta) - c) \mathrm{d}\theta + \int_t^{t+2\pi} (f(\theta) - c) \mathrm{d}\theta \\
&= g(t) + \int_0^{2\pi} (f(\theta) - c) \mathrm{d}\theta = g(t),
\end{aligned}$$

so we can think of g as a smooth function $g : \mathbb{S}^1 \to \mathbb{R}$. In $\mathbb{S}^1 - \{(1,0)\}$, we have that

$$\mathrm{d}g = f(\theta) \mathrm{d}\theta - c\, \mathrm{d}\theta = \alpha - c\omega.$$

Hence, we must have $\mathrm{d}g = \alpha - c\omega$ in $\mathbb{S}^1$ so that $[\alpha] = c[\omega]$. This shows that $[\omega]$ generates $H^1(\mathbb{S}^1)$ so we conclude that $H^1(\mathbb{S}^1) \simeq \mathbb{R}$.

The wedge product $\wedge : \Omega^k(M) \times \Omega^l(M) \to \Omega^{k+l}(M)$ induces the product in the de Rham cohomology of M called the **cup product** given by

$$[\alpha] \cup [\beta] := [\alpha \wedge \beta].$$

We leave it as an exercise to check that this definition is independent of the choice of representatives of the cohomology classes. With this product the space

$$H^\bullet(M) = \bigoplus_{k \in \mathbb{Z}} H^k(M)$$

becomes a $\mathbb{Z}$-graded ring (in fact, a $\mathbb{Z}$-graded algebra over $\mathbb{R}$).

Given a map $\Phi : M \to N$, pull-back gives a linear map $\Phi^* : \Omega^\bullet(N) \to \Omega^\bullet(M)$ which commutes with the differentials

$$\Phi^* d\omega = d(\Phi^* \omega).$$

Therefore, Φ^* takes closed (respectively, exact) forms to closed (respectively, exact) forms, and we have an induced map in cohomology

$$\Phi^* : H^\bullet(N) \to H^\bullet(M), \quad [\omega] \longmapsto [\Phi^* \omega].$$

This linear map satisfies:

(i) It is a ring homomorphism: $\Phi^*([\alpha] \cup [\beta]) = \Phi^*[\alpha] \cup (\Phi^*[\beta]$;
(ii) If $\Phi : M \to N$ and $\Psi : N \to Q$ are smooth maps, then the composition $(\Psi \circ \Phi)^* : H^\bullet(Q) \to H^\bullet(M)$ satisfies $(\Psi \circ \Phi)^* = \Phi^* \circ \Psi^*$;
(iii) The identity map Id $: M \to M$ induces the identity map $H^\bullet(M) \to H^\bullet(M)$.

In particular, when $\Phi : M \to N$ is a diffeomorphism, the induced linear map is an isomorphism in cohomology $\Phi^* : H^\bullet(N) \to H^\bullet(M)$. Hence, de Rham cohomology ring is an *invariant* of differentiable manifolds.

Corollary 22.1. *If M and N are diffeomorphic, then $H^\bullet(M)$ and $H^\bullet(N)$ are isomorphic rings.*

Note that the differential takes a compactly supported form to a compactly supported form, so we have another complex $(\Omega_c(M), d)$.

Definition 22.3. The **compactly supported de Rham cohomology space of degree** k is the vector space:

$$H_c^k(M) := Z_c^k(M)/B_c^k(M),$$

where $Z_c^k(M) \subset \Omega_c^k(M)$, respectively $B_c^k(M) \subset \Omega_c^k(M)$, denotes the subspaces of closed, respectively exact, compactly supported forms of degree k.

The inclusion $\Omega_c^k(M) \subset \Omega^k(M)$ gives a linear map in cohomology:

$$H_c^k(M) \to H^k(M).$$

When M is compact this is just the identity map, but for a general M it may be neither injective nor surjective:

- given a closed form $\omega \in \Omega^k(M)$ one may not be able to find a cohomologous form $\omega + d\eta$ with compact support, and
- given an exact form $\omega = d\eta \in \Omega_c^k(M)$ one may not be able to find a primitive η' with compact support.

Hence, $H_c^k(M)$ and $H^k(M)$ can be very different.

Proposition 22.2. *Let M be a smooth manifold. Then:*

(i) $H_c^0(M) = \mathbb{R}^s$, *where s is the number of compact connected components of M;*

(ii) $H_c^k(M) = \{0\}$, *if $k < 0$ or $k > \dim M$.*

Proof. If $f \in C_c^\infty(M)$ satisfies $df = 0$, then f is constant in the compact connected components of M and is zero in the non-compact connected components. Since $B_c^0(M) = \{0\}$, we conclude that

$$H_c^0(M) = \mathbb{R}^s,$$

where s is the number of compact connected components of M. $\square$

The wedge product of forms induces a cup product

$$\cup : H_c^k(M) \times H_c^l(M) \to H_c^{k+l}(M), \quad [\alpha] \cup [\beta] := [\alpha \wedge \beta],$$

so we have a $\mathbb{Z}$-graded ring (in fact, a $\mathbb{Z}$-graded algebra):

$$H_c^\bullet(M) = \bigoplus_{k \in \mathbb{Z}} H_c^k(M).$$

Pullback by a smooth map $\Phi : M \to N$ of a form ω with compact support is a form $\Phi^* \omega$ that may fail to have compact support. However, if Φ is a *proper map*, we do have an induced ring homomorphism

$$\Phi^* : H_c^\bullet(N) \to H_c^\bullet(M).$$

This satisfies properties analogous to (i)–(iii) above. Hence, compactly supported cohomology is also an invariant of differentiable manifolds.

Remark 22.1 (*A Crash Course in Homological Algebra — Part I*). The de Rham complex $(\Omega^\bullet(M), \mathrm{d})$ and the compactly supported de Rham complex $(\Omega_c^\bullet(M), \mathrm{d})$ are examples of *cochain complexes*. In general, a **cochain complex** is a pair (C, d) where:

(a) C is a $\mathbb{Z}$-graded vector space, i.e., $C = \oplus_{k \in \mathbb{Z}} C^k$ is the direct sum of vector spaces[1];
(b) $\mathrm{d} : C \to C$ is a linear transformation of degree 1, i.e., $\mathrm{d}(C^k) \subset C^{k+1}$, and such that $\mathrm{d}^2 = 0$.

One represents a complex (C, d) by a diagram

$$\cdots \longrightarrow C^{k-1} \xrightarrow{\ \mathrm{d}\ } C^k \xrightarrow{\ \mathrm{d}\ } C^{k+1} \longrightarrow \cdots$$

The transformation d is called the **differential** of the complex.

For any cochain complex, (C, d) one defines the subspace of all **cocycles** to be the kernel of d, denoted

$$Z^k(C) := \{z \in C^k : \mathrm{d}z = 0\},$$

and the subspace of all **coboundaries** to be the image of d, denoted

$$B^k(C) := \{\mathrm{d}z : z \in C^{k-1}\}.$$

The relation $\mathrm{d}^2 = 0$ implies that $B^k(C) \subset Z^k(C)$, so one can define the **cohomology of the complex** (C, d) by

$$H^\bullet(C) := \bigoplus_{k \in \mathbb{Z}} H^k(C), \quad \text{where } H^k(C) := \frac{Z^k(C)}{B^k(C)}.$$

---

[1] More generally, one can consider complexes formed by $\mathbb{Z}$-graded modules over commutative rings with unit (e.g., abelian groups).

Given two cochain complexes (A, d_A) and (B, d_B), a **cochain map** of degree d is a linear map $f : A \to B$ such that

(a) f shifts the grading by d, i.e., $f(A^k) \subset B^{k+d}$;
(b) f commutes with the differentials, i.e., $f\mathrm{d}_A = \mathrm{d}_B f$.

One can represent such a cochain map by the commutative diagram

$$
\begin{array}{ccccccc}
\cdots \longrightarrow & A^{k-1} & \xrightarrow{\;\mathrm{d}_A\;} & A^k & \xrightarrow{\;\mathrm{d}_A\;} & A^{k+1} & \longrightarrow \cdots \\
& \downarrow{\scriptstyle f} & & \downarrow{\scriptstyle f} & & \downarrow{\scriptstyle f} & \\
\cdots \longrightarrow & B^{k+d-1} & \xrightarrow[\;\mathrm{d}_B\;]{} & B^{k+d} & \xrightarrow[\;\mathrm{d}_B\;]{} & B^{k+d+1} & \longrightarrow \cdots
\end{array}
$$

A cochain map $f : A \to B$ takes cocycles to cocycles and coboundaries to coboundaries. Hence, it induces a linear map in cohomology, denoted by the same letter

$$ f : H^\bullet(A) \to H^{\bullet+d}(B). $$

Most often we consider cochain maps of degree 0, and omit mentioning the degree.

The study of cochain complexes and cochain maps is one of the central themes of *Homological Algebra* (see, e.g., Weibel, 1994).

Exercises

Exercise 22.1
Show that $H^1(\mathbb{R}) = 0$ and $H_c^1(\mathbb{R}) \neq 0$.

Exercise 22.2
Show that $H^1(\mathbb{R}^2) = H^2(\mathbb{R}^2) = 0$.

Exercise 22.3
Show that $H^1(\mathbb{T}^d) = \mathbb{R}^d$.
Hint: Show that a basis for $H^1(\mathbb{T}^d)$ is given by $\{[\mathrm{d}\theta_1], \dots, [\mathrm{d}\theta_d]\}$, where $(\theta_1, \dots, \theta_d)$ are the angles on each $\mathbb{S}^1$ factor.

Exercise 22.4
Consider the 2-sphere $\mathbb{S}^2 = \{(x, y, z) \in \mathbb{R}^3 : x^2 + y^2 + z^2 = 1\}$.

(a) Show that $H^1(\mathbb{S}^2) = 0$.
(b) Show that the 2-form in $\mathbb{R}^3$ given by

$$ \omega = x\mathrm{d}y \wedge \mathrm{d}z + y\mathrm{d}z \wedge \mathrm{d}x + z\mathrm{d}x \wedge \mathrm{d}y. $$

induces by restriction to $\mathbb{S}^2$ a non-zero cohomology class $[\omega] \in H^2(\mathbb{S}^2)$.

Hint: For (a), use the fact that a closed 1-form in $\mathbb{R}^2$ is always exact.

Exercise 22.5
Applying de Rham cohomology, prove that $\mathbb{T}^2$ and $\mathbb{S}^2$ are not diffeomorphic manifolds.

Exercise 22.6
Show that if M is a compact, orientable, d-manifold, then $H^d(M) \neq 0$.

Exercise 22.7
Prove that $H^2(\mathbb{RP}^2) = 0$.

Hint: Consider the projection $\pi : \mathbb{S}^2 \to \mathbb{RP}^2$.

Exercise 22.8
Show that the wedge product $\wedge : \Omega^k(M) \times \Omega^l(M) \to \Omega^{k+l}(M)$ induces a well-defined product $\cup$ in the de Rham cohomology of M, which makes $H(M) = \oplus_k H^k(M)$ into a ring.

Exercise 22.9
A symplectic form on a manifold M of dimension $2n$ is a 2-form $\omega \in \Omega^2(M)$ such that $d\omega = 0$ and $\wedge^n \omega$ is a volume form. Show that if M is compact and admits some symplectic form, then $H^{2k}(M) \neq 0$ for $k = 0, \ldots, n$.

Hint: Use the ring structure of $H^\bullet(M)$.

Lecture 23

The de Rham Theorem

We saw in the previous lecture that de Rham cohomology is an invariant of differentiable manifolds. Actually, it is a topological invariant: two smooth manifolds whose underlying topological spaces are homeomorphic have the same de Rham cohomology. This is a consequence of the famous de Rham Theorem, which identifies de Rham cohomology with singular cohomology. In this lecture, we will present the ingredients and the statement of this result. We will not give a complete proof, since it requires more advanced material going beyond the scope of this lecture.

Singular Homology

We recall the definition of the singular homology of a topological space M. Although we will continue to use the letter M, the following discussion only uses the topology of M. We denote by $\Delta^k \subset \mathbb{R}^{k+1}$ the **standard k-simplex**

$$\Delta^k := \left\{ (t_0, \ldots, t_k) \in \mathbb{R}^{k+1} : \sum_{i=0}^{k} t_i = 1, t_i \geq 0 \right\}.$$

Note that $\Delta^0 = \{1\}$ has only one element (see Figure 23.1).

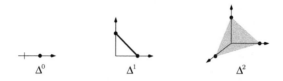

Fig. 23.1.   The standard k-simplex for $k = 0, 1, 2$.

Definition 23.1. A **singular k-simplex** in M is a continuous map $\sigma : \Delta^k \to M$. A **singular k-chain** is a formal linear combination

$$c = \sum_{i=1}^{p} a_i \sigma_i,$$

where $a_i \in \mathbb{R}$ and the σ_i are singular k-simplices.

We will denote by $S_k(M; \mathbb{R})$ the set of all singular k-chains. It is a real vector space. In fact, formally, $S_k(M; \mathbb{R})$ is the free vector space generated by the set of all singular k-simplices. One can also consider other abelian rings as coefficients besides $\mathbb{R}$, but here we will consider only real coefficients, since this is the case of interest to relate to differential forms.

We define the i-**face map** of the standard k-simplex, where $0 \le i \le k$, to be the map $\varepsilon^i : \Delta^{k-1} \to \Delta^k$ defined by

$$\varepsilon^i(t_0, \ldots, t_{k-1}) := (t_0, \ldots, t_{i-1}, 0, t_i, \ldots, t_{k-1}).$$

These face maps of the standard k-simplex induce face maps ε_i of any singular k-simplex $\sigma : \Delta^k \to M$ by setting

$$\varepsilon_i(\sigma) := \sigma \circ \varepsilon^i.$$

These clearly extend by linearity to any k-chain, yielding linear maps

$$\varepsilon_i : S_k(M; \mathbb{R}) \to S_{k-1}(M; \mathbb{R}).$$

Definition 23.2. The **boundary of a k-chain** c is the $(k-1)$-chain

$$\partial c := \sum_{i=0}^{k} (-1)^i \varepsilon_i(c).$$

The geometric meaning of this definition is that we consider the faces of each simplex with a certain choice of signs, which one should view as a kind of orientation of the faces (see Figure 23.2).

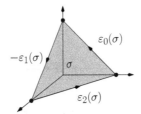

Fig. 23.2.   The boundary of a 2-simplex.

Example 23.1. The boundary of the standard 2-simplex $\sigma = $ id: $\Delta^2 \to \mathbb{R}^3$ is the chain

$$\partial\sigma = \varepsilon_0(\sigma) - \varepsilon_1(\sigma) + \varepsilon_2(\sigma),$$

where ε_0, ε_1 and ε_2 are the 1-simplices (faces) given by

$$\varepsilon_0(\sigma)(t_0, t_1) = (0, t_0, t_1), \quad \varepsilon_1(\sigma)(t_0, t_1) = (t_0, 0, t_1),$$
$$\varepsilon_2(\sigma)(t_0, t_1) = (t_0, t_1, 0).$$

The choice of signs can be thought of as orientations of the faces of the simplex.

Also, the 1-simplices ε_0, ε_1, and ε_2 have boundaries the 0-chains

$$\partial\varepsilon_0(\sigma)(1) = \varepsilon_0(\sigma)(0, 1) - \varepsilon_0(\sigma)(1, 0) = (0, 0, 1) - (0, 1, 0),$$
$$\partial\varepsilon_1(\sigma)(1) = \varepsilon_1(\sigma)(0, 1) - \varepsilon_1(\sigma)(1, 0) = (0, 0, 1) - (1, 0, 0),$$
$$\partial\varepsilon_2(\sigma)(1) = \varepsilon_2(\sigma)(0, 1) - \varepsilon_2(\sigma)(1, 0) = (0, 1, 0) - (1, 0, 0).$$

Note that

$$\partial^2\sigma = \partial(\partial\sigma) = \partial\varepsilon_0(\sigma) - \partial\varepsilon_1(\sigma) + \partial\varepsilon_2(\sigma) = 0.$$

In this example, we have $\partial^2\sigma = 0$. This is actually a general fact which is a consequence of the judicious choice of signs and parameterizations of the faces. We leave its proof as an exercise.

Lemma 23.1. *For every singular chain c, $\partial(\partial c) = 0$.*

In this way, we obtain a complex $(S(M; \mathbb{R}), \partial)$ represented by the diagram

$$\cdots \longleftarrow S_{k-1}(M; \mathbb{R}) \xleftarrow{\;\partial\;} S_k(M; \mathbb{R}) \xleftarrow{\;\partial\;} S_{k+1}(M; \mathbb{R}) \longleftarrow \cdots$$

One calls $(S(M;\mathbb{R}),\partial)$ the **complex of singular chains** in M. The homology of the complex $(S(M;\mathbb{R}),\partial)$ is called the **singular homology** of M with real coefficients, and is denoted

$$H_k(M;\mathbb{R}) := \frac{Z_k(M;\mathbb{R})}{B_k(M;\mathbb{R})}.$$

Remark 23.1 (*A Crash Course in Homological Algebra — Part II*). For a cochain complex, the differential *increases* the degree, while for the singular chains above the differential *decreases* the degree. In general, we call a complex $C = \oplus_{k\in\mathbb{Z}} C_k$ where the differential decreases the degree

$$\cdots \longleftarrow C_{k-1} \xleftarrow{\ \partial\ } C_k \xleftarrow{\ \partial\ } C_{k+1} \longleftarrow \cdots$$

a **chain complex**. We say that $z \in C_k$ is a **cycle** if $\partial z = 0$ and we say that z is a **boundary** if $z = \partial b$. In this case, one defines the **homology** of the complex (C,∂) to be vector space

$$H(C) := \oplus_{k\in\mathbb{Z}} H_k(C), \quad \text{with } H_k(C) := \frac{Z_k(C)}{B_k(C)},$$

where $Z_k(C)$ is the subspace of all cycles and $B_k(C)$ is the subspace of all boundaries. Note the position of the indices as subscripts.

Given a complex (C,∂) where the differential decreases degrees, one can define a new complex $(\bar{C},\mathrm{d})$ by setting $\bar{C}^k := C_{-k}$ and $\mathrm{d} = \partial$, obtaining a complex where the differential increases degrees. Therefore, these conventions are somewhat arbitrary.

If $\Phi : M \to N$ continuous map, then for any singular simplex $\sigma : \Delta^k \to M$, we have that $\Phi_*(\sigma) := \Phi \circ \sigma : \Delta^k \to N$ is a singular simplex in N. We extend this map to any chain $c = \sum_j a_j \sigma_j$ by requiring linearity to hold

$$\Phi_*(c) := \sum_j a_j (\Phi \circ \sigma_j).$$

It follows that $\Phi_* : S(M;\mathbb{R}) \to S(N;\mathbb{R})$ is a chain map

$$
\begin{array}{ccccccc}
\cdots \longleftarrow & S_{k-1}(M;\mathbb{R}) & \xleftarrow{\ \partial\ } & S_k(M;\mathbb{R}) & \xleftarrow{\ \partial\ } & S_{k+1}(M;\mathbb{R}) & \longleftarrow \cdots \\
& \downarrow{\scriptstyle\Phi_*} & & \downarrow{\scriptstyle\Phi_*} & & \downarrow{\scriptstyle\Phi_*} & \\
\cdots \longleftarrow & S_{k-1}(N;\mathbb{R}) & \xleftarrow{\ \partial\ } & S_k(N;\mathbb{R}) & \xleftarrow{\ \partial\ } & S_{k+1}(N;\mathbb{R}) & \longleftarrow \cdots
\end{array}
$$

Therefore, Φ_* induces a linear map in singular homology:

$$\Phi_* : H_\bullet(M;\mathbb{R}) \to H_\bullet(N;\mathbb{R}).$$

One checks easily that this assignment satisfies the following two properties:

(i) If $\Phi : M \to N$ and $\Psi : N \to Q$ are continuous maps, then

$$(\Psi \circ \Phi)_* = \Psi_* \circ \Phi_*;$$

(ii) The identity map id: $M \to M$ induces the identity map in homology

$$\mathrm{id}_* = \mathrm{id} : H_\bullet(M;\mathbb{R}) \to H_\bullet(M;\mathbb{R}).$$

From these, it follows that singular homology is a *topological invariant*.

Corollary 23.1. *If M and N are homeomorphic then $H_\bullet(M,\mathbb{R}) \simeq H_\bullet(N,\mathbb{R})$.*

Smooth Singular Homology

Assume now that M is a smooth manifold. The chain complex $(S_\bullet(M;\mathbb{R}),\partial)$ has a subcomplex $(S_\bullet^\infty(M;\mathbb{R}),\partial)$ consisting of the **smooth singular k-chains**:

$$S_k^\infty(M;\mathbb{R}) = \left\{ \sum_{i=1}^p a_i\sigma_i : \sigma_i : \Delta^k \to M \text{ is smooth} \right\}$$

This is a sub complex because if $c \in S_k^\infty(M;\mathbb{R})$ is a smooth k-chain, then so is $\partial c \in S_k^\infty(M;\mathbb{R})$.

Remark 23.2. Even when c is smooth, the use of the term "singular" is justified by the absence of any assumption on the differentials of the maps σ_i: in general, a smooth k-simplex *does not* parameterize any submanifold and its image may be a rather pathological subset of M.

One has the following important fact:

Proposition 23.1. *The inclusion* $S_\bullet^\infty(M, \mathbb{R}) \hookrightarrow S_\bullet(M, \mathbb{R})$ *induces an isomorphism in homology:*

$$H(S_\bullet^\infty(M, \mathbb{R})) \simeq H(S_\bullet(M, \mathbb{R})).$$

This proposition says that

(i) every class in $H_\bullet(M; \mathbb{R})$ has a representative c which is a C^∞ cycle, and
(ii) if c and c' are C^∞ cycles such that $c - c' = \partial b$ for some C^0 chain b, then $c - c' = \partial b'$ for some C^∞ chain b'.

Hence, smooth singular homology and singular homology coincide. The proof of the previous proposition is beyond the scope of these notes (see, e.g., Warner, 1983).

Singular Cohomology

One defines the singular cohomology of M by dualizing. First, one defines the space of **singular k-cochains** with real coefficients to be the vector space dual to $S_k(M; \mathbb{R})$

$$S^k(M; \mathbb{R}) := \operatorname{Hom}(S_k(M; \mathbb{R}), \mathbb{R}).$$

We have a **singular differential** obtained by transposing the singular boundary operator:

$$\mathrm{d} : S^k(M; \mathbb{R}) \to S^{k+1}(M; \mathbb{R}), \quad (\mathrm{d}l)(c) = l(\partial c), \quad \forall c \in S_k(M; \mathbb{R}).$$

It follows that $\mathrm{d}^2 = 0$, so we have a cochain complex $(S^\bullet(M; \mathbb{R}), \mathrm{d})$. The corresponding cohomology is called the **singular cohomology** of M with real coefficients and is denoted by $H^\bullet(M; \mathbb{R})$.

Remark 23.3. A more explicit form of the singular differential is as follows. Since the k-simplices form a basis for the vector space $S_k(M, \mathbb{R})$ a linear map $l : S_k(M, \mathbb{R}) \to \mathbb{R}$ amounts to a collection of real numbers $l = (l_\sigma)$, indexed by all singular simplices (so $l_\sigma = l(\sigma)$).

Then the singular differential $\mathrm{d}l \in S^{k+1}(M;\mathbb{R})$ is given by the collection $((\mathrm{d}l)_\sigma)$ indexed by $k+1$-simplices defined by

$$(\mathrm{d}l)_\sigma = \sum_{i=0}^{k+1}(-1)^i l_{\varepsilon_i(\sigma)}.$$

If $\Phi : M \to N$ we can transpose the map $\Phi_* : S_k(M;\mathbb{R}) \to S_k(N;\mathbb{R})$, obtaining a linear map $\Phi^* : S^k(N;\mathbb{R}) \to S^k(M;\mathbb{R})$ which is a cochain map, i.e., that satisfies

$$\Phi^*\mathrm{d} = \mathrm{d}\Phi^*.$$

This yields a linear map in singular cohomology $\Phi^* : H^\bullet(N;\mathbb{R}) \to H^\bullet(M;\mathbb{R})$, which satisfies the obvious functorial properties, and hence cohomology is a topological invariant.

Proposition 23.2. *If M and N are homeomorphic then $H^\bullet(M,\mathbb{R}) \simeq H^\bullet(N,\mathbb{R})$.*

Of course, one can also consider smooth singular k-cochains

$$S^k_\infty(M;\mathbb{R}) := \mathrm{Hom}(S^\infty_k(M;\mathbb{R}),\mathbb{R}).$$

which form a complex $(S^\bullet_\infty(M;\mathbb{R}),\mathrm{d})$. There is an obvious restriction map

$$S^k(M;\mathbb{R}) \to S^k_\infty(M;\mathbb{R}), \quad l \mapsto l|_{S^k_\infty(M;\mathbb{R})},$$

which is easily checked to be a cochain map. The induced map in cohomology is an isomorphism (see Warner, 1983).

Proposition 23.3. *The restriction map $S^k(M;\mathbb{R}) \to S^k_\infty(M;\mathbb{R})$ yields an isomorphism in cohomology*

$$H(S^\bullet(M;\mathbb{R}),\mathrm{d}) \simeq H(S^\bullet_\infty(M;\mathbb{R}),\mathrm{d}).$$

For this reason, in the sequel, we will not distinguish between these cohomologies.

Singular Cohomology vs. de Rham Cohomology

We now take advantage of the fact that singular cohomology and differentiable singular cohomology coincide to relate it with the de Rham cohomology. For that, we start by explaining how one can integrate differential forms over singular chains.

First, observe that one can parameterize the standard k-simplex Δ^k by the map $\phi : \Delta_0^k \to \Delta^k$, where

$$\Delta_0^k := \left\{ (x^1, \ldots, x^k) : x^i \geq 0, \sum_{i=1}^k x^i \leq 1 \right\}$$

$$\phi(x^1, \ldots, x^k) := \left(1 - \sum_{i=1}^k x^i, x^1, \ldots, x^k \right),$$

Hence, if $\omega \in \Omega^k(U)$ is a k-form which is defined in some open set $U \subset \mathbb{R}^{k+1}$ containing the standard k-simplex Δ^k, one can write

$$\phi^* \omega = f(x^1, \ldots, x^k) dx^1 \wedge \cdots \wedge dx^k,$$

and define

$$\int_{\Delta^k} \omega := \int_{\Delta_0^k} f dx^1 \cdots dx^k.$$

Next, given any differential form $\omega \in \Omega^k(M)$ on a smooth manifold M, we define the **integral of ω over a smooth simplex** $\sigma : \Delta^k \to M$ to be the real number

$$\int_\sigma \omega := \int_{\Delta^k} \sigma^* \omega.$$

We extend this definition to any smooth singular k-chain $c = \sum_{j=1}^p a_j \sigma_j$ by linearity

$$\int_c \omega := \sum_{j=1}^p a_j \int_{\sigma_j} \omega.$$

Note that, unlike the case of integration over manifolds, there is now no assumption neither about the orientability of M nor about the support of ω.

We leave it to the exercises the proof of the following version of Stokes formula for chains.

Theorem 23.1 (Stokes II). *Let M be a smooth manifold, $\omega \in \Omega^{k-1}(M)$ and c a smooth singular k-chain. Then,*

$$\int_c d\omega = \int_{\partial c} \omega.$$

Now we can define an *integration map* $I : \Omega^\bullet(M) \to S_\infty^\bullet(M; \mathbb{R})$ by setting

$$I(\omega)(\sigma) := \int_\sigma \omega,$$

for any $\omega \in \Omega^k(M)$ and $\sigma \in S_k^\infty(M; \mathbb{R})$.

Proposition 23.4. *The integration map* $I : (\Omega^\bullet(M), \mathrm{d}) \to (S_\infty^\bullet(M; \mathbb{R}), \mathrm{d})$ *is a chain map*

$$I(\mathrm{d}\omega) = \mathrm{d}I(\omega).$$

Proof. This follows from the following computation

$$(I(\mathrm{d}\omega))(\sigma) = \int_\sigma \mathrm{d}\omega = \int_{\partial\sigma} \omega = I(\omega)(\partial\sigma) = (\mathrm{d}I(\omega))(\sigma),$$

where we used Stokes formula for chains and the fact that the singular differential is the transpose of the singular coboundary operator. $\square$

Therefore, the integration map descends to the level of cohomology.

Theorem 23.2 (de Rham). *For any smooth manifold the integration map*

$$I : H^k(M) \to H^k(M; \mathbb{R}), \quad I([\omega])([\sigma]) := \int_\sigma \omega,$$

is an isomorphism.

There is also a cup product in singular cohomology and one can show that the integration map is actually a ring isomorphism (see the exercises). For a proof of de Rham's theorem see, e.g., Warner (1983, Chapter 5).

Corollary 23.2. *Two homeomorphic manifolds have isomorphic de Rham cohomologies.*

For example, the different exotic smooth structures on the spheres all have the same de Rham cohomology.

Exercises

Exercise 23.1
Show that for every singular chain c one has $\partial(\partial c) = 0$.

Exercise 23.2
Give a proof of Stokes formula for singular chains, by showing the following:

(a) It is enough to prove the formula for chains consisting of a singular simplex.
(b) It is enough to prove the formula for the standard k-simplex $\Delta_0^k \subset \mathbb{R}^k$.
(c) It is enough to prove the formula for $(k-1)$-differential forms in $\mathbb{R}^k$ of the type:

$$w = f\, dx^1 \wedge \cdots \wedge \widehat{dx^i} \wedge \cdots \wedge dx^k.$$

(d) Show that

$$\int_{\Delta_0^k} dw = \int_{\partial \Delta_0^k} w,$$

where w is a differential form of the type (c).

Exercise 23.3
In the torus $\mathbb{T}^d = \mathbb{S}^1 \times \cdots \times \mathbb{S}^1$ consider the 1-chains $c_j : [0,1] \to \mathbb{T}^d$ given by

$$c_j(t) := (1, \ldots, e^{2\pi i t}, \ldots, 1) \quad (j = 1, \ldots, d).$$

Show that

(a) The c_j's are 1-cycles, i.e., $\partial c_j = 0$.
(b) The c_j's are not 1-boundaries.
(c) The classes $\{[c_1], \ldots, [c_d]\} \subset H_1(\mathbb{T}^d, \mathbb{R})$ form a linearly independent set.

Hint: Use the Stokes formula.

Exercise 23.4
By the de Rham's Theorem, exterior product induces a product

$$\cup : H^k(M; \mathbb{R}) \times H^l(M : \mathbb{R}) \to H^{k+l}(M; \mathbb{R}),$$

so that $H^\bullet(M; \mathbb{R})$ becomes a ring. This is also called the **cup product**. Here is one way of constructing it directly.

(a) Show that for $l < k$ and $0 \leq i_0 < \cdots < i_l \leq k$ one has maps $\varepsilon_{i_0,\ldots,i_l} : \Delta^l \to \Delta^k$, defined by

$$\varepsilon_{i_0,\ldots,i_l}(t_0,\ldots,t_l) := (s_0,\ldots,s_k),$$

$$\text{where} \quad \begin{cases} s_l = 0, & \text{if } l \notin \{i_0,\ldots,i_l\} \\ s_{i_j} = t_j, & \text{otherwise.} \end{cases}$$

(b) Show that if $c_1 \in S^k(M; \mathbb{R})$ and $c_2 \in S^l(M; \mathbb{R})$ the formula

$$(c_1 \cup c_2)(\sigma) := c_1(\sigma \circ \varepsilon_{1,\ldots,k}) c_2(\sigma \circ \varepsilon_{k+1,\ldots,k+l}),$$

defines an element $c_1 \cup c_2 \in S^{k+l}(M; \mathbb{R})$.

(c) Show that for any chains $c_1 \in S^k(M; \mathbb{R})$ and $c_2 \in S^l(M; \mathbb{R})$ one has

$$\mathrm{d}(c_1 \cup c_2) = (\mathrm{d}c_1) \cup c_2 + (-1)^k c_1 \cup (\mathrm{d}c_2).$$

It follows that one can define $\cup : H^k(M; \mathbb{R}) \times H^l(M; \mathbb{R}) \to H^{k+l}(M; \mathbb{R})$ by

$$[c_1] \cup [c_2] := [c_1 \cup c_2].$$

If $I : \Omega^k(M) \to S^k_\infty(M)$ is the integration map, in general, $I(\omega \wedge \eta) \neq I(\omega) \cup I(\eta)$. However, show that this equality holds at the level of cohomology, i.e.,

$$I([\omega] \wedge [\eta]) = I([\omega]) \cup I([\eta]), \quad [\omega] \in H^k(M), [\eta] \in H^l(M).$$

Exercise 23.5

Let $S_k(M, \mathbb{Z}) \subset S_k(M, \mathbb{R})$ be the subgroup consisting of all **integral singular k-simplex**, i.e., all formal linear combinations

$$c = \sum_{i=1}^p n_i \sigma_i,$$

where $n_i \in \mathbb{Z}$ and the σ_i are singular k-simplices. So $S_k(M, \mathbb{Z})$ is the free abelian group generated by the set of all singular k-simplices.

(a) Show that $S_k(M, \mathbb{Z}) \subset (S_k(M, \mathbb{R}), \partial)$ is a subcomplex of abelian groups. The corresponding homology groups are denoted $H_k(M, \mathbb{Z})$ and are called the **integral singular homology groups** of M.

(b) Dually, define the complex of singular integral cochains $(S^k(M, \mathbb{Z}), d)$ by

$$S^k(M, \mathbb{Z}) := \operatorname{Hom}(S_k(M, \mathbb{Z}), \mathbb{Z}), \quad dc(\sigma) := c(\partial\sigma).$$

Denoting by $H_k(M, \mathbb{Z})$ the corresponding **integral singular cohomology groups**, show that there is a group homomorphism:

$$i : H^k(M, \mathbb{Z}) \to H^k(M, \mathbb{R}).$$

Exercise 23.6

Let $(S_k^\infty(M, \mathbb{Z}), \partial)$ be the complex of smooth integral singular chains and let $(S_\infty^k(M, \mathbb{Z}), d)$ be the complex of smooth integral singular cochains (see previous problem). It is a fact that these complexes still compute the integral singular homology and cohomology of M, i.e., we have

$$H_k(S_\bullet^\infty(M, \mathbb{Z}), \partial) = H_k(M, \mathbb{Z}), \quad H^k(S_\infty^\bullet(M, \mathbb{Z}), d) = H^k(M, \mathbb{Z}).$$

Assuming this, show that

(a) There is a homomorphism of abelian groups

$$I : H^k(M, \mathbb{Z}) \to H^k(M).$$

(b) For a closed-form $\omega \in \Omega^k(M)$, the set

$$\operatorname{Per}(\omega) := \left\{ \int_c \omega : [c] \in H_k(S_\bullet^\infty(M, \mathbb{Z}), \partial) \right\} \subset \mathbb{R},$$

is an additive subgroup. It is called the **group of periods** of ω.

(c) A cohomology class $[\omega] \in H^k(M)$ belongs to the image of the homomorphism $I : H^k(M, \mathbb{Z}) \to H^k(M)$ if and only if $\operatorname{Per}(\omega) \subset \mathbb{Z}$.

Lecture 24

Homotopy Invariance and Mayer–Vietoris Sequence

We shall now study some properties of de Rham cohomology which are very useful in the computation of these rings in specific examples.

The Poincaré Lemma

We start with the simplest example of manifold, namely $M = \mathbb{R}^d$. In order to compute its cohomology, we compare the cohomologies of M and of $M \times \mathbb{R}$, for an arbitrary smooth manifold M.

Proposition 24.1. *For a manifold M, let $\pi : M \times \mathbb{R} \to M$ be the projection and $i : M \to M \times \mathbb{R}$ the inclusion map given by*

$$
\begin{array}{cc}
M \times \mathbb{R} & \\
{\scriptstyle i}\big\uparrow \ \big\downarrow{\scriptstyle \pi} & \quad i(p) = (p,0), \\
M & \quad \pi(p,t) = p.
\end{array}
$$

The induced maps $i^* : H^\bullet(M \times \mathbb{R}) \to H^\bullet(M)$ and $\pi^* : H^\bullet(M) \to H^\bullet(M \times \mathbb{R})$ are inverse to each other.

Since $H^0(\mathbb{R}^0) = \mathbb{R}$ and $H^k(\mathbb{R}^0) = 0$ if $k \neq 0$, repeated use of the proposition gives the cohomology of Euclidean space.

Corollary 24.1 (Poincaré Lemma).

$$H^k(\mathbb{R}^d) = H^k(\mathbb{R}^0) = \begin{cases} \mathbb{R} & \text{if } k = 0, \\ 0 & \text{if } k \neq 0. \end{cases}$$

In other words, in $\mathbb{R}^d$ every closed form of positive degree is exact.

We now turn to the proof of Proposition 24.1. For that, we will need a bit more of homological algebra and the notion of of *homotopy operator*.

Remark 24.1 (*A Crash Course in Homological Algebra — Part III*). Given two cochain complexes (A, d) and (B, d), a **homotopy operator** between cochain maps $f, g : A \to B$ is a linear map $h : A \to B$ of degree -1, such that

$$f - g = \pm(\mathrm{d}h \pm h\mathrm{d})$$

(the choice of signs is irrelevant). In this case, we also say that f and g are **homotopic cochain maps** and we express it by the diagram

Since $\pm(\mathrm{d}h \pm h\mathrm{d})$ maps closed forms to exact forms, it induces the zero map in cohomology. Hence, homotopic chain maps f and g induce the same map in cohomology

$$f_* = g_* : H^\bullet(A) \to H^\bullet(B).$$

Proof of Proposition 24.1. Note that $\pi \circ i = \mathrm{Id}$, hence $i^* \circ \pi^* = \mathrm{Id}$. To complete the proof, we need to check that $\pi^* \circ i^* = \mathrm{Id}$. For this, we construct a homotopy operator $h : \Omega^\bullet(M \times \mathbb{R}) \to \Omega^{\bullet-1}(M \times \mathbb{R})$ such that

$$\mathrm{Id} - \pi^* \circ i^* = \mathrm{d}h + h\mathrm{d}.$$

For that we observe that differential k-form $\theta \in \Omega^k(M \times \mathbb{R})$ can be expressed as a locally finite sum

$$\theta = \sum_i \theta_i,$$

where each $\theta_i \in \Omega^k(M \times \mathbb{R})$ falls into one of the following two types

$$f_1 \, \pi^* \omega_1, \quad f_2 \, \mathrm{d}t \wedge \pi^* \omega_2,$$

with ω_1 and ω_2 differential forms in M of degree k and $k-1$, respectively, and $f_1, f_2 : M \times \mathbb{R} \to \mathbb{R}$ smooth functions. So, we define the homotopy operator h for each of these types of forms by setting

$$f_1 \, \pi^* \omega_1 \longmapsto 0, \quad f_2 \, \mathrm{d}t \wedge \pi^* \omega_2 \longmapsto \int_0^t f_2(x,s)\mathrm{d}s \; \pi^* \omega_2.$$

Then we extend it by linearity to all forms. We need to check that h is indeed a homotopy operator, i.e., that we have

$$(\mathrm{Id} - \pi^* \circ i^*)\theta = (\mathrm{d}h + \mathrm{d}h)\theta. \tag{24.1}$$

In order to check this, we show that it holds for each of the two types of the form above. Since h extends by linearity, it follows that (24.1) holds for any form θ.

- Let $\theta = \theta_1 := f_1 \, \pi^* \omega_1 \in \Omega^k(M \times \mathbb{R})$. Then

$$(\mathrm{Id} - \pi^* \circ i^*)\theta_1 = \theta_1 - \pi^*(f_1(x,0)\omega_1) = (f_1(x,t) - f_1(x,0))\pi^* \omega_1.$$

On the other hand,

$$
\begin{aligned}
(\mathrm{d}h + h\mathrm{d})\theta_1 &= h\mathrm{d}\theta_1 \\
&= h\left((\mathrm{d}f_1 \wedge \pi^* \omega_1 + f_1 \pi^* \mathrm{d}\omega_1)\right) \\
&= h\left(\frac{\partial f_1}{\partial t}\mathrm{d}t_1 \wedge \pi^* \omega_1\right) \\
&= \int_0^t \frac{\partial f_1}{\partial t}(x,s)\mathrm{d}s \; \pi^* \omega_1 = (f_1(x,t) - f_1(x,0))\pi^* \omega_1.
\end{aligned}
$$

Hence, (24.1) holds for the first type of forms.
- Let $\theta = \theta_2 := f_2 \, \mathrm{d}t \wedge \pi^* \omega_2$. On the one hand,

$$(\mathrm{Id} - \pi^* \circ i^*)\theta_2 = \theta_2.$$

On the other hand, in any local coordinates (U, x^i) for M, we find

$$(\mathrm{d}h + h\mathrm{d})\theta_2 = \mathrm{d}\left(\int_0^t f_2(x, s)\mathrm{d}s \; \pi^* \omega_2\right)$$

$$+ h\left(\sum_i \frac{\partial f_2}{\partial x^i}\mathrm{d}x^i \wedge \mathrm{d}t \wedge \pi^* \omega_2 - f_2 \mathrm{d}t \wedge \pi^* \mathrm{d}\omega_2\right)$$

$$= f_2(x, t)\mathrm{d}t \wedge \pi^* \omega_2 + \sum_i \int_0^t \frac{\partial f_2}{\partial x^i}\mathrm{d}s \; \mathrm{d}x^i \wedge \pi^* \omega_2$$

$$+ \int_0^t f_2(x, s)\mathrm{d}s \; \mathrm{d}\pi^* \omega_2$$

$$- \sum_i \int_0^t \frac{\partial f_2}{\partial x^i}\mathrm{d}s \; \mathrm{d}x^i \wedge \pi^* \omega_2 - \int_0^t f_2(x, s)\mathrm{d}s \; \pi^* \mathrm{d}\omega_2$$

$$= f_2(x, t)\mathrm{d}t \wedge \pi^* \omega_2 = \theta_2.$$

Therefore, (24.1) also holds for the second type of forms. $\square$

Homotopy Invariance

Proposition 24.1 is actually a very special case of a general property of cohomology, which loosely says that if a manifold can be continuously deformed into another manifold then their cohomologies are isomorphic. In order to formulate a precise statement, we make the following definition.

Definition 24.1. Let $\Phi, \Psi : M \to N$ be smooth maps. A **smooth homotopy** between Φ and Ψ is a smooth map $H : M \times \mathbb{R} \to N$ such that

$$H(p, t) = \begin{cases} \Phi(p) & \text{if } t \leq 0, \\ \Psi(p) & \text{if } t \geq 1. \end{cases}$$

Often, one defines a smooth homotopy between Φ and Ψ to be a smooth map $H : M \times [0, 1] \to N$ such that

$$H(p, 0) = \Phi(p), \quad H(1, p) = \Psi(p), \quad p \in M.$$

It is easy to see that the two definitions are equivalent. We also have the following less obvious facts (for a proof, see Hirsch, 1994):

(i) two smooth maps are smooth homotopic iff they are C^0-homotopic;

(ii) any continuous map between two smooth manifolds is C^0-homotopic to a smooth map.

Theorem 24.1 (Homotopy Invariance). *If $\Phi, \Psi : M \to N$ are smooth homotopic maps then $\Phi^* = \Psi^* : H^\bullet(N) \to H^\bullet(M)$.*

Proof. Denote by $\pi : M \times \mathbb{R} \to M$ the projection and $i_0, i_1 : M \to M \times \mathbb{R}$ the sections

$$i_0(p) = (p, 0) \quad \text{and} \quad i_1(p) = (p, 1).$$

By Proposition 24.1, i_0^* and i_1^* are linear maps which both invert π^*, so they must coincide: $i_0^* = i_1^*$.

Now, let $H : M \times \mathbb{R} \to N$ be a homotopy between Φ and Ψ. Then $\Phi = H \circ i_0$ and $\Psi = H \circ i_1$. At the level of cohomology, we find

$$\Phi^* = (H \circ i_0)^* = i_0^* \, H^*,$$

$$\Psi^* = (H \circ i_1)^* = i_1^* \, H^*.$$

Since $i_0^* = i_1^*$, we conclude that $\Phi^* = \Psi^*$. □

We say that two manifolds M and N have the same **homotopy type** if there exist smooth maps $\Phi : M \to N$ and $\Psi : N \to M$ such that $\Psi \circ \Phi$ and $\Phi \circ \Psi$ are homotopic to id_M and id_N, respectively. A manifold is said to be **contractible** if it has the same homotopy type as a point (i.e., $\mathbb{R}^0$).

Corollary 24.2. *If M and N have the same homotopy type, then $H^\bullet(M) \simeq H^\bullet(N)$. In particular, if M is a contractible manifold, then*

$$H^k(M) = \begin{cases} \mathbb{R} & \text{if } k = 0, \\ 0 & \text{if } k \neq 0. \end{cases}$$

Example 24.1. An open set $U \subset \mathbb{R}^d$ is called **star-shaped** if there exists some $x_0 \in U$ such that for any $x \in U$, the segment $tx + (1-t)x_0$

lies in U. We leave it as an exercise to show that a star-shaped open set U is contractible, so that

$$H^k(U) = \begin{cases} \mathbb{R} & \text{if } k = 0, \\ 0 & \text{if } k \neq 0. \end{cases}$$

Example 24.2. The manifold $M = \mathbb{R}^{d+1} \setminus \{0\}$ has the same homotopy type as $\mathbb{S}^d$. This follows because the inclusion $i : \mathbb{S}^d \hookrightarrow \mathbb{R}^{d+1} \setminus \{0\}$ and the projection $\pi : \mathbb{R}^{d+1} \setminus \{0\} \to \mathbb{S}^d$, $x \mapsto \frac{x}{\|x\|}$, are homotopic inverses to each other. Hence,

$$H^\bullet(\mathbb{S}^d) = H^\bullet(\mathbb{R}^{d+1} \setminus \{0\}).$$

Note that we don't know yet how to compute $H^\bullet(\mathbb{R}^{d+1} \setminus \{0\})$!

Mayer–Vietoris Sequence

Let us discuss now another important property of cohomology, which allows to compute the cohomology of a space by decomposing it into more elementary pieces of which we already know the cohomology.

Theorem 24.2 (Mayer–Vietoris Sequence). *Let M be a smooth manifold and let $U, V \subset M$ be open subsets such that $M = U \cup V$. There exists a long exact sequence:*

$$\longrightarrow H^k(M) \longrightarrow H^k(U) \oplus H^k(V) \longrightarrow H^k(U \cap V) \xrightarrow{\delta} H^{k+1}(M) \longrightarrow$$

The proof will require a bit more of homological algebra.

Remark 24.2 (*A Crash Course in Homological Algebra — Part IV*). A sequence of vector spaces and linear maps

$$\cdots \longrightarrow C^{k-1} \xrightarrow{f_{k-1}} C^k \xrightarrow{f_k} C^{k+1} \longrightarrow \cdots$$

is called **exact** if $\operatorname{Im} f_{k-1} = \operatorname{Ker} f_k$. An exact sequence of the form

$$0 \longrightarrow A \xrightarrow{f} B \xrightarrow{g} C \longrightarrow 0$$

is called a **short exact sequence**. This means that

(i) f is injective,
(ii) Im $f = \text{Ker } g$, and
(iii) g is surjective.

Given any exact sequence ending in trivial vector spaces

$$0 \longrightarrow C^0 \longrightarrow \cdots \longrightarrow C^k \longrightarrow \cdots \longrightarrow C^d \longrightarrow 0$$

the alternating sum of the dimensions of the spaces in the sequence is zero:

$$\sum_{i=0}^{d} (-1)^i \dim C^i = 0.$$

We leave the (easy) proof for the exercises.

A short exact sequence of complexes

$$0 \longrightarrow (A^\bullet, d) \xrightarrow{f} (B^\bullet, d) \xrightarrow{g} (C^\bullet, d) \longrightarrow 0$$

can be represented by a large commutative diagram where all rows are exact

An important basic fact about short exact sequence of complexes is that they possess an *associated long exact sequence* in cohomology

$$\cdots \longrightarrow H^k(A) \xrightarrow{f} H^k(B) \xrightarrow{g} H^k(C) \xrightarrow{\delta} H^{k+1}(A) \longrightarrow \cdots$$

where $\delta : H^k(C) \to H^{k+1}(A)$ is called the *connecting homomorphism*. The fact that Im $f = \text{Ker } g$ follows immediately from the definition of a short exact sequence. On the other hand, the identities $\operatorname{Im} g = \operatorname{Ker} \delta$ and $\operatorname{Im} \delta = \operatorname{Ker} f$ follow from the way δ is constructed, which we will now describe.

For the construction of δ one should keep in mind the large commutative diagram above. Given a cocycle $c \in C^k$ so that $\mathrm{d}c = 0$, it follows from the fact that the rows are exact that there exists $b \in B^k$ such that $g(b) = c$. Since the diagram commutes, we have

$$g(\mathrm{d}b) = \mathrm{d}g(b) = \mathrm{d}c = 0.$$

Using again that the rows are exact, we conclude that there exists a unique $a \in A^{k+1}$ such that $f(a) = \mathrm{d}b$. Note that

$$f(\mathrm{d}a) = \mathrm{d}f(a) = \mathrm{d}^2 b = 0,$$

and since f is injective, we have $\mathrm{d}a = 0$, i.e., a is cocycle. In this way, we have associated to a cocycle $c \in C^k$ a cocycle $a \in A^{k+1}$.

This association depends on a choice of an intermediate element $b \in C^k$. If we choose a different $b' \in C^k$ such $g(b') = c$, we obtain a different element $a' \in A^{k+1}$. However, noting that

$$g(b - b') = g(b') - g(b) = c - c = 0,$$

we see that there exist $\bar{a} \in A^k$ such that $f(\bar{a}) = b - b'$. Hence, we find

$$f(a - a') = f(a) - f(a') = \mathrm{d}b - \mathrm{d}b' = \mathrm{d}f(\bar{a}) = f(\mathrm{d}\bar{a}).$$

Since f is injective, we conclude that $a - a' = \mathrm{d}\bar{a}$. This shows that different intermediate choices lead to elements in the same cohomology class.

Finally, note that this assignment associates a coboundary to a coboundary. In fact, if $c \in C^k$ is a coboundary, i.e., $c = \mathrm{d}c'$, then there exists $b' \in C^{k-1}$ such that $g(b') = c'$. Moreover,

$$g(b - \mathrm{d}b') = g(b) - \mathrm{d}g(b') = c - \mathrm{d}c' = 0.$$

Therefore, there exists $a' \in A^k$ such that $f(a') = b - \mathrm{d}b'$, and

$$f(a - \mathrm{d}a') = f(a) - \mathrm{d}f(a') = \mathrm{d}b - \mathrm{d}b + \mathrm{d}^2 b' = 0.$$

Since f is injective, we conclude that $a = \mathrm{d}a'$ is a coboundary, as claimed.

In conclusion, there is a well-defined map in cohomology

$$\delta : H^k(C) \to H^{k+1}(A), [c] \mapsto [a].$$

We leave it as an exercise to check, using this definition, that $\operatorname{Im} g = \operatorname{Ker} \delta$ and $\operatorname{Im} \delta = \ker f$.

Proof of Theorem 24.2. We claim that we have a short exact sequence

$$0 \longrightarrow \Omega^\bullet(M) \longrightarrow \Omega^\bullet(U) \oplus \Omega^\bullet(V) \longrightarrow \Omega^\bullet(U \cap V) \longrightarrow 0$$

where the first map is given by

$$\omega \mapsto (\omega|_U, \omega|_V),$$

while the second map is defined by

$$(\theta, \eta) \mapsto \theta|_{U \cap V} - \eta|_{U \cap V}.$$

The corresponding long exact sequence in cohomology yields the statement of the theorem. It remains to prove the claim:

- Since $M = U \cup V$, the first map is injective.
- It is clear from the definitions that the image of the first map is contained in the kernel of the second map. On the other hand, if $(\theta, \eta) \in \Omega^\bullet(U) \oplus \Omega^\bullet(V)$ belongs to the kernel of the second map, then

$$\theta|_{U \cap V} = \eta|_{U \cap V}.$$

Hence, we can define a smooth differential form in M by

$$\omega_p = \begin{cases} \theta_p & \text{if } p \in U, \\ \eta_p & \text{if } p \in V. \end{cases}$$

Therefore, the image of the first map coincides with the kernel of the second map.

- Finally, let $\alpha \in \Omega^\bullet(U \cap V)$ and choose a partition of unity $\{\rho_U, \rho_V\}$ subordinated to the cover $\{U, V\}$. Then $\rho_V \alpha \in \Omega^\bullet(U)$ and $\rho_U \alpha \in \Omega^\bullet(V)$ and the effect of the second map on this pair of forms is

$$(\rho_V \alpha, -\rho_U \alpha) \mapsto \rho_V \alpha + \rho_U \alpha = \alpha.$$

Therefore, the second map is surjective. $\qquad \square$

Example 24.3. We saw in Example 22.1 how to find $H^\bullet(\mathbb{S}^1)$. One can use the Mayer–Vietoris sequence to compute the cohomology of $\mathbb{S}^d$ for $d \geq 2$.

Let $U = \mathbb{S}^d \setminus \{p_N\}$ and $V = \mathbb{S}^d \setminus \{p_S\}$, where $p_N, p_S \in \mathbb{S}^d$ are the north and south poles. Note that $\mathbb{S}^d = U \cup V$ and:

(i) U and V are contractible, since the stereographic projections $\pi_N : U \to \mathbb{R}^d$ and $\pi_S : V \to \mathbb{R}^d$ are diffeomorphism;
(ii) $U \cap V$ is diffeomorphic to $\mathbb{R}^d \setminus \{0\}$ (via any of the stereographic projections) and by Example 24.2 $\mathbb{R}^d \setminus \{0\}$ as the same homotopy type as $\mathbb{S}^{d-1}$.

Therefore,

- If $k \geq 1$, the Mayer–Vietoris sequence gives

$$\cdots \longrightarrow 0 \oplus 0 \longrightarrow H^k(\mathbb{S}^{d-1}) \overset{\delta}{\longrightarrow} H^{k+1}(\mathbb{S}^d) \longrightarrow 0 \oplus 0 \longrightarrow \cdots$$

Hence, $H^{k+1}(\mathbb{S}^d) \simeq H^k(\mathbb{S}^{d-1})$. By induction, we conclude that

$$H^k(\mathbb{S}^d) \simeq H^{k-1}(\mathbb{S}^{d-1}) \simeq \cdots \simeq H^1(\mathbb{S}^{d-k+1}).$$

- Since U, V, and $U \cap V$ are connected, the first terms of this sequence are

$$0 \longrightarrow \mathbb{R} \longrightarrow \mathbb{R} \oplus \mathbb{R} \longrightarrow \mathbb{R} \overset{\delta}{\longrightarrow} H^1(\mathbb{S}^d) \longrightarrow 0 \longrightarrow \cdots$$

Since the alternating sum of the dimensions must be zero, it follows that $\dim H^1(\mathbb{S}^d) = 0$, if $d \geq 2$.

Since $H^1(\mathbb{S}^1) = \mathbb{R}$, we conclude that

$$H^k(\mathbb{S}^d) = \begin{cases} \mathbb{R} & \text{if } k = 0 \text{ or } d, \\ 0 & \text{otherwise.} \end{cases}$$

Compactly Supported Cohomology

As we saw in the previous lecture, compactly supported cohomology does not behave functorially under smooth maps. Still this cohomology behaves functorially under proper maps and, because of this, compactly supported cohomology still satisfies properties analogous, but distinct, to the properties we have studied for de Rham cohomology.

Proposition 24.2. *Let M be a smooth manifold. Then,*

$$H_c^{\bullet}(M \times \mathbb{R}) \simeq H_c^{\bullet - 1}(M).$$

Proof. Note that if $\pi : M \times \mathbb{R} \to M$ is the projection and $\omega \neq 0$, then $\pi^* \omega$ does not have compact support. Instead, we claim that one has "push-forward" maps

$$\pi_* : \Omega_c^{\bullet + 1}(M \times \mathbb{R}) \to \Omega_c^{\bullet}(M), \quad e_* : \Omega_c^{\bullet}(M) \to \Omega_c^{\bullet + 1}(M \times \mathbb{R}).$$

which are cochain maps, homotopic inverse to each other.

We start by constructing π_*, which is known as *integration along the fibers*. For that note that every compactly supported k-form in $M \times \mathbb{R}$ is a locally finite sum of forms of the types

$$f_1 \, \pi^* \omega_1, \quad f_2 \, \pi^* \omega_2 \wedge \mathrm{d}t,$$

where $\omega_1 \in \Omega_c^k(M)$, $\omega_1 \in \Omega_c^{k-1}(M)$ and $f_1, f_2 : M \times \mathbb{R} \to \mathbb{R}$ are compactly supported smooth functions. The map π_* is defined on these two types of forms by

$$f_1 \, \pi^* \omega_1 \longmapsto 0, \quad f_2 \, \pi^* \omega_2 \wedge \mathrm{d}t \longmapsto \int_{-\infty}^{+\infty} f_2(x, t) \mathrm{d}t \, \omega_2,$$

and extended by linearity to arbitrary forms.

On the other hand, in order to construct e_* one chooses some compactly supported 1-form $\theta = g(t)\mathrm{d}t \in \Omega_c^1(\mathbb{R})$ with $\int_{\mathbb{R}} \theta = 1$ and sets

$$e_* : \omega \to \pi^* \omega \wedge \theta.$$

It follows from these definitions of π_* and e_* that

$$\pi_* \circ e_* = \mathrm{Id}, \quad \mathrm{d}\pi_* = \pi_* \mathrm{d}, \quad e_* \mathrm{d} = \mathrm{d}e_*.$$

To finish the proof, we check that $e_* \circ \pi_*$ is homotopic to the identity. We leave it as an exercise to check that the map $h : \Omega_c^{\bullet}(M \times \mathbb{R}) \to \Omega_c^{\bullet - 1}(M \times \mathbb{R})$ defined on forms of the two types by

$$f_1 \, \pi^* \omega_1 \longmapsto 0,$$

$$f_2 \, \pi^* \omega_2 \wedge \mathrm{d}t \longmapsto \left(\int_{-\infty}^{t} f_2(x, s)\mathrm{d}s - \int_{-\infty}^{+\infty} f_2(x, s)\mathrm{d}s \int_{-\infty}^{t} g(s)\mathrm{d}s \right) \pi^* \omega_2,$$

is indeed a homotopy from $e_* \circ \pi_*$ to the identity. $\qquad\square$

The proposition shows that compactly supported cohomology *is not* invariant under homotopy. On the other hand, the repeated use of the proposition shows that the Poincaré Lemma must be modified as follows.

Corollary 24.3 (Poincaré Lemma for compactly supported cohomology).

$$H_c^k(\mathbb{R}^d) = \begin{cases} \mathbb{R} & \text{if } k = d, \\ 0 & \text{if } k \neq d. \end{cases}$$

Next, we construct the Mayer–Vietoris sequence for compactly supported cohomology. Note that if $U, V \subset M$ are open sets with $U \cup V = M$, the inclusions $U, V \hookrightarrow M$, $U \cap V \hookrightarrow U$ and $U \cap V \hookrightarrow V$ give a short exact sequence

$$0 \longleftarrow \Omega_c^\bullet(M) \longleftarrow \Omega_c^\bullet(U) \oplus \Omega_c^\bullet(V) \longleftarrow \Omega_c^\bullet(U \cap V) \longleftarrow 0$$

where the two maps are

$$(\theta, \eta) \mapsto \theta + \eta, \quad \omega \mapsto (-\omega, \omega).$$

Hence, it follows that

Theorem 24.3 (Mayer–Vietoris sequence for compactly supported cohomology). *Let M be a smooth manifold and $U, V \hookrightarrow M$ open subsets such that $M = U \cup V$. There exists a long exact sequence*

$$\longleftarrow H_c^k(M) \longleftarrow H_c^k(U) \oplus H_c^k(V) \longleftarrow H_c^k(U \cap V)$$

$$\overset{\delta}{\longleftarrow} H_c^{k-1}(M) \longleftarrow$$

Note that in the Mayer–Vietoris sequence for compact supported cohomology the inclusions $U, V \hookrightarrow M$, $U \cap V \hookrightarrow U$ and $U \cap V \hookrightarrow V$ induce maps in the same direction, while for the ordinary de Rham cohomology, the inclusions are reversed in the sequence. In the next lecture, we will relate these two cohomology theories, and this will explain all the differences of behavior that we have just discussed.

Exercises

Exercise 24.1

Let $h : \Omega^k(M \times \mathbb{R}) \to \Omega^{k-1}(M \times \mathbb{R})$ be the homotopy operator used in the proof of Proposition 24.1. Show that h can be written in either of the following more invariant forms:

(a) If $E = t\frac{\partial}{\partial t} \in \mathfrak{X}(M \times \mathbb{R})$ is the Euler vector field and $\psi^s : M \times \mathbb{R} \to M \times \mathbb{R}$ the family of maps $\psi^s(x, t) = (x, st)$, then

$$h(\theta) = \int_0^1 \frac{1}{s} (\psi^s)^* i_E \theta \, \mathrm{d}s,$$

(b) If E is the Euler vector field and $\phi_E^s(x, t) = (x, e^s t)$ its flow, then

$$h(\theta) = \int_{-\infty}^0 (\phi_E^s)^* i_E \theta \, \mathrm{d}s,$$

(c) Use the second expression and Cartan Calculus to prove that h verifies (24.1).

Hint: The flow ϕ_X^s of a vector field X satisfies $\frac{\mathrm{d}}{\mathrm{d}s}(\phi_X^s)^* \omega = (\phi_X^s)^* \mathcal{L}_X \omega$.

Exercise 24.2

Show that a star-shaped open set is contractible.

Exercise 24.3

Let $i : N \hookrightarrow M$ be a submanifold. We say that a map $r : M \to N$ is a **retraction** of M in N if $r \circ i = \mathrm{Id}_N$ and that N is a **deformation retract** of M if there exists a retraction $r : M \to N$ such that $i \circ r$ is homotopic to Id_M. Show the following:

(a) If N is a deformation retract of M, then $H^\bullet(N) \simeq H^\bullet(M)$.
(b) Show that $\mathbb{S}^2$ is a deformation retract of $\mathbb{R}^3 \setminus \{0\}$.
(c) Show that $\mathbb{T}^2$, viewed as a submanifold of $\mathbb{R}^3$ as in Example 8.4, is a deformation retract of $\mathbb{R}^3 \setminus \{L \cup S\}$ where L is the z-axis and S is the circle in the xy-plane of radius R and center the origin.

Exercise 24.4

In Remark 24.2, show that the connecting homomorphism in the long exact sequence satisfies $\mathrm{Im}\, g = \mathrm{Ker}\, \delta$ and $\mathrm{Im}\, \delta = \ker f$.

Exercise 24.5

Given a long exact sequence of vector spaces

$$0 \longrightarrow C^0 \longrightarrow \cdots \longrightarrow C^k \longrightarrow \cdots \longrightarrow C^d \longrightarrow 0$$

show that

$$\sum_{i=0}^{d} (-1)^i \dim C^i = 0.$$

Exercise 24.6

Use the Mayer–Vietoris sequence to compute the cohomology of $\mathbb{T}^2$ and $\mathbb{RP}^2$.

Exercise 24.7

Complete the construction of the Mayer–Vietoris sequence for compactly supported cohomology, by showing that

$$0 \longleftarrow \Omega_c^\bullet(M) \longleftarrow \Omega_c^\bullet(U) \oplus \Omega_c^\bullet(V) \longleftarrow \Omega_c^\bullet(U \cap V) \longleftarrow 0$$

is a short exact sequence of complexes.

Exercise 24.8

Find $H_c^\bullet(\mathbb{R}^d \setminus \{0\})$.

Hint: Apply Mayer–Vietoris with $M = \mathbb{S}^d$, $U = \mathbb{S}^d \setminus \{p_N\}$ and $V = \mathbb{S}^d \setminus \{p_S\}$.

Exercise 24.9

Let M be the Möbius strip. Find $H_c^\bullet(M)$.

Exercise 24.10

Let M be an orientable manifold of dimension d. Show that $H_c^d(M) \neq 0$.

Hint: Consider a form $f\omega$ where ω is a volume form and $f \in C_c^\infty(M)$ is a non-negative compactly supported function.

Lecture 25

Computations in Cohomology and Applications

The Mayer–Vietoris sequence yields a very useful technique to compute cohomology by induction. In this lecture, we will see that it also allows to extract many properties of cohomology. In order to apply it, it is useful to cover M by open sets whose intersections have trivial cohomology.

Definition 25.1. An open cover $\{U_\alpha\}$ of a smooth manifold M is called a **good cover** if all finite intersections $U_{\alpha_1} \cap \cdots \cap U_{\alpha_k}$ are diffeomorphic to $\mathbb{R}^d$. We say that M is a **manifold of finite type** if it admits a finite good cover.

Proposition 25.1. *Every smooth manifold M admits a good cover. If M is compact then it admits a finite good cover.*

Sketch of Proof. Let g be a Riemannian metric for M.[1] A classical result in Riemannian geometry shows that each point $p \in M$ has a strong geodesically convex neighborhood U_p, i.e., a neighborhood such that for any two points $q, q' \in U_p$, there exists a unique length minimizing geodesic in U_p which connects q and q'. One checks that

---

[1]This proof requires some knowledge of Riemannian geometry. If you are not familiar with the notion of geodesics, you may wish to skip the proof and admit the result as valid.

(i) a strong geodesically convex open set is diffeomorphic to $\mathbb{R}^d$, and
(ii) the intersection of two strong geodesically convex open sets is a strong geodesically convex open set.

It follows that a cover $\{U_p\}_{p\in M}$ by strong geodesically convex open neighborhoods is a good cover of M.

If M is compact, then any good cover has a finite subcover which is also good. □

Finite-Dimensional Cohomology

As a first application of good covers and the Mayer–Vietoris sequence, we show that the cohomology is finite-dimensional under mild assumptions.

Theorem 25.1. *If M is a manifold of finite type, then the cohomology spaces $H^k(M)$ and $H^k_c(M)$ have finite dimension.*

Proof. For any two open sets U and V, the Mayer–Vietoris sequence:

$$\cdots \longrightarrow H^{k-1}(U\cap V) \xrightarrow{\delta} H^k(U\cup V) \xrightarrow{r} H^k(U)\oplus H^k(V) \longrightarrow \cdots$$

shows that

$$H^k(U\cup V) \simeq \operatorname{Im}\delta \oplus \operatorname{Im} r.$$

Hence, if the cohomologies of U, V, and $U\cap V$ are finite-dimensional, then so is the cohomology of $U\cup V$.

Now we can use induction on the number of open sets in a cover, to show that manifolds which admit a finite good cover have finite-dimensional cohomology:

- If M is diffeomorphic to $\mathbb{R}^d$ the Poincaré Lemma shows that M has finite-dimensional cohomology.
- Now assume that all manifolds admitting a good cover with at most n open sets have finite-dimensional cohomology. Let M be manifold which admits a good cover with $n+1$ open sets $\{U_1,\ldots,U_{n+1}\}$.

We observe that the open sets:

$$U_{n+1},$$

$$U_1 \cup \cdots \cup U_n, \text{ and}$$

$$(U_1 \cup \cdots \cup U_n) \cap U_{n+1} = (U_1 \cap U_{n+1}) \cup \cdots \cup (U_n \cap U_{n+1}),$$

all have finite-dimensional cohomology, since they all admit a good cover with at most n open sets. Hence, the cohomology of $M = U_1 \cup \cdots \cup U_{n+1}$ is also finite-dimensional.

The proof for compactly supported cohomology is similar. $\qquad\square$

Poincaré Duality

We saw in Lecture 22 that the exterior product induces a ring multiplication in cohomology, denoted

$$\cup : H^k(M) \times H^l(M) \to H^{k+l}(M), \ [\omega] \cup [\eta] \equiv [\omega \wedge \eta].$$

Obviously, if η has compact support then $\omega \wedge \eta$ also has compact support, hence we also obtain also a "product"

$$\cup : H^k(M) \times H_c^l(M) \to H_c^{k+l}(M).$$

Now, Stokes formula shows that the integral of differential forms descends to the level of cohomology. Hence, if M is an oriented manifold of dimension d we obtain a bilinear form

$$H^k(M) \times H_c^{d-k}(M) \to \mathbb{R}, \ ([\omega], [\eta]) \mapsto \int_M \omega \wedge \eta. \qquad (25.1)$$

Theorem 25.2 (Poincaré duality). *If M is an oriented manifold of finite type the bilinear form (25.1) is non-degenerate. In particular, one has*

$$H^k(M) \simeq H_c^{d-k}(M)^*.$$

Remark 25.1 (*A Crash Course in Homological Algebra — Part V*). For the proof, we recall the following useful fact from Homological Algebra.

Lemma 25.1 (Five Lemma). *Consider a commutative diagram of homomorphisms of vector spaces where the rows are exact*

$$
\begin{array}{ccccccccc}
A & \xrightarrow{f_1} & B & \xrightarrow{f_2} & C & \xrightarrow{f_3} & D & \xrightarrow{f_4} & E \\
\downarrow{\alpha} & & \downarrow{\beta} & & \downarrow{\gamma} & & \downarrow{\delta} & & \downarrow{\varepsilon} \\
A' & \xrightarrow{f_1'} & B' & \xrightarrow{f_2'} & C' & \xrightarrow{f_3'} & D' & \xrightarrow{f_4'} & E'
\end{array}
$$

If α, β, δ, and ε are isomorphisms, then γ is also an isomorphism.

The proof of this lemma is by diagram chasing and is left as an easy exercise.

Proof of Theorem 25.2. The bilinear form (25.1) yields a linear map $H^k(M) \to H^{d-k}_c(M)^*$. Them, if U and V are open sets, one checks easily that the Mayer–Vietoris sequences for $\Omega^\bullet$ and $\Omega^\bullet_c$ give a diagram of long exact sequences

$$
\begin{array}{ccccccccc}
\longrightarrow & H^k(U \cup V) & \longrightarrow & H^k(U) \oplus H^k(V) & \longrightarrow & H^k(U \cap V) & \xrightarrow{\delta} & H^{k+1}(U \cup V) & \longrightarrow \\
& \downarrow & & \downarrow & & \downarrow & & \downarrow & \\
\longrightarrow & H^{d-k}_c(U \cup V)^* & \to & H^{d-k}_c(U)^* \oplus H^{d-k}_c(V)^* & \to & H^{d-k}_c(U \cap V)^* & \xrightarrow{\delta^*} & H^{d-k-1}_c(U \cup V)^* & \to
\end{array}
$$

which is commutative up to signs. For example, we have

$$
\int_{U \cap V} \omega \wedge \delta\theta = \pm \int_{U \cup V} \delta\omega \wedge \tau.
$$

If we apply the Five Lemma to this diagram, we conclude that if Poincaré duality holds for U, V, and $U \cap V$, then it also holds for $U \cup V$.

Now let M be a manifold with a finite good cover. We show that Poincaré duality holds using induction on the cardinality of the cover.

- If the cover has only one element, i.e., if $M \simeq \mathbb{R}^d$, the Poincaré Lemmas give

$$
H^k(\mathbb{R}^d) = \begin{cases} \mathbb{R} & \text{if } k = 0, \\ 0 & \text{if } k \neq 0. \end{cases} \qquad H^k_c(\mathbb{R}^d) = \begin{cases} \mathbb{R} & \text{if } k = d, \\ 0 & \text{if } k \neq d. \end{cases}
$$

Since $(\cdot, \cdot) : H^0(\mathbb{R}^d) \times H_c^d(\mathbb{R}^d) \to \mathbb{R}$ is non-zero, it is non-degenerate.

- Now assume that Poincaré duality holds for any manifold admitting a good cover with at most n open sets. If M is a manifold which admits an open cover $\{U_1, \ldots, U_{n+1}\}$ with $n + 1$ open sets, we note that the open sets:

$$U_{n+1}, \ U_1 \cup \cdots \cup U_n, \ \text{and}$$

$$(U_1 \cup \cdots \cup U_n) \cap U_{n+1} = (U_1 \cap U_{n+1}) \cup \cdots \cup (U_n \cap U_{n+1}),$$

all satisfy Poincaré duality, since they all admit a good cover with at most n open sets. It follows that $M = U_1 \cup \cdots \cup U_{n+1}$ also satisfies Poincaré duality.

$\square$

Corollary 25.1. *If M is a compact oriented manifold, then*

$$H^k(M) \simeq H^{d-k}(M).$$

Remark 25.2. If M does have a finite good cover, so the cohomology of M may be infinite-dimensional, it is possible to show that one still has an isomorphism

$$H^k(M) \simeq (H_c^{d-k}(M))^*.$$

However, in general, *one does not have* a dual isomorphism $H_c^{d-k}(M) \simeq H^k(M)^*$. The reason is that while the dual of direct product is a direct sum, the dual of an infinite direct sum is not a direct product. We discuss an example in the exercises.

Corollary 25.2. *Let M be a connected manifold of dimension d. Then,*

$$H_c^d(M) \simeq \begin{cases} \mathbb{R} & \text{if } M \text{ is orientable,} \\ 0 & \text{if } M \text{ is not orientable.} \end{cases}$$

In particular, if M is compact and connected of dimension d, then M is orientable if and only if $H^d(M) \simeq \mathbb{R}$.

Proof. By Poincaré duality, if M is a connected orientable manifold of dimension d, then $(H_c^d(M))^* \simeq H^0(M) \simeq \mathbb{R}$. The proof of the converse is left as an exercise. $\square$

Triangulations and Euler's Formula

As another application of the Mayer–Vietoris sequence, we show how the familiar Euler's formula for regular polygons can be extended to any compact manifold M admitting a *triangulation*.[2]

A **regular simplex** is a simplex $\sigma : \Delta^d \to M$ which can be extended to a diffeomorphism $\tilde{\sigma} : U \to \tilde{\sigma}(U) \subset M$, where U is some open neighborhood of Δ^d. We have defined before the $(d-1)$-dimensional faces of a simplex $\sigma : \Delta^d \to M$. For a regular simplex, these are regular $(d-1)$-simplices $\varepsilon_i(\sigma) : \Delta^{d-1} \to M$ of dimension $(d-1)$. By iterating this construction we obtain the $(d-k)$-**dimensional faces** of a simplex, which are regular $(d-k)$-simplices $\varepsilon_{i_1,i_2,\ldots,i_{d-k}}(\sigma) : \Delta^{d-k} \to M$.

Definition 25.2. A **triangulation** of a compact manifold M of dimension d is a finite collection $\{\sigma_i\}$ of regular d-simplices such that

(i) the collection $\{\sigma_i\}$ covers M, and
(ii) if two simplices in $\{\sigma_i\}$ have non-empty intersection, then there intersection $\sigma_i \cap \sigma_j$ is a face of both simplices σ_i and σ_j.

Figure 25.1 illustrates condition (ii) for dimensions 2 and 3. Note that on the top subdivisions the condition is satisfied while on the bottom the condition fails.

If M is a manifold with finite-dimensional cohomology (e.g., if M is compact) one defines the **Euler characteristic** of M to be the integer $\chi(M)$ given by

$$\chi(M) := \dim H^0(M) - \dim H^1(M) + \cdots + (-1)^d \dim H^d(M).$$

Applying Poincaré duality we conclude this integer must be zero for odd-dimensional manifolds.

Corollary 25.3. *If M is a compact oriented odd-dimensional manifold then $\chi(M) = 0$.*

---

[2]Actually, one can show that every smooth compact manifold can be triangulated. This result is very technical and we will not be discuss in these lectures. See Manolescu (2014) for an historical account and references.

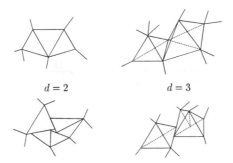

$d = 2$ $d = 3$

Fig. 25.1. Triangulations vs. non-triangulations in dimensions 2 and 3.

On the other hand, for even-dimensional manifolds the Euler characteristic is, in general, non-zero and can be computed using a triangulation.

Theorem 25.3 (Euler's Formula). *If M is a compact manifold of dimension d, then for any triangulation*

$$(-1)^d \chi(M) = r_0 - r_1 + \cdots + (-1)^d r_d,$$

where r_i denotes the number of faces of dimension i of the triangulation.

Proof. Fix a triangulation $\{\sigma_1, \sigma_2, \ldots, \sigma_{r_d}\}$ of M and define open sets

$$V_k := M \setminus \{k\text{-faces of the triangulation}\}.$$

We claim that for $0 \le k \le d - 1$ we have

$$\chi(M) = \chi(V_k) + (-1)^d (r_0 - r_1 + \cdots + (-1)^k r_k). \qquad (25.2)$$

Assuming this claim, since

$$V_{d-1} = \bigcup_{j=1}^{r_d} \text{int}(\sigma_j),$$

and each open set $\text{int}(\sigma_j)$ is contractible, we have $H^k(V_{d-1}) = 0$, for $k > 0$. Hence

$$\chi(V_{d-1}) = \dim H^0(V_{d-1}) = r_d.$$

This identity and (25.2) show that Euler's formula holds.

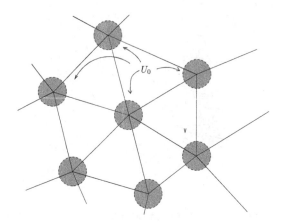

Fig. 25.2.   The open set U_0.

It remains to prove (25.2). We first verify it for $k = 0$. For each 0-dimensional face we can choose disjoint open neighborhoods $U_{0,1}, \ldots, U_{0,r_0}$, each diffeomorphic to the open ball $B_1^d = \{x \in \mathbb{R}^d : \|x\| < 1\}$. We set (see Figure 25.2)

$$U_0 := \bigcup_{i=0}^{r_0} U_{0,i}.$$

Note that $V_0 \cup U_0 = M$. Since each $U_{0,i}$ is contractible, we have

$$\dim H^k(U_0) = \begin{cases} r_0, & \text{if } k = 0, \\ 0, & \text{if } k \neq 0. \end{cases}$$

On the other hand, the intersection $V_0 \cap U_{0,i}$ deformation retracts in $\mathbb{S}^{d-1}$, hence

$$\dim H^k(V_0 \cap U_0) = \begin{cases} r_0, & \text{if } k = 0, d-1, \\ 0, & \text{if } k \neq 0, d-1. \end{cases}$$

We can apply the Mayer–Vietoris argument to the pair (U_0, V_0). Assuming $d > 2$, this sequence gives the following information.

(i) The lowest degree terms in the sequence are

$$0 \longrightarrow H^0(M) \longrightarrow H^0(U_0) \oplus H^0(V_0) \longrightarrow H^0(U_0 \cap V_0) \longrightarrow$$

$$\longrightarrow H^1(M) \longrightarrow 0 \oplus H^1(V_0) \longrightarrow 0$$

so it follows that

$$\dim H^0(M) - \dim H^0(U_0) - \dim H^0(V_0)$$
$$+ \dim H^0(U_0 \cap V_0) - \dim H^1(M) + \dim H^1(V_0) = 0.$$

Since M and V_0 have the same number of connected components we find

$$\dim H^0(M) = \dim H^0(V_0).$$

On the other hand, the number of connected components of U_0 and $V_0 \cap U_0$ are also the same, hence we conclude that

$$\dim H^1(M) = \dim H^1(V_0).$$

(ii) For $1 < k < d - 1$, the Mayer–Vietoris sequence gives

$$0 \longrightarrow H^k(M) \longrightarrow 0 \oplus H^k(V_0) \longrightarrow 0$$

Hence

$$\dim H^k(M) = \dim H^k(V_0).$$

(iii) Finally, the last terms in the sequence give

$$0 \longrightarrow H^{d-1}(M) \longrightarrow 0 \oplus H^{d-1}(V_0) \longrightarrow H^{d-1}(U_0 \cap V_0) \longrightarrow$$
$$\longrightarrow H^d(M) \longrightarrow 0 \oplus H^d(V_0) \longrightarrow 0$$

Since $\dim H^{d-1}(U_0 \cap V_0) = r_0$, we conclude that

$$\dim H^{d-1}(M) - \dim H^{d-1}(V_0) + \dim H^{d-1}(V_0) - \dim H^d(M) = -r_0.$$

When $d = 2$, we obtain exactly the same results except that we can consider the whole sequence at once.

In any case, we conclude that

$$\chi(M) = \sum_{i=0}^{d} (-1)^i \dim H^i(M)$$

$$= \sum_{i=0}^{d} (-1)^i \dim H^i(V_0) + (-1)^d r_0 = \chi(V_0) + (-1)^d r_0,$$

which yields (25.2) if $k = 0$.

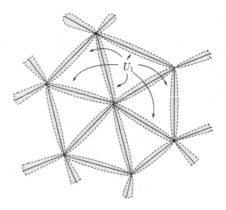

Fig. 25.3.   The open set U_1.

In order to prove (25.2) when $k = 1$, we can proceed as follows. For each 1-face, we choose open disjoint neighborhoods $U_{1,1}, \ldots, U_{1,r_1}$ of the (1-faces)-(0-faces), diffeomorphic to (int Δ^1) $\times B_1^{d-1}$, and we define the open set (see Figure 25.3)

$$U_1 := \bigcup_{i=0}^{r_1} U_{1,i}.$$

We have that $V_0 = U_1 \cup V_1$. Moreover, U_1 is a disjoint union of r_1 contractible open sets, while $U_1 \cap V_1$ as the same homotopy type as the disjoint union of $(d-2)$-spheres. Using the Mayer–Vietoris sequence, exactly like in the case $k = 0$. one shows that

$$\chi(V_0) = \chi(V_1) + (-1)^{d-1} r_1.$$

In general, for each k, we choose open disjoint neighborhoods $U_{k,1}, \ldots, U_{k,r_k}$ of $\{k\text{-faces}\} - \{(k-1)\text{-faces}\}$, diffeomorphic to (int Δ^k) $\times B_1^{d-k}$, and define

$$U_k := \bigcup_{i=0}^{r_k} U_{k,i}.$$

We have that $V_{k-1} = U_k \cup V_k$, where U_k is a union of r_k contractible open sets, while $U_k \cap V_k$ as the same homotopy type as the disjoint union of $(d-k-1)$-spheres. Applying Mayer–Vietoris sequence one then shows that

$$\chi(V_{k-1}) = \chi(V_k) + (-1)^{d-k} r_k.$$

This proves (25.2) and finishes the proof of Euler's formula. $\qquad\square$

Exercises

Exercise 25.1
Give an example of a connected manifold which is not of finite type.

Exercise 25.2
Prove the Five Lemma. Are there weaker conditions on the maps α, β, ε and δ, so that the conclusion still holds?

Exercise 25.3
Check the commutativity, up to signs, of the diagram of long exact sequences that appears in the proof of Poincaré duality.

Exercise 25.4
Show that

(a) $\dim H^k(\mathbb{T}^d) = \binom{d}{k}$;
(b) $\dim H^{2k}(\mathbb{CP}^d) = 1$ if $2k \leq d$, and 0 otherwise.

Exercise 25.5
Let M be a connected manifold of dimension d, which is not orientable. Prove that

$$H_c^d(M) = 0,$$

by proceeding as follows. Let $\widetilde{M}_{\mathrm{ori}}$ denote the orientation cover of M (see Exercise 19.4). Show the following:

(a) $\widetilde{M}$ is a connected orientable manifold of dimension d;
(b) The map $\Phi : \widetilde{M} \to \widetilde{M}$, $(p, [\mu_p]) \mapsto (p, -[\mu_p])$ is a diffeomorphism that changes orientation and satisfies:

$$\pi = \pi \circ \Phi, \qquad \Phi \circ \Phi = \mathrm{Id};$$

(c) Given $\widetilde{\omega} \in \Omega^k(\widetilde{M})$, there exists $\omega \in \Omega^k(M)$ such that $\widetilde{\omega} = \pi^*\omega$ iff $\Phi^*\widetilde{\omega} = \widetilde{\omega}$;

(d) Conclude that one must have $H_c^d(M) = 0$.

Exercise 25.6

Let $M_1, M_2, \ldots$, be orientable d-dimensional manifolds of finite type and consider the disjoint union of the M_i:

$$M = \bigcup_{i=1}^{+\infty} M_i.$$

Show that

(a) $H^k(M) = \prod_{i=1}^{+\infty} H^k(M_i)$;

(b) $H_c^k(M) = \bigoplus_{i=1}^{+\infty} H_c^k(M_i)$;

(c) Conclude that there exists an isomorphism: $H^k(M) \simeq (H_c^{d-k}(M))^*$;

(d) Give an example of an orientable M with $H_c^{d-k}(M)$ not isomorphic to $H^k(M)^*$.

Exercise 25.7

Consider the two subdivisions of the square $[0,1] \times [0,1]$ in Figure 25.4.

(a) Verify that only one of these subdivisions induces a triangulation of $\mathbb{T}^2$;

(b) Compute r_0, r_1 and r_2 for this triangulation.

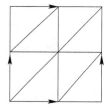 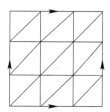

Fig. 25.4.   Subdivisions of $[0,1] \times [0,1]$.

Exercise 25.8

Let M and N be connected compact manifolds of dimension d. The **connected sum** of M and N (see Figure 25.5) is the manifold $M \# N$ obtained by gluing M and N along the boundary of open sets $U \subset M$ and $V \subset N$ both diffeomorphic to the ball $\{x \in \mathbb{R}^d : \|x\| < 1\}$. Show that the Euler characteristics satisfy

$$\chi(M \# N) = \chi(M) + \chi(N) - \chi(\mathbb{S}^d).$$

Conclude that the Euler characteristic of a compact, oriented, surface of genus g (i.e., with g holes) is $2 - 2g$.

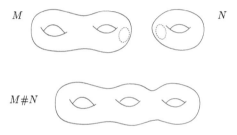

Fig. 25.5.   Connected sum of M and N.

Lecture 26

The Degree and the Index

We saw in the previous lecture that a connected manifold M of dimension d is orientable if and only if $H_c^d(M) \simeq \mathbb{R}$. Note that a choice of orientation for M determines a generator of $H_c^d(M)$. In fact, in this case, integration

$$H_c^d(M) \to \mathbb{R}, \quad [\omega] \mapsto \int_M \omega$$

gives an isomorphism $H_c^d(M) \simeq \mathbb{R}$. This isomorphism is really just Poincaré duality, since M being connected $H^0(M)$ is the space of constant functions in M. In the sequel, we will often use the same symbol μ_M to denote the orientation of M and the generator $\mu_M \in H_c^d(M)$ that corresponds to the constant function 1.

Now let $\Phi : M \to N$ be a proper map between connected, oriented manifolds of the same dimension d. The induced isomorphism in cohomology

$$\Phi^* : H_c^d(N) \to H_c^d(M),$$

together with the canonical isomorphisms $H_c^d(M) \simeq \mathbb{R}$ and $H_c^d(N) \simeq \mathbb{R}$, allow to associate to Φ a real number called the *degree of the map*. We formulate this more explicitly as follows.

Definition 26.1. Let $\Phi : M \to N$ be a proper map between connected, oriented manifolds of the same dimension d. The **degree** of

Φ is the unique real number deg Φ such that

$$\int_M \Phi^*\omega = \deg\Phi \int_N \omega,$$

for every differential form $\omega \in \Omega_c^d(N)$.

Our next aim is to give a geometric characterization of the degree which allows also for its computation. For simplicity, we consider only the case where both manifolds are compact, but you may wish to try to extend these results to any proper map. First, we observe the following property of the degree.

Proposition 26.1. *Let* $\Phi : M \to N$ *be a smooth map between compact, connected, oriented manifolds of the same dimension d. If* Φ *is not surjective then* deg $\Phi = 0$.

Proof. Let $q_0 \in N \setminus \Phi(M)$. Since $\Phi(M)$ is closed, there is an open neighborhood of q_0 such that $U \subset N \setminus \Phi(M)$. Choose $\omega \in \Omega^d(N)$ with support in U such that $\int_N \omega \neq 0$. Then

$$0 = \int_M \Phi^*\omega = \deg\Phi \int_N \omega,$$

hence deg $\Phi = 0$. $\qquad\square$

The following geometric interpretation of the degree shows that the degree is always an integer, something which is not obvious from its definition.

Theorem 26.1. *Let* $\Phi : M \to N$ *be a smooth map between compact, connected, oriented manifolds of the same dimension d. Let* $q \in N$ *be a regular value of* Φ *and for each* $p \in \Phi^{-1}(q)$ *define*

$$\mathrm{sgn}_p \Phi := \begin{cases} 1 & \text{if } d_p\Phi : T_pM \to T_qN \text{ preserves orientations,} \\ -1 & \text{if } d_p\Phi : T_pM \to T_qN \text{ switches orientations.} \end{cases}$$

Then

$$\deg\Phi = \sum_{p\in\Phi^{-1}(q)} \mathrm{sgn}_p \Phi.$$

In particular, the degree is an integer.

Remark 26.1. If $\Phi^{-1}(q)$ is empty then q is a regular value and we convention that the sum is zero.

Proof. Let q be a regular value of Φ. If $\Phi^{-1}(q)$ is empty, then Φ is not surjective and the result follows from the previous proposition. On the other hand, if $\Phi^{-1}(q)$ is non-empty then it is a discrete subset of M which, by compactness, must be finite, so $\Phi^{-1}(q) = \{p_1, \ldots, p_N\}$. We apply the following lemma.

Lemma 26.1. *There exists a neighborhood V of q and disjoint neighborhoods $U_1, \ldots, U_N$ of $p_1, \ldots, p_N$ such that*

$$\Phi^{-1}(V) = U_1 \cup \cdots \cup U_N.$$

Assuming that this lemma holds, since each p_i is a regular point, we can further assume that V is the domain of a chart $(y^1, \ldots, y^d)$ for N and that the restrictions $\Phi|_{U_i} : U_i \to V$ are diffeomorphisms. Let

$$\omega := f dy^1 \wedge \cdots \wedge dy^d \in \Omega^d(N),$$

where $f \geq 0$ has $\operatorname{supp} f \subset V$. Obviously, we have

$$\operatorname{supp} \Phi^* \omega \subset U_1 \cup \cdots \cup U_N,$$

so we find

$$\int_M \Phi^* \omega = \sum_{i=1}^N \int_{U_i} \Phi^* \omega.$$

Since each $\Phi|_{U_i}$ is a diffeomorphism, the change of variables formula gives

$$\int_{U_i} \Phi^* \omega = \pm \int_V \omega = \pm \int_N \omega,$$

where the sign is positive if $\Phi|_{U_i}$ preserves orientations and negative otherwise. Since $\Phi|_{U_i}$ preserves orientations if $\operatorname{sgn}_{p_i} \Phi > 0$ and switches orientations if $\operatorname{sgn}_{p_i} \Phi < 0$, we conclude that

$$\int_M \Phi^* \omega = \sum_{i=1}^N \operatorname{sgn}_{p_i} \Phi \int_N \omega.$$

This yields the formula for the degree of Φ in the statement.

To finish the proof it remains to prove the lemma. Let $O_1, \ldots, O_N$ be any disjoint open neighborhoods of $p_1, \ldots, p_N$, and W a compact neighborhood of q. The set $\widetilde{W} \subset M$ defined by

$$\widetilde{W} := \Phi^{-1}(W) \setminus (O_1 \cup \cdots \cup O_N),$$

is compact. Hence, $\Phi(\widetilde{W})$ is a compact set which does not contain q. Therefore, there exists an open set $V \subset W \setminus \Phi(\widetilde{W})$ containing q and we have

$$\Phi^{-1}(V) \subset O_1 \cup \cdots \cup O_N.$$

If we let $U_i := O_i \cap \Phi^{-1}(V)$ the lemma follows. $\qquad\square$

The degrees of two homotopic maps coincide since such maps induce the same map in cohomology. This is a very useful fact in computing degrees and can be explored to deduce the global properties of manifolds. A classic illustration of this is given in the next example.

Example 26.1. Consider the antipodal map $\Phi : \mathbb{S}^d \to \mathbb{S}^d$, $p \mapsto -p$. For the canonical orientation of the sphere $\mathbb{S}^d$ defined by the form

$$\omega = \sum_{i=1}^{d+1} (-1)^{i+1} x^i dx^1 \wedge \cdots \wedge \widehat{dx^i} \wedge \cdots \wedge dx^{d+1}.$$

we see that Φ preserves or switches orientations if d is odd or even, respectively. Since $\Phi^{-1}(q)$ contains only one point, we conclude that

$$\deg \Phi = (-1)^{d-1}.$$

By the way, one can also compute the degree directly from the definition, since $\Phi^*\omega = (-1)^{d-1}\omega$ so

$$\int_{\mathbb{S}^d} \Phi^*\omega = (-1)^{d-1} \int_{\mathbb{S}^d} \omega.$$

One can use this fact to show that every vector field on a even-dimensional sphere vanishes at some point. Let $X \in \mathfrak{X}(\mathbb{S}^{2d})$ be a nowhere vanishing vector field. Then for each $p \in \mathbb{S}^{2d}$ there exists a unique semi-circle γ_p joining p to $-p$ with $\gamma_p'(0) = X|_p$. It follows that the map

$$H : \mathbb{S}^{2d} \times [0,1] \to \mathbb{S}^{2d}, \quad (p,t) \mapsto \gamma_p(t),$$

is a homotopy between Φ and the identity map. Hence, one would have

$$-1 = \deg \Phi = \deg(\mathrm{id}) = 1,$$

a contradiction.

Note that, in contrast, any odd degree $\mathbb{S}^{2d-1} \subset \mathbb{R}^{2d}$ admits the nowhere vanishing vector field

$$X = x^2 \frac{\partial}{\partial x^1} - x^1 \frac{\partial}{\partial x^2} + \cdots + x^{2d} \frac{\partial}{\partial x^{2d-1}} - x^{2d-1} \frac{\partial}{\partial x^{2d}}.$$

As another application of degree theory, we will introduce now the index of a vector field at a zero. This will lead us to a famous formula for the Euler characteristic of a manifold known as the Poincaré–Hopf Theorem.

Consider first a vector field X defined in an open set $U \subset \mathbb{R}^d$ which has an isolated zero at $x_0 \in U$. We can view it as a map $X : U \to \mathbb{R}^d$ which vanishes at x_0 and is non-zero in a deleted neighborhood $V \setminus \{x_0\}$. Let $D_\varepsilon(x_0) \subset U$ be a closed disk of radius ε centered at x_0, which does not contain any other zero of X, and let $\mathbb{S}_\varepsilon := \partial D_\varepsilon(x_0)$ be the sphere of radius ε centered at x_0. We introduce the **Gauss map**

$$G_\varepsilon : \mathbb{S}_\varepsilon \to \mathbb{S}^{d-1}, \quad x \mapsto \frac{X(x)}{\|X(x)\|},$$

and define the **index** of X at x_0 to be the degree of the Gauss map

$$\mathrm{ind}_{x_0} X := \deg G_\varepsilon.$$

Here, on both domain and target spheres, one considers the induced orientation from $\mathbb{R}^d$. The following result states that the degree is independent of ε and is a diffeomorphism invariant.

Proposition 26.2. *Let $U \subset \mathbb{R}^d$ be open and let $X \in \mathfrak{X}(U)$ a vector field with an isolated zero at x_0.*

(i) *Any two Gauss maps G_{ε_0} and G_{ε_1} have the same degree.*
(ii) *If $\Phi : U \to U'$ a diffeomorphism and $X' = \Phi_* X$ then*

$$\mathrm{ind}_{x_0} X = \mathrm{ind}_{\Phi(x_0)} X'.$$

Proof. We leave (i) as an exercise. To prove (ii), we assume that $\Phi(x_0) = x_0 = 0$ and that U is star-shaped with center 0.

Assume first that Φ preserves orientations. The map

$$H(t,x) := \begin{cases} \frac{1}{t}\Phi(tx), & \text{if } t > 0, \\ d_0\Phi(x), & \text{if } t = 0. \end{cases}$$

is a homotopy between Φ and $d_0\Phi$, consisting of diffeomorphisms that fix the origin. Since $d_0\Phi$ preserves orientation it is homotopic to the identity via linear isomorphisms. Hence, we see that there exists a homotopy, via diffeomorphisms that fix the origin, between Φ and the identity. It follows that the Gauss maps of X and $X' = \Phi_*X$ are homotopic, so the indices of X and X' coincide.

If Φ switches orientations, composing with a reflection we obtain a orientation preserving diffeomorphism. So to finish the proof it remains to conside the case where Φ is a reflection. In this case, Φ is a linear map, so

$$X' = \Phi_*X = \Phi \circ X \circ \Phi^{-1}.$$

The corresponding Gauss maps are then related by

$$G'_\varepsilon = \Phi \circ G_\varepsilon \circ \Phi^{-1},$$

and so their degrees coincide. $\qquad\qquad\qquad\qquad\qquad\qquad\square$

Proposition 26.2 allows us to define the index for an arbitrary vector field.

Definition 26.2. The **index of a vector field** $X \in \mathfrak{X}(M)$ at an isolated zero $p_0 \in M$ is the integer

$$\operatorname{ind}_{p_0} X := \operatorname{ind}_0 \phi_*(X|_U),$$

where (U, ϕ) is any coordinate system centered at p_0.

In some simple cases, it is possible to determine the index of a vector field from its phase portrait. The pictures in Figure 26.1 give some examples of planar vector fields with a zero and the value of its index. You should check that the degree of the corresponding Gauss maps is indeed the integer in each figure.

In general, it maybe hard to compute the index, but for *non-degenerate* zeros of vector fields, there is a simple formula, as we will now explain.

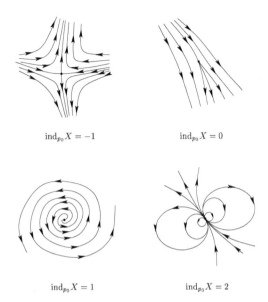

Fig. 26.1.   Index of vector fields.

Let $X \in \mathfrak{X}(M)$ be a vector field and let $p_0 \in M$ be a zero of X. Note that the zero section $Z \subset TM$ and the fiber $T_{p_0}M \subset TM$ intersect transversely at $0 \in T_{p_0}M$, since we have

$$T_0(TM) = T_{p_0}Z \oplus T_{p_0}(T_{p_0}M) \simeq T_{p_0}M \oplus T_{p_0}M,$$

where the isomorphism in the second factor is given by the differential $\mathrm{d}_{p_0}\pi$ of the projection $\pi : TM \to M$. Under this decomposition, the differential of the vector field

$$\mathrm{d}_{p_0}X : T_{p_0}M \to T_0(TM).$$

has first component the identity, since $\pi \circ X = \mathrm{id}_M$, while the second component is a linear map $T_{p_0}M \to T_{p_0}M$. This linear map will be denoted also by $\mathrm{d}_{p_0}X$, and is called the **linear approximation** to X at the zero p_0. One can also view $\mathrm{d}_{p_0}X$ as a *linear vector field* on the tangent space $T_{p_0}M$.

In a chart $(U, x^1, \ldots, x^d)$ centered at p_0 the vector field can be written

$$X = \sum_{i=1}^{d} X^i \frac{\partial}{\partial x^i}, \quad \text{where } X^i(0) = 0.$$

The linear approximation is the linear vector field on $T_{p_0}M$

$$\mathrm{d}_{p_0}X = \sum_{i,j=1}^{d} \frac{\partial X^i}{\partial x^j}(0)x^j \left.\frac{\partial}{\partial x^i}\right|_{p_0},$$

and can be viewed as a linear map $\mathrm{d}_{p_0}X : T_{p_0}M \to T_{p_0}M$ which relative to the basis $\{\left.\frac{\partial}{\partial x^1}\right|_{p_0}, \ldots, \left.\frac{\partial}{\partial x^d}\right|_{p_0}\}$ represented by the matrix

$$\left[\frac{\partial X^i}{\partial x^j}(0)\right]_{i,j=1}^{d}.$$

Definition 26.3. A zero p_0 of $X \in \mathfrak{X}(M)$ is called **non-degenerate** if the linear approximation $\mathrm{d}_{p_0}X : T_{p_0}M \to T_{p_0}M$ is an invertible linear transformation.

Non-degenerate zeros are always isolated and their indices can be found easily.

Proposition 26.3. *Let $p_0 \in M$ be a non-degenerate zero of $X \in \mathfrak{X}(M)$. Then p_0 is an isolated zero and*

$$\mathrm{ind}_{p_0}X = \begin{cases} +1, & \text{if } \det \mathrm{d}_{p_0}X > 0, \\ -1, & \text{if } \det \mathrm{d}_{p_0}X < 0. \end{cases}$$

Proof. Choose a local chart (U, ϕ) centered at p_0. The vector field $\phi_*(X|_U)$ has an associated Gauss map $G : \mathbb{S}_\varepsilon \to \mathbb{S}^{d-1}$ which is a diffeomorphism. This diffeomorphism preserves (respectively, switches) orientations if and only if $\det \mathrm{d}_{p_0}X > 0$ (respectively, < 0). Hence the result follows from Theorem 26.1. $\qquad\square$

Example 26.2. Consider $\mathbb{R}^3$ with coordinates (x, y, z). The vector field

$$X = y\frac{\partial}{\partial x} - x\frac{\partial}{\partial y} \in \mathfrak{X}(\mathbb{R}^3),$$

is tangent to the sphere $\mathbb{S}^2 = \{(x, y, z) : x^2 + y^2 + z^2 = 1\}$ and hence defines a vector field $X \in \mathfrak{X}(\mathbb{S}^2)$, with exactly two zeros (the north pole p_N and the south pole p_S; see Figure 26.2).

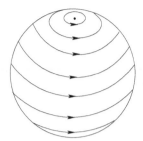

Fig. 26.2.   A vector field on $\mathbb{S}^2$ with two zeros.

The projection $\phi = (u, v) : (x, y, z) \mapsto (x, y)$ restricts on the upper and lower hemispheres to local charts on $\mathbb{S}^2$ centered at p_N and p_S. We have

$$\phi_* X = v\frac{\partial}{\partial u} - u\frac{\partial}{\partial v}.$$

The matrix representation of the linear approximation to X at p_N and p_S relative to the basis $\{\frac{\partial}{\partial u}, \frac{\partial}{\partial v}\}$ is then given in both cases by

$$\mathrm{d}_{p_N} X = \mathrm{d}_{p_S} X = \begin{bmatrix} 0 & 1 \\ -1 & 0 \end{bmatrix}.$$

We conclude that p_N and p_S are non-degenerate zeros and:

$$\mathrm{ind}_{p_N} X = \mathrm{ind}_{p_S} X = 1.$$

In the previous example, the sum of the indices of the zeros the vector field $X \in \mathfrak{X}(\mathbb{S}^2)$ equals 2, so it coincides with the value of the Euler characteristic of $\mathbb{S}^2$. This is an illustration of the following famous result.

Theorem 26.2 (Poincaré–Hopf). *Let $X \in \mathfrak{X}(M)$ is a vector field on a compact manifold with a finite number of zeros $\{p_1, \ldots, p_N\}$. Then,*

$$\chi(M) = \sum_{i=1}^{N} \mathrm{ind}_{p_i} X.$$

This beautiful theorem connects the topology of M with its smooth structure, i.e., its tangent bundle. The proof will be given in the last part of these notes where we will study bundle theory.

Exercises

Exercise 26.1
Identify $M = \mathbb{R}^2$ with the field of complex numbers $\mathbb{C}$. If $\Phi : \mathbb{C} \to \mathbb{C}$ is a polynomial map of degree d, find $\deg \Phi$.

Exercise 26.2
Show that for a manifold M of dimension $d > 0$ the identity map $\mathrm{Id} : M \to M$ is never homotopic to a constant map. Use this fact to prove that there is no retraction of the closed unit disk $D^d \subset \mathbb{R}^d$ on its boundary $\mathbb{S}^{d-1} = \partial D^d$.

Hint: If there was a retraction $r : D^d \to \mathbb{S}^{d-1}$ consider the map $H(x,t) = r(rx)$.

Exercise 26.3
Fix a 2×2 matrix with integer entries

$$A = \begin{pmatrix} a & b \\ c & d \end{pmatrix}.$$

Identifying $\mathbb{T}^2 = \mathbb{R}^2/\mathbb{Z}^2$, consider the map $\Phi : \mathbb{T}^2 \to \mathbb{T}^2$ defined by

$$\Phi([x, y]) := [ax + by, cx + dy].$$

Determine $\deg \Phi$.

Exercise 26.4
Let $X : \mathbb{R}^d \to \mathbb{R}^d$ be a vector field with an isolated zero at $x = 0$ and associated Gauss map G_ε. Define $\widetilde{G}_\varepsilon : \mathbb{S}^{d-1} \to \mathbb{S}^{d-1}$ to be the composition of $G_\varepsilon : \mathbb{S}_\varepsilon \to \mathbb{S}^{d-1}$ with the map $\mathbb{S}^{d-1} \to \mathbb{S}_\varepsilon$, $x \mapsto \varepsilon x$.

(a) Show that $\deg \widetilde{G}_\varepsilon = \deg G_\varepsilon$.
(b) Show that for any $\varepsilon_0, \varepsilon_1 > 0$ the maps $\widetilde{G}_{\varepsilon_0}$ and $\widetilde{G}_{\varepsilon_1}$ are homotopic.
(c) Conclude that the degree of the Gauss map G_ε is independent of ε.

Exercise 26.5
Identify $M = \mathbb{R}^2$ with the field of complex numbers $\mathbb{C}$. Show that the polynomial map $z \mapsto z^k$ defines a vector field in $\mathbb{R}^2$ which has

a zero at the origin of index k. How would you change $z \mapsto z^k$ to obtain a vector field with a zero of index $-k$?

Exercise 26.6
Find the index of the zeros of the following vector fields in $\mathbb{R}^2$:

(a) $x\frac{\partial}{\partial x} \pm y\frac{\partial}{\partial y}$;

(b) $(x^2y + y^3)\frac{\partial}{\partial x} - (x^3 + xy^2)\frac{\partial}{\partial y}$;

Exercise 26.7
Show that a vector field on a compact, oriented, surface of genus g must have at least one zero if $g \neq 1$.

Exercise 26.8
Consider the vector field $X \in \mathfrak{X}(\mathbb{S}^{2d})$ obtained by restriction of the vector field

$$X = x^2\frac{\partial}{\partial x^1} - x^1\frac{\partial}{\partial x^2} + \cdots + x^{2d}\frac{\partial}{\partial x^{2d-1}} - x^{2d-1}\frac{\partial}{\partial x^{2d}} \in \mathfrak{X}(\mathbb{R}^{2d+1}).$$

Show that there is a vector field Y in $\mathbb{RP}^{2d}$ such that $\pi_* X = Y$ and apply the Poincaré–Hopf theorem to compute the Euler characteristic of $\mathbb{RP}^{2d}$. What can you say about the Euler characteristic of $\mathbb{RP}^{2d+1}$?

PART 4
Fiber Bundles

Lecture 27

Vector Bundles

A vector bundle is a collection $\{E_p\}_{p\in M}$ of vector spaces parameterized by a manifold M as in Figure 27.1. The union of these vector spaces is a manifold E and the map

$$\pi : E \to M, \quad \pi(E_p) = p,$$

must satisfy a local trivialization condition. These properties should be familiar from our study of the tangent and cotangent bundles of a manifold.

In order to formalize this concept properly, let $\pi : E \to M$ be a smooth map between differentiable manifolds. A **trivializing chart** of dimension r for π is a pair (U, ϕ), where $U \subset M$ is open and $\phi : \pi^{-1}(U) \to U \times \mathbb{R}^r$ is a diffeomorphism, such that we have a commutative diagram

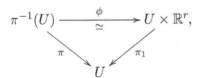

In this diagram, $\pi_1 : U \times \mathbb{R}^r \to U$ denotes the projection in the first factor. Note that an arbitrary map $\pi : E \to M$ may not admit trivializing charts. For example, if (U, ϕ) is a trivializing chart then the restriction $\pi : \pi^{-1}(U) \to U$ must be a submersion.

Let $E_p = \pi^{-1}(p)$ be the fiber over $p \in U$. Given a trivializing chart (U, ϕ) containing this fiber, we define a diffeomorphism $\phi^p : E_p \to \mathbb{R}^r$

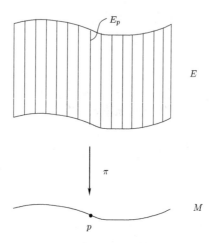

Fig. 27.1. Fiber bundle.

as the composition

$$\phi^p : E_p \xrightarrow{\phi} \{p\} \times \mathbb{R}^r \longrightarrow \mathbb{R}^r.$$

Hence, if $\mathbf{v} \in E_p$, we have

$$\phi(\mathbf{v}) = (p, \phi^p(\mathbf{v})).$$

Since each ϕ^p is a diffeomorphism, we can use it to transport the vector space structure of $\mathbb{R}^r$ to E_p. Given two trivializing charts whose domains intersect we would like that the induced vector space structures on the fibers to coincide. This leads to the following definition.

Definition 27.1. A **vector bundle** of rank r over a manifold M is a triple $\xi = (\pi, E, M)$, where $\pi : E \to M$ is a smooth map admitting a collection of trivializing charts $\mathcal{C} = \{(U_\alpha, \phi_\alpha) : \alpha \in A\}$ of dimension r, satisfying the following properties:

(i) $\{U_\alpha : \alpha \in A\}$ is an open cover of M: $\bigcup_{\alpha \in A} U_\alpha = M$.
(ii) The charts are compatible: for any $\alpha, \beta \in A$ and every $p \in U_\alpha \cap U_\beta$, the **transition functions** $g_{\alpha\beta}(p) \equiv \phi_\alpha^p \circ (\phi_\beta^p)^{-1} : \mathbb{R}^r \to \mathbb{R}^r$ are linear isomorphisms.
(iii) The collection $\mathcal{C}$ is maximal: if (U, ϕ) is a trivializing chart of dimension r with the property that for every $\alpha \in A$, the maps $\phi^p \circ (\phi_\alpha^p)^{-1}$ and $\phi_\alpha^p \circ (\phi^p)^{-1}$ are linear isomorphisms, then $(U, \phi) \in \mathcal{C}$.

We will call E the **total space**, M the **basis space**, and π the **projection** of the bundle ξ. The projection is a surjective submersion.

A collection of charts satisfying (i) and (ii) is called a **vector bundle atlas** or a **trivialization** of ξ. A vector bundle atlas defines a vector bundle since every atlas is contained in a unique maximal atlas. As we have already remarked, (ii) implies that the fiber E_p has a vector space structure such that for any trivializing chart (U, ϕ) the map $\phi^p : E_p \to \mathbb{R}^r$ is a linear isomorphism.

In the definition above all maps are C^∞. Of course, one can also define C^k-vector bundles over C^k-manifold or even topological manifolds. Also, one can define complex vector bundles over smooth manifolds by replacing $\mathbb{R}^r$ by $\mathbb{C}^r$ and where the base is still a *real* smooth manifold. In these notes, we will consider mainly real C^∞ vector bundles, but complex vector bundles are also important and will appear occasionally.

Let $\xi = (\pi, E, M)$ be a vector bundle and $U \subset M$ an open set. A map $s : U \to E$ is called a **section** over U if $\pi \circ s = \mathrm{Id}_U$. The sections over U form a real vector space which we denote by $\Gamma_U(E)$. When $U = M$ we call a section over M a **global section** of E and we write $\Gamma(E)$ instead of $\Gamma_M(E)$. If rank $\xi = r$, a collection $s_1, \ldots, s_r$ of sections over U such that $\{s_1(p), \ldots, s_r(p)\}$ form a basis for E_p for each $p \in U$ is called is called a **frame** over U.

Definition 27.2. Let $\xi_1 = (\pi_1, E_1, M_1)$ and $\xi_2 = (\pi_2, E_2, M_2)$ be two vector bundles. A **morphism of vector bundles** is a smooth map $\Psi : E_1 \to E_2$ which maps the fibers of ξ_1 linearly in the fibers of ξ_2, i.e., Ψ covers a smooth map $\psi : M_1 \to M_2$

$$
\begin{array}{ccc}
E_1 & \xrightarrow{\;\Psi\;} & E_2 \\
\pi_1 \downarrow & & \downarrow \pi_2 \\
M_1 & \xrightarrow{\;\psi\;} & M_2
\end{array}
$$

and the map of the fibers $\Psi^p \equiv \Psi|_{(E_1)_p} : (E_1)_p \to (E_2)_{\psi(p)}$ is a linear transformation for each $p \in M_1$.

In this way, one has the **category of all vector bundles**. Often one considers vector bundles over a *fixed* base manifold M and morphisms over the identity $\psi = \mathrm{Id}_M : M \to M$. These form the **category of vector bundles over** M. Two vector bundles $\xi_1 = (\pi_1, E_1, M_1)$ and $\xi_2 = (\pi_2, E_2, M_2)$ are called:

- **equivalent** if there exist morphisms $\Psi : \xi_1 \to \xi_2$ and $\Psi' : \xi_2 \to \xi_1$ which are inverse to each other. This means that Ψ is an isomorphism in the category of vector bundles — it covers a diffeomorphism $\psi : M_1 \to M_2$ and each fiber map $\Psi^p : (E_1)_p \to (E_2)_{\psi(p)}$ is a linear isomorphism.
- **isomorphic** if $M_1 = M_2 = M$ and there exist morphisms $\Psi : \xi_1 \to \xi_2$ and $\Psi' : \xi_2 \to \xi_1$, covering the identity which are inverse to each other. This means that Ψ is an isomorphism in the category of vector bundles over M — it covers the identity $\psi = \mathrm{Id}_M$ and each fiber map $\Psi^p : (E_1)_p \to (E_2)_p$ is a linear isomorphism.

Example 27.1.

(1) Obviously, for any smooth manifold M, we have the associated vector bundles TM, T^*M, $\wedge^k T^*M$, $\otimes^r TM \otimes^s T^*M$, etc. The sections of these bundles are the vector fields, the differential forms and general tensor fields, that we have studied before. If $\Psi : M \to N$ is a smooth map, its differential $d\Psi : TM \to TN$ is a morphism of vector bundles (note, however, that the transpose $(d_x\Psi)^*$, in general, *is not* a vector bundle morphism).

(2) The **trivial vector bundle** of rank r over M is the vector bundle $\varepsilon_M^r = (\pi, M \times \mathbb{R}^r, M)$, where $\pi : M \times \mathbb{R}^r \to M$ is the projection in the first factor. The global sections of ε_M^r can be identified with $C^\infty(M; \mathbb{R}^r)$. In general, a vector bundle ξ over M of rank r is said to be **trivial** if it is isomorphic to ε_M^r. A vector bundle is trivial if and only if it admits a global frame.

　　A **parallelizable manifold** is a manifold M for which TM is a trivial vector bundle. For example, any Lie group G is parallelizable, while $\mathbb{S}^2$ is not parallelizable. Actually, one can show that $\mathbb{S}^d$ is parallelizable if and only if $d = 0, 1, 3$, and 7.

(3) A r-dimensional distribution D in a manifold M, defines a vector bundle over M of rank r. The fibers are the subspaces $D_p \subset T_pM$. A section of this vector bundle is simply a vector field tangent to the distribution.

(4) A vector bundle of rank 1 is usually refer to as a **line bundle**. For example, any non-vanishing vector field $X \in \mathfrak{X}(M)$ defines a line bundle which is always trivial. More generally, a rank 1

distribution defines a line bundle which is trivial if and only if the distribution is generated by a single vector field.

(5) Consider the manifold E formed by pairs $([x], \mathbf{v})$, where $[x]$ is a line through the origin in $\mathbb{R}^{d+1}$ and $\mathbf{v}$ is a point in this line, i.e.,

$$E = \{([x], \mathbf{v}) \in \mathbb{RP}^d \times \mathbb{R}^{d+1} : \mathbf{v} = \lambda x, \text{ for some } \lambda \in \mathbb{R}\}.$$

The map $\pi : E \to \mathbb{RP}^d$ given by $\pi([x], \mathbf{v}) = [x]$ satisfies the local triviality condition. To see this, given an open set $V \subset \mathbb{S}^d$ such that if $x \in V$ then $-x \notin V$, denoted by $U = \{[x] : x \in V\} \subset \mathbb{RP}^d$ the corresponding open set in real projective space. Then the map defined by

$$\psi : U \times \mathbb{R} \to \pi^{-1}(U), \quad \psi([x], t) = ([x], tx), \forall x \in V,$$

is a diffeomorphism, and its inverse $\phi = \psi^{-1}$ defines a trivializing chart over U. The family of all such charts (U, ϕ) is a vector bundle atlas over $\mathbb{RP}^d$. This vector bundle is called the **canonical line bundle** over $\mathbb{RP}^d$ and denoted γ_d^1.

Let $\xi = (\pi, E, M)$ be a rank r vector bundle. If (U_α, ϕ_α) and (U_β, ϕ_β) are trivializing charts, the corresponding **transition function** is the map

$$g_{\alpha\beta} : U_\alpha \cap U_\beta \to GL(r), \quad p \mapsto g_{\alpha\beta}(p) \equiv \phi_\alpha^p \circ (\phi_\beta^p)^{-1},$$

so that

$$\phi_\alpha \circ (\phi_\beta)^{-1}(p, \mathbf{v}) = (p, g_{\alpha\beta}(p) \cdot \mathbf{v}).$$

The transition function satisfies the following fundamental identity

$$g_{\alpha\beta}(p) g_{\beta\gamma}(p) = g_{\alpha\gamma}(p), \quad (p \in U_\alpha \cap U_\beta \cap U_\gamma). \tag{27.1}$$

If $\alpha = \beta = \gamma$, this condition reduces to

$$g_{\alpha\alpha}(p) = I, \quad (p \in U_\alpha),$$

and when $\gamma = \alpha$ we obtain

$$g_{\beta\alpha}(p) = g_{\alpha\beta}(p)^{-1}, \quad (p \in U_\alpha \cap U_\beta).$$

The family of transition functions $\{g_{\alpha\beta}\}$ depends on the choice of trivializing charts. However, they can be used to describe vector bundles if one notices the following.

Lemma 27.1. *Let ξ and η be vector bundles over M with trivializations $\{\phi_\alpha\}$ and $\{\phi'_\alpha\}$ subordinated to the same open cover $\{U_\alpha\}$. Denote by $\{g_{\alpha\beta}\}$ and $\{g'_{\alpha\beta}\}$ the corresponding collections of transition functions. If ξ is isomorphic to η, then there exist smooth maps $\lambda_\alpha : U_\alpha \to GL(r)$ such that*

$$g'_{\alpha\beta}(p) = \lambda_\alpha(p) \cdot g_{\alpha\beta}(p) \cdot \lambda_\beta^{-1}(p), \quad (p \in U_\alpha \cap U_\beta). \qquad (27.2)$$

Proof. Let $\Psi : \xi \to \eta$ be an isomorphism. For each U_α we define smooth maps $\lambda_\alpha : U_\alpha \to GL(r)$ by

$$\lambda_\alpha(p) = \phi'^p_\alpha \circ \Psi \circ (\phi^p_\alpha)^{-1}.$$

If $p \in U_\alpha \cap U_\beta$, we have

$$g'_{\alpha\beta}(p) = \phi'^p_\alpha \circ (\phi'^p_\beta)^{-1} = \lambda_\alpha(p) \circ \phi^p_\alpha \circ (\phi^p_\beta)^{-1} \circ (\lambda_\beta(p))^{-1}$$
$$= \lambda_\alpha(p) \circ g_{\alpha\beta}(p) \circ \lambda_\beta(p)^{-1}.$$

$\square$

Given a manifold M and an open cover $\{U_\alpha\}_{\alpha \in A}$ we call a family of maps $g_{\alpha\beta} : U_\alpha \cap U_\beta \to GL(r)$ satisfying (27.1) a **cocycle** subordinated to the cover. Two cocycles $\{g_{\alpha\beta}\}$ and $\{g'_{\alpha\beta}\}$ subordinated to the same cover are said to be **equivalent** if they are related by (27.2) for some family of smooth maps $\lambda_\alpha : U_\alpha \to GL(r)$.

We saw above that (i) a trivialization of a vector bundle determines a cocycle and that (ii) two trivializations of isomorphic vector bundles subordinated to the *same* cover determine equivalent cocycles. Moreover, we have the following converse:

Proposition 27.1. *Let $\{g_{\alpha\beta}\}$ be a cocycle subordinated to an open cover $\{U_\alpha\}$ of M. There exists a vector bundle $\xi = (\pi, E, M)$ admitting a trivialization $\{\phi_\alpha\}$ with collection of transition functions $\{g_{\alpha\beta}\}$. Two equivalent cocycles $\{g_{\alpha\beta}\}$ and $\{g'_{\alpha\beta}\}$ determine isomorphic vector bundles.*

Proof. Given a cocycle $\{g_{\alpha\beta}\}$, subordinated to the cover $\{U_\alpha\}$ of M, we construct the manifold E as the quotient

$$E = \bigsqcup_{\alpha \in A} (U_\alpha \times \mathbb{R}^r) \Big/ \sim$$

where $\sim$ is the equivalence relation defined by

$$(p, \mathbf{v}) \sim (q, \mathbf{w}) \text{ iff } \begin{cases} p = q \quad \text{and} \\ \exists \, \alpha, \beta \in A : \ g_{\alpha\beta}(p) \cdot \mathbf{v} = \mathbf{w}. \end{cases}$$

We leave it as an exercise to check that this equivalence relation satisfies the conditions of Godement's Criterion (Theorem 10.1) so E is a smooth manifold.

The obvious projection $\pi : E \to M$, $\pi([p, \mathbf{v}]) = p$ is a smooth map. Also, the maps

$$\phi_\alpha : \pi^{-1}(U_\alpha) \to U_\alpha \times \mathbb{R}^r, \quad \phi_\alpha([p, \mathbf{v}]) = (p, \mathbf{v}),$$

give trivializing charts for $\pi : E \to M$ and the corresponding transition functions are exactly the $\{g_{\alpha\beta}\}$. Hence, we have a vector bundle $\xi = (\pi, E, M)$ as in the statement.

If $\{g'_{\alpha\beta}\}$ is another cocycle equivalent to $\{g_{\alpha\beta}\}$ through the family $\{\lambda_\alpha\}$ and $\xi' = (\pi', E', M)$ denotes the vector bundle associated with $\{g'_{\alpha\beta}\}$, we have a vector bundle isomorphism $\Psi : \xi \to \xi'$ defined on each open set $\pi^{-1}(U_\alpha)$ by

$$\Psi([p, \mathbf{v}]) = [p, \lambda_\alpha(p) \cdot \mathbf{v}].$$

$\square$

Remark 27.1. If two cocycles are subordinated to *different* covers we can refine the covers and obtain cocycles subordinated to the same cover. Hence, one obtains a 1:1 correspondence

$$\left\{ \begin{array}{c} \text{vector bundles } \xi = (\pi, E, M) \\ \text{up to isomorphism} \end{array} \right\}$$

$$\overset{\sim}{\longleftrightarrow} \left\{ \begin{array}{c} \text{cocycles } \{g_{\alpha\beta} : U_\alpha \cap U_\beta \to GL(r)\} \\ \text{up to equivalence and refinement} \end{array} \right\}.$$

We will see later that a vector bundle over a contractible space is always trivial. Hence, if one fixes an open cover of M consisting of contractible open set $\{U_\alpha\}$, one does not need refinements in this correspondence.

We now turn to constructions with vector bundles. We have the following general principle:

> For every functorial construction with vector spaces there is a
> similar construction with vector bundles.

This principle can be made precise. However, instead of following the abstract route, we will describe explicitly the constructions that are most relevant for us.

Subbundles and Quotients

Every vector bundle $\xi = (\pi, E, M)$ can be restricted to a submanifold $N \subset M$. The **restriction** ξ_N is the vector bundle with total space

$$E_N = \{E_p : p \in N\},$$

and projection $\pi_N : E_N \to N$ the restriction of π to E_N. The restriction is an example of a *vector subbundle*.

Definition 27.3. A vector bundle $\eta = (\tau, F, N)$ is called a **vector subbundle** of a vector bundle $\xi = (\pi, E, M)$ if F is a submanifold of E, and the inclusion $F \hookrightarrow E$ is a morphism of vector bundles.

If $\Psi : \eta \to \xi$ is a morphism of vector bundles covering the identity, in general, its image and its kernel *are not* vector subbundles: these are made of vector spaces of varying dimension. This issue disappears if we assume that Ψ has **constant rank**, i.e., if all linear maps $\Psi^p : E_p \to F_p$ have the same rank. In this case, we have

- The **kernel** of Ψ is the vector subbundle $\operatorname{Ker} \Psi := \{\mathbf{v} \in E : \Psi(\mathbf{v}) = 0\} \subset E$.
- The **image** of Ψ is the vector subbundle $\operatorname{Im} \Psi := \{\Psi(\mathbf{v}) \in F : \mathbf{v} \in E\} \subset F$.
- The **co-kernel** of Ψ is the vector bundle $\operatorname{coKer} \Psi := F/\sim$, where $\sim$ the equivalence relation $\mathbf{w}_1 \sim \mathbf{w}_2$ if and only if $\mathbf{w}_1 - \mathbf{w}_2 = \Psi(\mathbf{v})$, for some $\mathbf{v} \in E$.

Note that if Ψ is a *monomorphism* (i.e., each Ψ^p is injective) or if Ψ is an *epimorphism* (i.e., each Ψ^p is surjective) then Ψ has constant rank. Therefore, the kernel, image, and cokernel of monomorphisms and epimorphisms are vector subbundles.

The notions associated with exact sequences can be easily extended to vector bundles and morphisms of constant rank. For example, a short exact sequence of vector bundles is a sequence of vector bundle morphisms

$$0 \longrightarrow \xi \overset{\Phi}{\longrightarrow} \eta \overset{\Psi}{\longrightarrow} \theta \longrightarrow 0$$

where Φ is a monomorphism, Ψ is an epimorphism, and $\operatorname{Im}\Phi = \operatorname{Ker}\Psi$. In this case, we have vector bundle isomorphisms $\xi \simeq \operatorname{Ker}\Psi$ and $\theta \simeq \operatorname{coKer}\Psi$. We say that θ is the **quotient vector bundle** of the monomorphism Φ and we will denote it by η/ξ. Note that the fibers of η/ξ are the quotient vector spaces E_p/F_p.

Example 27.2. Let M be a manifold and $N \subset M$ a submanifold. The restriction of TM to N is denoted $T_N M$. The tangent bundle TN is a vector subbundle of $T_N M$ and the quotient bundle $\nu(N) \equiv T_N M/TN$ is usually called the **normal bundle to** N in M.

Similarly, given a foliation $\mathcal{F}$ of M, we have the vector subbundle $T\mathcal{F} \subset TM$. The quotient bundle $\nu(\mathcal{F}) \equiv TM/T\mathcal{F}$ is usually called the **normal bundle of** $\mathcal{F}$ in M. If L is a leaf of $\mathcal{F}$, the restriction of $\nu(\mathcal{F})$ to L is the normal bundle $\nu(L)$.

Direct Sums and Tensor Products

Let $\xi = (\pi, E, M)$ and $\eta = (\tau, F, M)$ be vector bundles over the same manifold M. The **Whitney sum** or **direct sum** of ξ and η is the vector bundle $\xi \oplus \eta$ whose total space is

$$E \oplus F := E \times_M F = \{(\mathbf{v}, \mathbf{w}) \in E \times F : \pi(\mathbf{v}) = \tau(\mathbf{w})\},$$

and whose projection is

$$E \oplus F \to M, \quad (\mathbf{v}, \mathbf{w}) \mapsto \pi(\mathbf{v}) = \tau(\mathbf{w}).$$

Note that the fiber of $\xi \oplus \eta$ over $p \in M$ is the direct sum $E_p \oplus F_p$. The local triviality condition is easily verified: if $\{\phi_\alpha\}$ and $\{\psi_\alpha\}$ are trivializations of ξ and η, subordinated to the same covering, with corresponding cocycles $\{g_{\alpha\beta}\}$ and $\{h_{\alpha\beta}\}$, then we have the trivialization of $\xi \oplus \eta$ given by $\{(\phi_\alpha \times \psi_\alpha)|_{E \oplus F}\}$, to which corresponds the cocycle defined by

$$g_{\alpha\beta} \oplus h_{\alpha\beta} = \begin{bmatrix} g_{\alpha\beta} & 0 \\ 0 & h_{\alpha\beta} \end{bmatrix}.$$

Similarly, we can define:

- The **tensor product $\xi \otimes \eta$:** the fibers are the tensor products $E_p \otimes F_p$ and the transition functions are $g_{\alpha\beta} \otimes h_{\alpha\beta}$.
- The **dual vector bundle ξ^*:** the fibers are the dual vector spaces E_p^* and the transition functions are the inverse transpose maps $(g_{\alpha\beta})^{-T}$.
- The **exterior product $\wedge^k \xi$:** the fibers are the exterior products $\wedge^k E_p$ and the transition functions are the exterior powers $\wedge^k g_{\alpha\beta}$.
- The **Hom(ξ, η)-bundles:** the fibers are the space of all linear morphisms $\text{Hom}(E_x, F_x)$. We leave as an exercise to show that there is a natural isomorphism $\text{Hom}(\xi, \eta) \simeq \xi^* \otimes \eta$.

Orientations

A vector bundle $\xi = (\pi, E, M)$ of rank r is called an **orientable vector bundle** if the exterior product $\wedge^r \xi$ has a section which never vanishes. Note that this section corresponds to a smooth choice of an orientation in each vector space E_p. We call an **orientation** for ξ an equivalence class $[s]$, where two non-vanishing sections $s_1, s_2 \in \Gamma(\wedge^r \xi)$ are equivalent if and only if $s_2 = f s_1$ for some smooth positive function $f \in C^\infty(M)$. We leave as an exercise to check that ξ is orientable if and only if it admits a trivialization $\{\phi_\alpha\}$ for which the associated cocycle $\{g_{\alpha\beta}\}$ takes values in $GL^+(r)$, the group of invertible $r \times r$ matrices with positive determinant

$$g_{\alpha\beta} : U_\alpha \cap U_\beta \to GL^+(r) \subset GL(r).$$

For a manifold M, the notion of orientation that we studied before corresponds to the notion of orientation of the vector bundle TM. For a vector bundle $\xi = (\pi, E, M)$, the possible orientations for ξ, E and M are related as follows:

Lemma 27.2. *Let $\xi = (\pi, E, M)$ be a vector bundle. If two among the vector bundles TM, TE and ξ are orientable so is the third one.*

The proof is left as an exercise.

Riemmanian Structures

A **Riemmanian structure** in a vector bundle $\xi = (\pi, E, M)$ is a choice of an inner product $\langle \ , \ \rangle : E_p \times E_p \to \mathbb{R}$ in each fiber which varies smoothly, i.e., for any sections $s_1, s_2 \in \Gamma(E)$ the map $p \mapsto \langle s_1(p), s_2(p) \rangle$ is smooth. This condition is equivalent to say that the section of the vector bundle $\otimes^2 \xi^*$ defined by $\langle \ , \ \rangle$ is smooth.

It is easy to see that a vector bundle always admits a Riemann structure. Given a trivialization $\{(U_\alpha, \phi_\alpha)\}$, one chooses a partition of unity $\{\rho_\alpha\}$ subordinated to the cover $\{U_\alpha\}$ and defines a Riemmanian structure by

$$\langle \mathbf{v}, \mathbf{w} \rangle := \sum_\alpha \rho_\alpha(p)(\phi_\alpha^p(\mathbf{v}), \phi_\alpha^p(\mathbf{w}))_{\mathbb{R}^r} \ (\mathbf{v}, \mathbf{w} \in E_p).$$

For this reason, any vector bundle admits a trivialization $\{\phi_\alpha\}$ whose associated cocycle $\{g_{\alpha\beta}\}$ take values in the orthogonal group $O(r)$

$$g_{\alpha\beta} : U_\alpha \cap U_\beta \to O(r) \subset GL(r).$$

This follows also from the polar decomposition

$$GL(r) = O(r) \times \{\text{positive definite symmetric matrices}\}.$$

If $\xi = (\pi, E, M)$ is a vector bundle and $\langle \ , \ \rangle$ is a Riemann structure in ξ, then for any vector subbundle $\eta = (\tau, F, N)$ we can define the **orthogonal vector bundle** $\eta^\perp$ over N as the subbundle of ξ with total space $F^\perp$, where

$$F_p^\perp \equiv \{\mathbf{v} \in E_p : \langle \mathbf{v}, \mathbf{w} \rangle = 0, \forall \mathbf{w} \in F_p\}.$$

When $M = N$, we obtain

$$\xi = \eta \oplus \eta^\perp.$$

It follows that $\eta^\perp \simeq \xi/\eta$, since the natural projection $\xi \to \xi/\eta$ restricts to an isomorphism on $\eta^\perp$.

Exercises

Exercise 27.1

Show that a vector bundle is trivial if and only it admits a global frame.

Exercise 27.2

Let $G_r(\mathbb{R}^d)$ be the Grassmannian manifold of r-planes in $\mathbb{R}^d$. Consider the submanifold $E \subset G_r(\mathbb{R}^d) \times \mathbb{R}^d$ defined by

$$E = \{(S, x) : S \text{ is a subspace of } \mathbb{R}^d \text{ and } x \in S\},$$

and the smooth map

$$\pi : E \to G_r(\mathbb{R}^d), \quad (S, x) \mapsto S.$$

Show that $\gamma_d^r = (\pi, E, G_r(\mathbb{R}^d))$ is a vector bundle of rank r. It is called the **canonical bundle** over $G_r(\mathbb{R}^d)$.

Exercise 27.3

Let $\Psi : \eta \to \xi$ be a morphism of vector bundles which covers the identity. Show that the kernel and the image of Ψ are vector subbundles if the rank of the linear maps Ψ^p is constant. Give counterexamples when the rank is not constant.

Exercise 27.4

Let $\xi = (\pi, E, M)$ and $\eta = (\tau, F, M)$ be vector bundles.

(a) Show that there exists a vector bundle $\mathrm{Hom}(\xi, \eta)$ whose fibers are the vector spaces $\mathrm{Hom}(E_x, F_x)$.
(b) Find the transition function of $\mathrm{Hom}(\xi, \eta)$ in terms of the transition functions of ξ and η.
(c) Find an isomorphism $\mathrm{Hom}(\xi, \eta) \simeq \xi^* \otimes \eta$.

Exercise 27.5

Given a vector bundle, ξ show that there exists a trivialization of ξ for which the transition functions take values in $O(r)$.

Exercise 27.6

Consider a short exact sequence of vector bundles

$$0 \longrightarrow \xi_1 \longrightarrow \xi_2 \overset{\Psi}{\longrightarrow} \xi_3 \longrightarrow 0$$

Show that

(a) Such a short exact sequence always *splits*, i.e., there exists a morphism of vector bundles $\Phi : \xi_3 \to \xi_2$ such that $\Psi \circ \Phi = \mathrm{Id}$.

(b) There is an isomorphism of vector bundles: $\xi_2 \simeq \xi_1 \oplus \xi_3$.
(c) If two among the vector bundles ξ_1, ξ_2, and ξ_3 are orientable, so is the third one.

Exercise 27.7

Let $\xi = (\pi, E, M)$ be a vector bundle. Show the following:

(a) There exists a natural isomorphism of vector bundles $T_M E \simeq \xi \oplus TM$.
(b) If two among the vector bundles TM, TE, and ξ are orientable so is the third one.

Exercise 27.8

For a vector bundle, ξ show that the following statements are equivalent:

(a) ξ is orientable.
(b) There exists a trivialization of ξ for which the transition functions take values in $GL^+(r)$.
(c) There exists a trivialization of ξ for which the transition functions take values in $SO(r)$.

Exercise 27.9

Let M be a manifold with fundamental group $\pi_1(M)$, $q : \widetilde{M} \to M$ its universal covering space and $\rho : \pi_1(M) \to GL(V)$ a representation. Consider the action of $\pi_1(M)$ on $\widetilde{M} \times V$ given by

$$\gamma \cdot (p, v) := (\gamma \cdot p, \rho(\gamma)(v)).$$

Show that there is a vector bundle $\xi = (\pi, E, M)$ with total space and projection

$$E := (\widetilde{M} \times V)/\pi_1(M), \quad \pi([p, v]) = q(p).$$

Exercise 27.10

Let M be a manifold with fundamental group $\pi_1(M)$, $q : \widetilde{M} \to M$ its universal covering space and $f : \pi_1(M) \times \widetilde{M} \to \mathbb{C}$ a smooth function.

(a) Show that one has a group action of $\pi_1(M)$ on $\tilde{M} \times \mathbb{C}$ defined by

$$\gamma \cdot (p, z) := (\gamma \cdot p, f(\gamma, p)z),$$

if and only if

$$f(\gamma_1 \cdot \gamma_2, p) = f(\gamma_1, \gamma_2 \cdot p)f(\gamma_2, p), \qquad (27.3)$$

for all $\gamma_i \in \pi_1(M)$ and $p \in \widetilde{M}$.

(b) Show that if (27.3) holds, then one has a complex line bundle $\xi = (\pi, L, M)$ with total space and projection

$$L := (\tilde{M} \times \mathbb{C})/\pi_1(M), \quad \pi(p, z) := q(p).$$

(c) Let $M = \mathbb{T}^2$, so $\pi_1(M) = \mathbb{Z}^2$ and $\widetilde{M} = \mathbb{R}^2$. Show that

$$f : \mathbb{Z}^2 \times \mathbb{R}^2 \to \mathbb{C}, \quad f((n_1, n_2), (x, y)) := e^{i2\pi n_2 x},$$

satisfies (27.3) and that the resulting line bundle from (b) is non-trivial.

The Thom Class and the Euler Class

The homotopy invariance of de Rham cohomology relied crucially on the isomorphism

$$H^\bullet(M \times \mathbb{R}^r) \simeq H^\bullet(M).$$

One can interpret this isomorphism as relating the cohomology of the total space of the trivial bundle $\mathrm{pr}_1 : M \times \mathbb{R}^r \to M$ with the cohomology of its base. Indeed, this holds for any vector bundle.

Proposition 28.1. *For any vector bundle* $\xi = (\pi, E, M)$,

$$H^\bullet(E) \simeq H^\bullet(M).$$

Proof. Let $s : M \to E$ be the zero section. Scalar multiplication on the fibers, gives a deformation retract of E on its image. Therefore, by homotopy invariance we see that $s^* : H^\bullet(E) \to H^\bullet(M)$ is an isomorphism. $\qquad\qquad\square$

One may guess that the corresponding statement for compactly supported cohomology, namely

$$H_c^\bullet(E) \simeq H_c^{\bullet-r}(M),$$

should also hold. The following example shows that one must be careful.

Example 28.1. Consider the canonical line bundle γ_1^1 over $\mathbb{RP}^1 = \mathbb{S}^1$ (see Exercise 27.2). The total space E of this bundle is the Möbius band, a non-oriented manifold of dimension 2, so we have $H_c^2(E) = 0$. On the other hand, for the base we find

$$H_c^{2-1}(\mathbb{S}^1) = H^1(\mathbb{S}^1) \simeq \mathbb{R} \neq 0.$$

The issue with this example is exactly the lack of orientability.

Proposition 28.2 (Thom Isomorphism — First version). *Let* $\xi = (\pi, E, M)$ *be a vector bundle of rank* r, *where* E *is orientable and* M *is of finite type. Then*

$$H_c^\bullet(E) \simeq H^{\bullet - r}(M).$$

Proof. Since M is of finite type, so is E. Hence, E is both orientable and of finite type and we can apply Poincaré duality to conclude that

$$H_c^\bullet(E) \simeq H^{d+r-\bullet}(E) \quad \text{(by Poincaré duality for } E\text{)},$$

$$\simeq H^{d+r-\bullet}(M) \quad \text{(by Proposition 28.1).}$$

$\square$

The isomorphism behind the Thom isomorphism in the previous proposition can be described explicitly. It relies on a push-forward map

$$\pi_* : \Omega_c^\bullet(E) \to \Omega^{\bullet - r}(M)$$

called **integration along the fibers**. For the case of the trivial line bundle, this map appeared in the proof of Proposition 24.2. Using a local trivialization, we can extend the description given in that proof to any vector bundle. We start by covering M by trivializing oriented charts (U_α, ϕ_α) for the vector bundle ξ, where each U_α is the domain of a chart $(x^1, \ldots, x^d)$ of the base M. This yields a chart $(x^1, \ldots, x^d, t^1, \ldots, t^r)$ for the total space E with domain $\pi^{-1}(U_\alpha)$, where $(t^1, \ldots, t^r)$ are linear coordinates on the fibers. If $\omega \in \Omega_c^\bullet(E)$, then $\omega_\alpha = \omega|_{\pi^{-1}(U_\alpha)}$ is a linear combination of two kinds of forms

$$f_1(x,t)\pi^*\theta_1 \wedge dt^{i_1} \wedge \cdots \wedge dt^{i_k}, \quad \text{with } k < r$$

$$f_2(x,t)\pi^*\theta_2 \wedge dt^1 \wedge \cdots \wedge dt^r$$

where θ_i are differential forms in M and the functions $f_i(x,t)$ have compact support. Integration along the fibers $\pi_* : \Omega_c^\bullet(E) \to \Omega^{\bullet - r}(M)$

is the unique linear map defined by

$$f_1(x,t)\pi^*\theta_1 \wedge dt^{i_1} \wedge \cdots \wedge dt^{i_k} \longmapsto 0, \quad (k < r),$$

$$f_2(x,t)(\pi^*\theta_2) \wedge dt^1 \wedge \cdots \wedge dt^r \longmapsto \theta_2 \int_{\mathbb{R}^r} f_2(x, t^1, \ldots, t^r) \, dt^1 \cdots dt^r.$$

We leave it as an exercise to check that this definition is independent of the choices made.

Remark 28.1 (Differential forms with compact vertical support). The description above of fiber integration π_* shows that

(i) π_* is defined as long as the vector bundle ξ is oriented, so the base and/or the total space can be non-orientable.

(ii) π_* can be defined for any differential form ω in E with compact vertical support, i.e., such that $\mathrm{supp}\,\omega \cap \pi^{-1}(K)$ is compact for every compact set $K \subset M$.

We denote by $\Omega_{cv}^\bullet(E)$ the space of differential forms with compact vertical support so for any oriented vector bundle of rank r fiber integration is a linear map

$$\pi_* : \Omega_{cv}^\bullet(E) \to \Omega^{\bullet - r}(M).$$

Another exercise using the definition above shows that fiber integration satisfies the following properties.

Proposition 28.3. *Fiber integration* $\pi_* : \Omega_{cv}^\bullet(E) \to \Omega^{\bullet - r}(M)$ *satisfies:*

(i) π_* *is a cochain map:* $d\pi_*\omega = \pi_* d\omega$.

(ii) *Projection formula: for any* $\theta \in \Omega^\bullet(M)$ *and* $\omega \in \Omega_{cv}^\bullet(E)$:

$$\pi_*(\pi^*\theta \wedge \omega) = \theta \wedge \pi_*\omega. \qquad (28.1)$$

(iii) *If* $\Psi : \xi_1 \to \xi_2$ *is a vector bundle map covering a map* $\psi : M_1 \to M_2$, *which is a fiberwise isomorphism and preserves orientations, then for any* $\omega \in \Omega_{cv}^\bullet(E)$:

$$(\pi_1)_*\Psi^* = \psi^*(\pi_2)_*. \qquad (28.2)$$

The space $\Omega_{cv}^*(E)$ of differential forms with compact vertical support is a subcomplex of the de Rham complex and gives rise to a cohomology $H_{cv}^\bullet$. By property (i), fiber integration induces a map in cohomology

$$\pi_* : H_{cv}^\bullet(E) \to H^{\bullet-r}(M).$$

The general version of the Thom isomorphism can then be stated as follows.

Proposition 28.4 (Thom Isomorphism). *Let $\xi = (\pi, E, M)$ be an oriented vector bundle of rank r over a manifold of finite type. Then fiber integration gives an isomorphism*

$$H_{cv}^\bullet(E) \simeq H^{\bullet-r}(M).$$

Proof. For a trivial vector bundle, the proof is the same as Proposition 24.2.

Using a partition of unity argument, one sees that the cohomology $H_{cv}^\bullet(E)$ satisfies the Mayer–Vietoris sequence property. Then, given open sets $U, V \subset M$, one obtains a commutative diagram of Mayer–Vietoris sequences

$$\to H_{cv}^k(E|_{U\cup V}) \to H_{cv}^k(E|_U) \oplus H_{cv}^k(E|_V) \to H_{cv}^k(E|_{U\cap V}) \xrightarrow{\delta} H_{cv}^{k+1}(E|_{U\cup V}) \to$$

$$\downarrow \pi_* \qquad \downarrow \pi_* \qquad \downarrow \pi_* \qquad \downarrow \pi_*$$

$$\to H_c^{k-r}(U\cup V)^* \to H_c^{k-r}(U)^* \oplus H_c^{k-r}(V)^* \to H_c^{k-r}(U\cap V)^* \xrightarrow{\delta^*} H_c^{k+1-r}(U\cup V)^* \to$$

If the vector bundle ξ is trivial over U and V, then in the previous diagram one obtains that π_* is an isomorphism for U, V and $U\cap V$. By the Five Lemma, it follows that π_* is an isomorphism also over $U\cup V$. Then the proof proceeds by an induction argument over the number of elements of a good cover, as in the proof of Poincaré duality. $\square$

We can now introduce a cohomological invariant of a vector bundle.

Definition 28.1. The **Thom class** of an oriented vector bundle $\xi = (\pi, E, M)$ of rank r is the image of 1 under the Thom isomorphism $H^0(M) \simeq H_{cv}^r(E)$. We will denote this class by $U \in H_{cv}^r(E)$.

The Thom class allows one to write, in a more or less explicit way, the inverse to the fiber integration $\pi_* : H_{cv}^\bullet(E) \to H^{\bullet-r}(M)$. In fact, since $\pi_* U = 1$, the projection formula (28.1) shows that the inverse is the linear map

$$(\pi_*)^{-1} : H^\bullet(M) \to H_{cv}^{\bullet+r}(E), \quad [\omega] \mapsto [\pi^*\omega] \cup U.$$

The following result gives an alternative characterization of the Thom class.

Theorem 28.1. *The Thom class of an oriented vector bundle $\xi = (\pi, E, M)$ of rank r is the unique class $U \in H_{cv}^r(E)$ whose pullback to each fiber E_p is the canonical generator of $H_c^r(E_p)$, i.e., that satisfies*

$$\int_{E_p} i^* U = 1, \quad \forall p \in M,$$

where $i : E_p \hookrightarrow E$ is the inclusion.

Proof. Since $\pi_* U = 1$, we see that the restriction $i^* U$ to each fiber E_p is a compactly supported form with $\int_{E_p} i^* U = 1$.

Conversely, let $U' \in H_{cv}^r(E)$ be a class such for each $p \in M$ the restriction $i^* U' \in H_c^r(E_p)$ is the canonical generator. By the projection formula (28.1), we obtain

$$\pi_*(\pi^*[\theta] \cup U') = [\theta] \cup \pi_* U' = [\theta], \quad \forall [\theta] \in H^\bullet(M).$$

Hence, $[\theta] \mapsto \pi^*[\theta] \cup U'$ inverts π_*. The image of 1, which is U', must then coincide with the Thom class. $\square$

From now on, to simplify the presentation, we will assume that M is compact so that it is of finite type. Moreover, it follows that for a vector bundle $\xi = (\pi, E, M)$ we have $H_{cv}^\bullet(E) = H_c^\bullet(E)$.

The Thom class of a vector bundle $\xi = (\pi, E, M)$ is an invariant of the bundle, but it lies in the cohomology of the total space. We can use a global section to obtain an invariant which lies in the cohomology of the base. For that observe that given a section $s : M \to E$ we have an induced map in cohomology

$$s^* : H_c^\bullet(E) \to H^\bullet(M),$$

which we can view as the composition of two maps

$$H_c^\bullet(E) \longrightarrow H^\bullet(E) \xrightarrow{\ s^*\ } H^\bullet(M).$$

On the other hand, any two sections $s_0, s_1 : M \to E$ are homotopic via

$$H(p, t) := t s_1(p) + (1 - t) s_0(p)$$

From the homotopy invariance of cohomology, we conclude that the maps induced in cohomology by any two sections are identical

$$s_0^* = s_1^* : H_c^\bullet(E) \to H^\bullet(M).$$

Therefore, the following definition makes sense.

Definition 28.2. Let $\xi = (\pi, E, M)$ be an oriented vector bundle of rank r over a compact manifold M. The **Euler class** of ξ is the class $e(\xi) \in H^r(M)$ defined by

$$e(\xi) := s^* U,$$

where U is the Thom class of ξ and $s : M \to E$ is *any* global section of ξ.

Note that, in particular, we can define the Euler class by pulling back along the zero section. The following proposition lists some properties of the Euler class. We leave its proof for the exercises.

Proposition 28.5. *Let* $\xi = (\pi, E, M)$, $\xi' = (\pi', E', M)$ *and* $\eta = (\pi, F, N)$ *be oriented vector bundles of rank r over compact manifolds.*

  (i) *If the rank r is odd, then $e(\xi) = 0$.*
 (ii) *If $\bar{\xi}$ denotes ξ with the opposite orientation, then $e(\bar{\xi}) = -e(\xi)$.*
(iii) *If $\xi \oplus \xi'$ has the direct sum orientation, then $e(\xi \oplus \xi') = e(\xi) \cup e(\xi')$.*
 (iv) *If a vector bundle map $\Psi : \eta \to \xi$, covering a map $\psi : M \to N$, is a fiberwise isomorphism and preserves orientations, then $e(\eta) = \psi^* e(\xi)$.*

The Euler class of a vector bundle is an *obstruction* to the existence of a non-vanishing global section.

Theorem 28.2. *Let* $\xi = (\pi, E, M)$ *be an oriented vector bundle over a compact manifold M. If ξ admits a non-vanishing section then $e(\xi) = 0$.*

Proof. Let $s : M \to E$ be a non-vanishing section. If $\omega \in \Omega_c^r(E)$ is a compactly supported form representing the Thom class, then there exists $c \in \mathbb{R}$ such that the image of the section cs does not intersect $\mathrm{supp}\,\omega$. Hence:

$$e(\xi) = (cs)^*U = [(cs)^*\omega] = 0. \qquad \Box$$

Note that there are examples of oriented vector bundles ξ with $e(\xi) = 0$, for which every global section has a zero.

The next example gives some additional intuition to the meaning of the Euler class for rank 2 vector bundles.

Example 28.2. Consider an oriented vector bundle $\xi = (\pi, E, M)$ of rank 2. Fix some Riemannian metric on ξ and cover M by charts $\{(U_\alpha, x_\alpha^i)\}$ over which we have a positive orthonormal frame $\{s_1^\alpha, s_2^\alpha\}$. These define coordinates $(\pi^*x_\alpha^i, r_\alpha, \theta_\alpha)$ on

$$(E - \{0_M\})|_{U_\alpha} \simeq U_\alpha \times (\mathbb{R}^2 - \{0\})$$

where $(r_\alpha, \theta_\alpha)$ are polar coordinates on $\mathbb{R}^2 - \{0\}$. On an overlap $U_\alpha \cap U_\beta$ the radial functions coincide $r_\alpha = r_\beta$, while the angles differ by a rotation, i.e.,

$$\theta_\alpha - \theta_\beta = \pi^*\varphi_{\alpha\beta}, \qquad \varphi_{\alpha\beta} : U_\alpha \cap U_\beta \to \mathbb{S}^1.$$

Note that on triple intersections $U_\alpha \cap U_\beta \cap U_\gamma$ we have

$$\varphi_{\alpha\beta} + \varphi_{\beta\gamma} = \varphi_{\alpha\gamma}.$$

Let $\{\rho_\alpha\}$ be a partition of unity subordinated to the cover $\{U_\alpha\}$. We obtain a 1-form on each open U_α defined by

$$\varepsilon_\alpha := \sum_\gamma \rho_\gamma \mathrm{d}\varphi_{\alpha\gamma} \in \Omega^1(U_\alpha).$$

On a double intersection $U_\alpha \cap U_\beta$ we find

$$\varepsilon_\alpha - \varepsilon_\beta = \sum_\gamma \rho_\gamma(\mathrm{d}\varphi_{\alpha\gamma} - \mathrm{d}\varphi_{\beta\gamma}) = \sum_\gamma \rho_\gamma \mathrm{d}\varphi_{\alpha\beta} = \mathrm{d}\varphi_{\alpha\beta}.$$

Hence, it follows that there is a well-defined 2-form $\varepsilon \in \Omega^2(M)$ such that

$$\varepsilon|_{U_\alpha} = \mathrm{d}\varepsilon_\alpha.$$

On the other hand, on $(E - 0_M)|_{U_\alpha \cap U_\beta}$, we have

$$\mathrm{d}\theta_\alpha - \mathrm{d}\theta_\beta = \pi^*\mathrm{d}\varphi_{\alpha\beta} = \pi^*\varepsilon_\alpha - \pi^*\varepsilon_\beta.$$

Hence, we also have a global "angular form" $\phi \in \Omega^1(E - 0_M)$ such that

$$\phi = d\theta_\alpha - \pi^* \varepsilon_\alpha \quad \text{on } (E - 0_M)|_{U_\alpha}.$$

Note that

$$d\phi = -\pi^* \varepsilon.$$

Finally, let $\delta > 0$ and choose a smooth function $\rho : \mathbb{R} \to \mathbb{R}$ which is non-decreasing, $\rho(r) = -\frac{1}{2\pi}$ for $t < \delta$, $\rho(r) = 0$ for $t \geq 1$ and $\int_{\mathbb{R}} \rho'(r) dr = \frac{1}{2\pi}$. We can promote it to a function $\rho : E \to \mathbb{R}$ of the radius

$$\rho(\mathbf{v}) = \rho(||\mathbf{v}||) \ (\mathbf{v} \in E).$$

Then we define the 2-form

$$u := d(\rho\phi) = d\rho \wedge \phi - \rho \pi^* \varepsilon \in \Omega^2_{\mathrm{cv}}(E).$$

A priori this form is only defined outside the zero section, but the second expression shows that it extends smoothly to E, since ρ is constant in a neighborhood of 0_M. The restriction of u to a fiber E_p is the compactly supported 2-form $(d\rho \wedge \phi)|_{E_p}$, which is positively oriented and has integral 1. Hence, $U = [u]$ is the Thom class of the bundle ξ. Moreover, if we pullback u by the zero section $s_0 : M \to E$ we obtain

$$s_0^* u = -\rho(0) s_0^* \pi^* \varepsilon = \frac{1}{2\pi} \varepsilon.$$

So we conclude also that $e(\xi) = \frac{1}{2\pi}[\varepsilon]$.

The name *Euler class* is related with the special case where $\xi = TM$. Let M be an oriented, connected, manifold with $\dim M = d$ and denote the orientation by μ. The corresponding canonical generator in cohomology will also be denoted by $\mu \in H_c^d(M)$ and it can be represented by any top degree form $\omega \in \Omega_c^d(M)$ such that

$$\int_M \omega = 1.$$

If we assume that M is of finite type, then μ is the image of 1 under Poincaré duality $H^0(M) \simeq H_c^d(M)$. Recalling from Lecture 26, the notion of index of an isolated zero of a vector field one finds the following result.

Theorem 28.3. *Let M be a compact, oriented, connected manifold of dimension d. If $X \in \mathfrak{X}(M)$ is a vector field with a finite number of zeros $\{p_1, \ldots, p_N\}$, then*

$$e(TM) = \left(\sum_{i=1}^{N} \mathrm{ind}_{p_i} X\right) \mu \in H^d(M),$$

where $\mu \in H^d(M)$ is the class defined by the orientation of M.

Proof. Let $\omega \in \Omega_c^d(TM)$ be a compactly supported form representing the Thom class. We need to show that

$$\int_M X^* \omega = \left(\sum_{i=1}^{N} \mathrm{ind}_{p_i} X\right).$$

Choose charts (U_i, ϕ_i) centered at p_i and consider D_i the closed balls

$$D_i := \phi_i^{-1}(\{x \in \mathbb{R}^d : ||x|| \le 1\}).$$

We have identifications $TU_i \simeq U_i \times \mathbb{R}^d$ provided by the charts, and we denote by $p : TU_i \to \mathbb{R}^d$ the projection on the second factor. Using a partition of unity argument, it follows from Theorem 28.1 that we can choose the representative ω so that on each coordinate system U_i we have

$$\omega|_{TD_i} = p^* d\theta \quad \text{where} \quad \int_{\mathbb{S}^{d-1}} \theta = 1.$$

For any $c > 0$, the vector fields X and cX have the same zeros and the same indices. Hence, by choosing c sufficiently large we can assume that

$$X_p \notin \mathrm{supp}\,\omega, \quad \forall p \notin \bigcup_{i=1}^{N} D_i.$$

Therefore,

$$\int_M X^* \omega = \sum_{i=1}^{N} \int_{D_i} X^* \omega,$$

and so it is enough to verify that

$$\int_{D_i} X^* \omega = \mathrm{ind}_{p_i} X.$$

Recall that $\mathrm{ind}_{p_i}(X) = \deg G_i$ where the Gauss map G_i is obtained using the identification $TU_i \simeq U_i \times \mathbb{R}^d$ provided by the charts

$$X|_{U_i} : U_i \to TU_i, \quad p \mapsto (p, X_i(p)), \quad G_i = \frac{X_i}{\|X_i\|} : \partial D_i \to \mathbb{S}^{d-1} \subset \mathbb{R}^d.$$

Since the maps $X_i : \partial D_i \to \mathbb{R}^d$ and $G_i : \partial D_i \to \mathbb{R}^d$ are homotopic, we find

$$\int_{D_i} X^* \omega = \int_{D_i} \mathrm{d}(X^* p^* \theta) = \int_{\partial D_i} X_i^* \theta$$

$$= \int_{\partial D_i} G_i^* \theta = (\deg G_i) \int_{\mathbb{S}^{d-1}} \theta = \mathrm{ind}_{p_i}(X).$$

$\square$

Corollary 28.1. *Let X and Y be vector fields with a finite number of zeros on an oriented, compact, connected manifold M. The sum of the indices of the zeros of X coincides with the sum of the indices of the zeros of Y.*

Theorem 28.4 (Poincaré–Hopf). *Let M be an oriented, compact, connected manifold of dimension d. Then for any vector field $X \in \mathfrak{X}(M)$ with a finite number of zeros $\{p_1, \ldots, p_N\}$, one has*

$$\chi(M) = \sum_{i=1}^{N} \mathrm{ind}_{p_i} X.$$

In particular, $e(TM) = \chi(M)\mu$, where $\mu \in H^d(M)$ is the orientation class.

Remark 28.2. As we remarked before, there exist vector bundles with $e(\xi) = 0$, but where every section has a zero. However, in the case of the tangent bundle one can show that $e(TM) = 0$ if and only if there exists a non-vanishing vector field in M — see the Exercises at the end of this lecture. This result admits a "dual result" due to Thurston: a compact, oriented manifold admits a codimension 1 foliation if and only if $e(TM) = 0$. Thurston's Theorem is much harder to prove.

Proof. By the corollary above, it is enough to construct a vector field X in M, with a finite number of zeros, for which the equality holds. For that, we fix a triangulation $\{\sigma_1, \ldots, \sigma_l\}$ of M, and we construct a vector field X with the following properties:

(a) X has exactly one zero p_i in each face of the triangulation.
(b) The zero p_i is non-degenerate and $\mathrm{ind}_{p_i} X = (-1)^k$, where k is the dimension of the face containing p_i.

Hence, if r_k is the number of faces of dimension k, we have

$$\sum_{i=1}^{N} \mathrm{ind}_{p_i} X = r_0 - r_1 + \cdots + (-1)^d r_d,$$

so the result follows from Euler's Formula — see Theorem 25.3. To construct X we describe its phase portrait in each face of the triangulation:

- In each face of dimension 0, the vector field X has a zero.
- In each face of dimension 1, we put a zero in the center of the face and connect it by orbits to the zeros in the vertices.
- In each face of dimension 2, we put a zero in the center of the face and connect it by heteroclinic orbits to the zeros in the faces of dimension 1 (see Figure 28.1).
- Then, we complete the phase portrait of X in the face of dimension 2, so that the zero in its interior becomes an attractor of the vector field restricted to the face (see Figure 28.2).
- In general, once one has constructed the phase portrait in the faces of dimension $k - 1$, we construct the phase portrait in a face of dimension k, putting a zero in the center of the face and connecting it by heteroclinic orbits to the zeros in the faces of dimension $k-1$. We then complete the phase portrait so that the new zero is an attractor of the vector field restricted to the face of dimension k.

The vector field one constructs in this way has exactly one zero in each face. Moreover, we can assume that they are non-degenerate

Fig. 28.1.   Phase portrait in 2-dimensional faces.

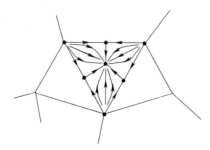

Fig. 28.2.   Phase portrait in 2-dimensional faces.

zeros. For a zero p_i in the face of dimension k, the linearization of the vector field at p_i is a real matrix with k eigenvalues with negative real part, corresponding to the directions along the face, and $n - k$ eigenvalues with positive real part, corresponding to the directions normal to the face. The sign of the determinant of this matrix is $(-1)^k$. Hence, we have that

$$\mathrm{ind}_{p_i} X = (-1)^k.$$

Hence, the vector field X satisfies (a) and (b) and this completes the proof of the Poincaré–Hopf Theorem.   □

Exercises

Exercise 28.1
Prove the properties of fiber integration given in Proposition 28.3.

Exercise 28.2
Let $E_1 \to M$ and $E_2 \to M$ be oriented vector bundles over a compact manifold M. Consider their Whitney sum with the direct sum of the orientations. Denoting the projections:

show that the Thom classes of E_1, E_2 and $E_1 \oplus E_2$ are related by

$$U_{E_1 \oplus E_2} = \pi_1^* U_{E_1} \cup \pi_2^* U_{E_2}.$$

Use this property to prove that

$$e(\xi \oplus \xi') = e(\xi) \cup e(\xi').$$

Exercise 28.3

Let $\xi = (\pi, E, M)$ and $\eta = (\tau, F, N)$ be oriented vector bundles of rank r over compact manifolds M and N. If $\Psi : \eta \to \xi$ is a morphism of vector bundles covering a map $\psi : N \to M$, which preserves orientations and is a fiberwise isomorphism, show that

$$e(\eta) = \psi^* e(\xi).$$

Use this property to conclude that

(a) $e(\bar{\xi}) = -e(\xi)$, where $\bar{\xi}$ denotes the vector bundle ξ with the opposite orientation.
(b) $e(\xi) = 0$ whenever rank ξ is odd.

Exercise 28.4

Let $M = \mathbb{CP}^1 \simeq \mathbb{S}^2$ embedded in $\mathbb{CP}^2$ as the submanifold:

$$\mathbb{CP}^1 \hookrightarrow \mathbb{CP}^2, \quad [x : y] \mapsto [x : y : 0].$$

Find the Euler class of the normal bundle $\nu(\mathbb{CP}^1)$ and conclude that this vector bundle is non-trivial.

Exercise 28.5

Consider the canonical complex line bundle $\gamma_d^1(\mathbb{C})$ over $\mathbb{CP}^d$, defined analogously to the canonical real line bundle γ_d^1 over $\mathbb{RP}^d$ — see Exercise 27.2. Show that it is orientable and that is Euler class is non-trivial.

Exercise 28.6

Let M be a compact manifold of dimension d. One can show that

(a) If $p_1, \ldots, p_N \in M$ there exists an open set $U \subset M$, diffeomorphic to the ball $\{x \in \mathbb{R}^d : ||x|| < 1\}$, such that $p_1, \ldots, p_n \in U$.

(b) If $\psi : \mathbb{S}^{d-1} \to \mathbb{S}^{d-1}$ is a map with degree zero, then it is homotopic to the constant map.

Use these facts to show that if $\chi(M) = 0$, then there exists a nowhere vanishing vector field in M.

Exercise 28.7

Sketch the portrait of a vector field, with a finite number of zeros, of your choice on a compact, oriented, surface Σ_g of genus g and use it to find $\chi(\Sigma_g)$.

Lecture 29

Pullbacks of Vector Bundles

The following *pullback* construction for vector bundles plays a crucial role.

Definition 29.1. Let $\psi : M \to N$ be a smooth map and $\xi = (\pi, E, N)$ a vector bundle over N of rank r. The **pullback** of ξ by ψ is the vector bundle $\psi^*\xi = (\hat{\pi}, \psi^*E, M)$ of rank r, with total space

$$\psi^*E := \{(p, \mathbf{v}) \in M \times E : \psi(p) = \pi(\mathbf{v})\},$$

and projection $\hat{\pi} : \psi^*E \to M$, $(p, \mathbf{v}) \mapsto p$.

Note that in a pullback vector bundle $\psi^*\xi$ for each point in the preimage $\psi^{-1}(q)$ one takes a copy of the fiber of ξ over q.

We still need to check that the construction in the definition above does indeed produce a vector bundle. First of all, note that

$$\psi^*E = (\psi \times \pi)^{-1}(\Delta),$$

where $\Delta \subset N \times N$ is the diagonal. Since $\pi : E \to N$ is a submersion, we have that $(\psi \times \pi) \pitchfork \Delta$, so $\psi^*E \subset M \times E$ is a submanifold. To cheek local triviality of $\psi^*\xi$, let $\{(U_\alpha, \phi_\alpha)\}$ be a trivialization of ξ. Then we obtain a trivialization $\{(\psi^{-1}(U_\alpha), \tilde{\phi}_\alpha)\}$ for $\psi^*\xi$, where

$$\tilde{\phi}_\alpha : \hat{\pi}^{-1}(\psi^{-1}(U_\alpha)) \to \psi^{-1}(U_\alpha) \times \mathbb{R}^r$$

$$(p, \mathbf{v}) \longmapsto (p, \phi_\alpha^{\psi(p)}(\mathbf{v})).$$

This proves the local triviality. Moreover, if $\{g_{\alpha\beta}\}$ is the cocycle of ξ associated with the trivialization $\{(U_\alpha, \phi_\alpha)\}$, then

$\{\psi^* g_{\alpha\beta}\} = \{g_{\alpha\beta} \circ \psi\}$ is the cocycle of $\psi^* \xi$ associated with the trivialization $\{(\psi^{-1}(U_\alpha), \tilde{\phi}_\alpha)\}$.

The map

$$\Psi : \psi^* \xi \to \xi \ (p, \mathbf{v}) \mapsto \mathbf{v},$$

is a morphism of vector bundles covering ψ. Hence, the pullback construction allows to complete the following commutative diagram of morphisms of vector bundles

$$
\begin{array}{ccc}
\psi^* E & \overset{\Psi}{- - - - \to} & E \\
{\scriptstyle \hat{\pi}} \downarrow & & \downarrow {\scriptstyle \pi} \\
M & \overset{\psi}{\longrightarrow} & N
\end{array}
$$

The following universal property characterizes the pullback up to isomorphism.

Proposition 29.1. *Let $\psi : M \to N$ be a smooth map, $\eta = (\tau, F, M)$ and $\xi = (\pi, E, N)$ vector bundles and $\Phi : \eta \to \xi$ a morphism of vector bundles covering ψ. Then there exists a unique morphism of vector bundles $\tilde{\Phi} : \eta \to \psi^* \xi$, covering the identity, which makes the following diagram commutative:*

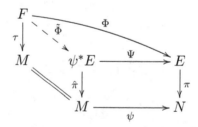

Moreover, $\tilde{\Phi}$ is a vector bundle isomorphism iff $\Phi^p : F_p \to E_{\psi(p)}$ is a linear isomorphism for all $p \in M$.

Proof. The map $\tilde{\Phi} : \eta \to \psi^* \xi$ is given by $\tilde{\Phi}(\mathbf{w}) := (\tau(\mathbf{w}), \Phi(\mathbf{w}))$. We leave the details as an (easy) exercise. $\qquad \square$

One can also **pullback morphisms** covering the identity: if $\xi = (\pi, E, N)$ and $\eta = (\tau, F, N)$ are vector bundles and $\Phi : \xi \to \eta$

is a morphism covering the identity, then for any smooth map $\psi : M \to N$, we define a morphism of vector bundles $\psi^*(\Phi) : \psi^*\xi \to \psi^*\eta$ by

$$\psi^*(\Phi)(p, \mathbf{v}) = (p, \Phi(\mathbf{v})).$$

Obviously, this morphism makes the following diagram commute

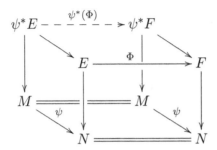

Proposition 29.2. *Let* $\psi : M \to N$ *and* $\phi : Q \to M$ *be smooth maps,* ξ, η *and* θ *vector bundles over* N, *and* $\Phi : \xi \to \eta$ *and* $\Psi : \eta \to \theta$ *morphisms of vector bundles over the identity. Then:*

  (i) $\psi^*(\mathrm{Id}_\xi) = \mathrm{Id}_{\psi^*\xi}$;
 (ii) $\psi^*(\Psi \circ \Phi) = \psi^*(\Psi) \circ \psi^*(\Phi)$;
(iii) $\psi^*(\varepsilon_N^r) = \varepsilon_M^r$;
(iv) $(\mathrm{Id})^*\xi = \xi$;
 (v) $(\psi \circ \phi)^*\xi = \phi^*(\psi^*\xi)$.

Remark 29.1. Some of the equalities in the previous proposition are actually isomorphisms. However, they are canonical, i.e., they do not depend on any choices. So we still use the symbol "=" instead of "≃" to ease the notation. This same remark applies to many of the "equalities" that follow.

    The previous result shows that if we fix a smooth map $\psi : M \to N$, then:

- Pullback defines a covariant functor from the category of vector bundles over N to the category of vector bundles over M.

On the other hand, if we denote by $\mathrm{Vect}_r(M)$ the set of isomorphism classes of vector bundles of rank r over a manifold M, there is a distinguished point in $\mathrm{Vect}_r(M)$, namely the class of the trivial vector bundle. Given a smooth map $\psi : M \to N$, pullback $\psi^* : \mathrm{Vect}_r(N) \to \mathrm{Vect}_r(M)$ preserves this distinguished point, so we also have:

- Pullback defines a contravariant functor from the category of smooth manifolds to the category of sets with a distinguished point.

All the functorial constructions with vector bundles are preserved under pullbacks. For example, one finds that

(i) $\psi^*(\xi \oplus \eta) = \psi^*\xi \oplus \psi^*\eta$;
(ii) $\psi^*(\xi^*) = (\psi^*\xi)^*$;
(iii) $\psi^*(\wedge^k \xi) = \wedge^k \psi^*\xi$.

One can also commute the operations of restriction and pullback, under a transversality assumption. Namely, if the map $\psi : M \to N$ is transverse to the submanifold $Q \subset N$, so that $\psi^{-1}(Q) \subset M$ is a submanifold, then

$$\psi^*(\xi|_Q) = \psi^*(\xi)|_{\psi^{-1}(Q)}.$$

There is also an operation of **pullback of sections**, taking sections of a vector bundle $\xi = (\pi, E, N)$ to sections of the pullback $\psi^*\xi = (\hat\pi, \psi^*E, M)$, namely

with $\psi^*s(p) := (p, s(\phi(p)))$.

In particular, if rank $\xi = r$ then $\psi^* \wedge^r \xi = \wedge^r \psi^*\xi$, and the pullback of a non-vanishing section of $\wedge^r \xi$ is a non-vanishing section of $\wedge^r \psi^*\xi$. Hence, the pullback $\psi^*\xi$ of an oriented vector bundle ξ has a natural **pullback orientation**.

Proposition 29.3. *Let* $\psi : M \to N$ *be a smooth map and let* $\xi = (\pi, E, N)$ *be an oriented vector bundle. For any form* $\omega \in \Omega^\bullet_{cv}(E)$,

$$\hat\pi_* \Psi^* \omega = \psi^* \pi_* \omega,$$

where $\psi^*\xi = (\hat\pi, \psi^*E, M)$ *has the pullback orientation and* $\Psi : \psi^*\xi \to \xi$ *is the canonical vector bundle map covering* ψ.

We leave the proof as an exercise. Another fundamental property of the pullback of vector bundles is the following.

Theorem 29.1 (Homotopy invariance). *If $\psi, \phi : M \to N$ are homotopic maps and ξ is a vector bundle over N, then $\psi^*\xi$ and $\phi^*\xi$ are isomorphic vector bundles.*

Proof. Let $H : M \times [0,1] \to N$ be a homotopy between ϕ and ψ. Then,

$$\phi^*\xi = H_0^*\xi = H^*\xi|_{M \times \{0\}},$$
$$\psi^*\xi = H_1^*\xi = H^*\xi|_{M \times \{1\}}.$$

Hence, it is enough to show that for any vector bundle η over $M \times [0,1]$, the restrictions $\eta|_{M \times \{0\}}$ and $\eta|_{M \times \{1\}}$ are isomorphic. Note that H is only C^0, but one can show that

(a) a vector bundle morphism of class C^0 covering a map of class C^∞ can be approximated by a morphism of classe C^∞ covering the same map.
(b) a vector bundle morphism which is close enough to an isomorphism is also an isomorphism.

Hence, it is enough to proof that for any vector bundle $\eta = (\pi, E, M \times [0,1])$, there exists a C^0-morphism of vector bundles $\Delta : \eta \to \eta$, covering the map

$$\delta : M \times [0,1] \to M \times [0,1], \quad (p,t) \mapsto (t,1),$$

and such that the induced maps in the fibers are isomorphisms. In order to construct Δ, we use the following lemma, whose proof is left as an exercise.

Lemma 29.1. *Let η be a vector bundle over $M \times [0,1]$. There exists an open cover $\{U_\alpha\}_{\alpha \in A}$ of M such that the restrictions $\eta|_{U_\alpha \times [0,1]}$ are trivial vector bundles.*

Now choose a locally finite countable open cover $\{U_k\}_{k \in \mathbb{N}}$ of M such that each $\eta|_{U_k \times [0,1]}$ is trivial, with trivialization

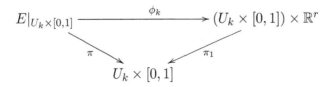

Denote by $\{\rho_k\}_{k\in\mathbb{N}}$ an *envelope of unity* subordinated to the cover $\{U_k\}_{k\in Nn}$, i.e., a collection of continuous maps $\rho_k : M \to \mathbb{R}$ such that $0 \le \rho_k \le 1$, $\operatorname{supp}\rho_k \subset U_k$ and, for all $p \in M$,

$$\max\{\rho_k(p) : k \in \mathbb{N}\} = 1.$$

Such an envelope of unity can be constructed starting with a partition of unity $\{\theta_k\}$ and defining

$$\rho_k(p) := \frac{\theta_k(p)}{\max\{\theta_k(p) : k \in \mathbb{N}\}}.$$

For each $k \in \mathbb{N}$, we define vector bundle morphisms $\Delta_k : \eta \to \eta$ by

(a) Δ_k cover the map $\delta_k : M \times [0,1] \to M \times [0,1]$ given by

$$\delta_k(p,t) = (p, \max(\rho_k(p),t));$$

(b) In $\pi^{-1}(U_k \times [0,1])$, Δ_k is defined by

$$\Delta_k(\phi_k^{-1}(p,t,\mathbf{v})) \equiv \phi_k^{-1}(p, \max(\rho_k(p),t), v),$$

and Δ_k is the identity outside $\pi^{-1}(U_k \times [0,1])$.

Finally, one defines $\Delta : \eta \to \eta$ by composition

$$\Delta = \cdots \circ \Delta_k \circ \cdots \circ \Delta_1.$$

Since each $p \in M$ has a neighborhood which intersects a finite number of open sets U_k, this is a well-defined vector bundle morphism $\Delta : \eta \to \eta$, which locally is the composition of vector bundle maps that are isomorphisms on the fibers. Hence, Δ is a vector bundle isomorphism which covers $\delta : M \times [0,1] \to M \times [0,1]$. $\qquad\square$

Corollary 29.1. *Any vector bundle over a contractible manifold is trivial.*

Proof. Let $\xi = (\pi, E, M)$ be a vector bundle and let $\phi : M \to \{*\}$ and $\psi : \{*\} \to M$ be smooth maps such that $\psi \circ \phi$ is homotopic to id_M. The theorem gives

$$\xi \simeq (\psi \circ \phi)^*\xi \simeq \phi^*(\psi^*\xi).$$

Since $\psi^*\xi$ is a vector bundle over a set which consist of a single point, it is a trivial vector bundle. Hence $\xi \simeq \phi^*(\psi^*\xi)$ is also a trivial vector bundle. □

Hence, when M is contractible the space $\text{Vect}_r(M)$ consisting of isomorphism classes of vector bundles of rank r over M has only one point.

Example 29.1. Given a line bundle $\xi = (\pi, E, \mathbb{S}^1)$, we can cover $\mathbb{S}^1$ by the two contractible open sets $U = \mathbb{S}^1 - \{p_N\}$ and $V = \mathbb{S}^1 - \{p_S\}$. By the corollary, over each open set U and V the vector bundle trivializes: $\phi_U : E|_U \simeq U \times \mathbb{R}$ and $\phi_V : E|_V \simeq V \times \mathbb{R}$. Therefore, the line bundle is completely characterized by the transition function $g_{UV} : U \cap V \to \mathbb{R}$, so that

$$\phi_V \circ \phi_U^{-1} : U \times \mathbb{R} \to V \times \mathbb{R}, \ (p, v) \mapsto (p, g_{UV}(p)v).$$

The intersection $U \cap V$ has two connected components, and we leave it as an exercise to check that if $g_{UV}(x)$ has the same sign in both components, then ξ is trivial, while if $g_{UV}(x)$ has the opposite signs in the two components then the line bundle is isomorphic to the line bundle whose total space is the Möbius band. In other words, the space $\text{Vect}_1(\mathbb{S}^1)$ consisting of isomorphism classes of line bundles over $\mathbb{S}^1$ has two elements.

This allows is to define the **first Stiefel–Whitney class** $w_1(\xi) \in H^1(M, \mathbb{Z}_2)$ of a line bundle ξ over a manifold M. Namely, if $[\gamma] \in H_1(M, \mathbb{Z})$ is represented by a loop $\gamma : \mathbb{S}^1 \to M$, then one sets

$$\langle \widetilde{w}_1(\xi), [\gamma] \rangle := \begin{cases} 0 & \text{if } \gamma^*\xi \text{ is trivial,} \\ 1 & \text{if } \gamma^*\xi \text{ is not trivial.} \end{cases}$$

We leave as an exercise to check that w_1 gives a bijection

$$w_1 : \text{Vect}_1(M) \simeq H^1(M, \mathbb{Z}_2).$$

so a line bundle ξ is uniquely determined, up to isomorphism, by its first Stiefel–Whitney class.

For example, one has $H^1(\mathbb{RP}^d, \mathbb{Z}_2) = \mathbb{Z}_2$. So $\text{Vect}_1(\mathbb{RP}^d)$ has two elements: the class of the trivial bundle and the class of the canonical line bundle γ_d^1.

Exercises

Exercise 29.1

Give a proof of the universal property of pullbacks in Proposition 29.1. Show that this property characterizes the pullback of vector bundles up to isomorphism.

Exercise 29.2

Verify the properties of the pullback of vector bundles given by Proposition 29.2.

Exercise 29.3

Let $\Psi : \eta \to \xi$ be a vector bundle map covering a map $\psi : M \to N$. Show that if Ψ is a fiberwise isomorphism then η is isomorphic to $\phi^*\eta$. Use this to conclude that Proposition 29.3 follows from Proposition 28.3.

Exercise 29.4

Let $\phi : M \to N$ be a submersion and denote by $\mathcal{F}$ the foliation of M by the fibers of ϕ. Show that the normal bundle $\nu(\mathcal{F})$ and the conormal bundle $\nu^*(\mathcal{F})$ are naturally isomorphic to the pullback bundles ϕ^*TN and ϕ^*T^*N, respectively.

Exercise 29.5

Let ξ be a vector bundle over $M \times [0,1]$. Show that there exists an open cover $\{U_\alpha\}_{\alpha \in A}$ of M such that the restrictions $\xi|_{U_\alpha \times [0,1]}$ are trivial.

Hint: If ξ is a vector bundle over $M \times [a,c]$ which is trivial when restricted to both $M \times [a,b]$ and $M \times [b,c]$, for some $a < b < c$, then ξ is a trivial vector bundle.

Exercise 29.6

Complete the details of Example 29.1, showing that $\mathrm{Vect}_1(\mathbb{S}^1) \simeq \mathbb{Z}_2$.

Exercise 29.7

For a line bundle ξ denote by $w_1(\xi) \in H^1(M, \mathbb{Z}_2)$ its first Stiefel–Whitney class.

(a) Given a class $c \in H^1(M, \mathbb{Z}_2)$ show that there exists a line bundle ξ whose first Stiefel–Whitney class is $w_1(\xi) = c$.

(b) Conclude that $w_1 : \text{Vect}_1(M) \to H^1(M, \mathbb{Z}_2)$ is a bijection.

(c) Show that the tensor product makes $\text{Vect}_1(M)$ into a group. What is the group structure induced on $H^1(M, \mathbb{Z}_2)$?

Exercise 29.8

Let Γ be a discrete group that acts properly and free on a manifold N and by linear transformations on $\mathbb{R}$. Then Γ acts diagonally on the trivial line bundle $\text{pr}_1 : N \times \mathbb{R} \to N$. Show that the induced map between the quotients

$$p : (N \times \mathbb{R})/\Gamma \to N/\Gamma,$$

defines a line bundle. Conversely, show that every line bundle over M is the quotient of a trivial line over the universal covering space $\widetilde{M}$.

Exercise 29.9

Let $\xi = (\pi, E, M)$ be a vector bundle and let $\phi : N \to M$ be a smooth map. Show that the Euler classes of ξ and $\phi^*\xi$ are related by

$$\phi^* e(\xi) = e(\phi^* \xi).$$

Exercise 29.10

Denote by $\text{Pic}(M)$ the space of isomorphism classes of complex line bundles over a manifold M. Show that the tensor product turns $\text{Pic}(M)$ into a group, called the **Picard group** of M. Find $\text{Pic}(\mathbb{S}^1)$.

Lecture 30

The Classification of Vector Bundles

The problem of determining $\mathrm{Vect}_k(M)$ can be reduced to a problem in homotopy theory. We will only sketch this briefly since this topic belongs to the realm of algebraic topology. For a detailed discussion, we refer, e.g., to Husemoller (1994) and Milnor and Stasheff (1974).

Recall from Exercise 27.2 that γ_n^r denotes the **canonical bundle** over the Grassmannian $G_r(\mathbb{R}^n)$. It has total space and projection

$$E := \{(S, x) : S \subset \mathbb{R}^n \text{ is } r\text{-dimensional subspace and } x \in S\},$$

$$\pi(S, x) := S.$$

This is a rank r subbundle of the trivial vector bundle $\varepsilon_{G_r(\mathbb{R}^n)}^n$.

Over the Grassmannian $G_{n-r}(\mathbb{R}^n)$ there is another rank r vector bundle, called the **universal quotient bundle** and denoted η_n^r, which can be defined as follows. It has total space

$$F := \{(S, x + S) : S \subset \mathbb{R}^n \text{ is } (n-r)\text{-dimensional subspace and } x \in \mathbb{R}^n\},$$

and projection $\pi : F \to G_{n-r}(\mathbb{R}^n)$, $(S, x + S) \mapsto S$. In other words, the fiber over S is the normal space $\mathbb{R}^n / S$.

These vector bundles are related via the short exact sequence of vector bundles

$$0 \longrightarrow \gamma_n^{n-r} \longrightarrow \varepsilon_{G_{n-r}(\mathbb{R}^n)}^n \longrightarrow \eta_n^r \longrightarrow 0$$

where the last map is the projection $(S, x) \mapsto (S, x + S)$. In particular, choosing n global sections of the trivial bundle $\varepsilon^n_{G_{n-r}(\mathbb{R}^n)}$ yields n global sections of η^r_n which at each point S generate the fiber $\mathbb{R}^n/S$.

The reason for the name *universal* is justified by the following proposition.

Proposition 30.1. *Let ξ be a rank r vector bundle over a manifold M. If ξ admits n global sections $s_1, \ldots, s_n$ which generate E_p for all $p \in M$, then there exists a smooth map $\psi : M \to G_{n-r}(\mathbb{R}^n)$ such that*

$$\xi \simeq \psi^*(\eta^r_n).$$

Proof. Let

$$V := \bigoplus_{i=1}^n \mathbb{R}s_i \simeq \mathbb{R}^n.$$

By assumption, the sections s_i generate E_p, for each $p \in M$, so evaluation of these sections at p give a linear surjective map

$$V \xrightarrow{\ \mathrm{ev}_p\ } E_p \longrightarrow 0 .$$

The kernel of this map is a subspace of V of codimension r.

Define a smooth map:

$$\psi : M \to G_{n-r}(V), \quad p \mapsto \mathrm{Ker}\,\mathrm{ev}_p.$$

Then we have a vector bundle map:

$$\xi \mapsto \psi^*\eta^r_n, \quad \mathbf{v} \mapsto \left(\pi(\mathbf{v}), \mathrm{ev}^{-1}_{\pi(\mathbf{v})}(\mathbf{v}) \right).$$

This is a fiberwise isomorphism covering the identity, so it is a vector bundle isomorphism. Choosing a basis for V, one obtains the desired map $\psi : M \to G_{n-r}(\mathbb{R}^n)$. □

A map $\psi : M \to G_{n-r}(\mathbb{R}^n)$ such that $\xi \simeq \psi^*\eta^r_n$ is called a **classifying map** for the vector bundle ξ. We leave as an exercise to check that any such classifying map arises from the choice of n global sections $s_1, \ldots, s_n \in \Gamma(\xi)$ generating each fiber E_p, as in the previous proof.

The next result shows that a rank r vector bundle over a manifold M of finite type always admit a classifying map $\psi : M \to G_{n-r}(\mathbb{R}^n)$, provide one takes n sufficient large.

Proposition 30.2. *Let ξ be a rank r vector bundle over a manifold M. If M admits a finite good cover with k open sets, then for $n \geq rk$:*

(i) *There exist classifying maps $\psi : M \to G_{n-r}(\mathbb{R}^n)$ for ξ.*
(ii) *Any two classifying maps are homotopic.*

Proof. (i) We claim that ξ admits global sections $s_1, \ldots, s_n$ which generate E_p, for all $p \in M$, so (i) follows from Proposition 30.1. To see this, let $U_1, \ldots, U_k$ be a finite good cover of M. Since each U_i is contractible, the restriction $\xi|_{U_i}$ is trivial. Hence, we can choose a basis of local sections $\{s_1^i, \ldots, s_r^i\}$ for $\Gamma(\xi|_{U_i})$. Note that there are open sets $V_1, \ldots, V_k$, with $\overline{V}_i \subset U_i$ which still cover M. If we choose smooth functions $f_i : U_i \to \mathbb{R}$ such that $f_i|_{V_i} = 1$ and $f_i = 0$ outside U_i, then $\{f_i s_1^i, \ldots, f_i s_r^i : i = 1, \ldots, k\}$ are the desired global sections.

(ii) Let $\psi : M \to G_{n-r}(V)$ and $\psi : M \to G_{n-r}(V')$ be two classifying maps constructed from two choices of global sections $\{s_1, \ldots, s_n\}$ and $\{s_1', \ldots, s_n'\}$, as in the proof of the previous proposition. Then we have a canonical identification between V and V' and also between $G_{n-r}(V)$ and $G_{n-r}(V')$. It follows that the classifying map is well-defined up to a choice of identification $V \simeq \mathbb{R}^n$. If we fix this choice, then we conclude that two classifying maps $\psi : M \to G_{n-r}(\mathbb{R}^n)$ and $\psi : M \to G_{n-r}(\mathbb{R}^n)$ differ by the action of an element $A \in GL(n)$, i.e., $\psi' = A \circ \psi$, where A is the matrix relating the two bases.

Note that the order of the basis is irrelevant to construct the classifying map, so we can assume that A has a positive determinant. Since $GL^+(n)$ is connected, we can choose a continuous path $A_t \in GL^+(n)$ with $A_1 = A$ and $A_0 = I$, so that the map

$$\psi_t := A_t \circ \psi : M \to G_{n-r}(\mathbb{R}^n),$$

is a homotopy between ψ and ψ'. $\square$

Denote by $[M, N]$ the set of homotopy classes of maps $\phi : M \to N$. The classification of vector bundles mentioned before can be stated as follows.

Theorem 30.1 (Classification of vector bundles). *Let M be a manifold which admits a good open cover with k open sets. For every $n \geq rk$, there exists a bijection*

$$\mathrm{Vect}_r(M) \simeq [M, G_{n-r}(\mathbb{R}^n)].$$

Proof. We saw above that the homotopy class of a classifying map for ξ is determined by the isomorphism class of ξ, so we have a well-defined map

$$\mathrm{Vect}_r(M) \to [M : G_{n-r}(\mathbb{R}^n)],$$

which to an isomorphism class of ξ associates the homotopy class of its classifying map.

On the other hand, the homotopy invariance of the pullbacks implies that pullback of the universal bundle yields a map

$$[M : G_{n-r}(\mathbb{R}^n)] \to \mathrm{Vect}_r(M), \quad \psi \mapsto \psi^* \eta_n^r.$$

We leave as an exercise to show that these maps are inverse to each other, so the result follows. $\qquad\square$

This result reduces the classification of vector bundles to a homotopy problem. We illustrate this in the next example, which assumes some knowledge of homotopy theory.

Example 30.1. Recall that if X is a path connected topological space then the free homotopies and the homotopies based at $x_0 \in X$ are related by

$$\pi_k(X, x)/\pi_1(X, x) \simeq [\mathbb{S}^k, X],$$

where the left-hand side is the orbit space for the natural action of $\pi_1(X, x)$ in $\pi_k(X, x)$. Therefore, we have

$$\mathrm{Vect}_r(\mathbb{S}^k) = [\mathbb{S}^k, G_{n-r}(\mathbb{R}^n)] \simeq \pi_k(G_{n-r}(\mathbb{R}^n))/\pi_1(G_{n-r}(\mathbb{R}^n)),$$

for n large enough. On the other, since the Grassmannian can be described as the homogeneous space

$$G_{n-r}(\mathbb{R}^n) = O(n)/(O(n-r) \times O(r)),$$

and $\pi_k(O(n)/O(n-r)) = 0$, if n is large enough, the long exact sequence in homotopy yields

$$\pi_k(G_{n-r}(\mathbb{R}^n)) = \pi_{k-1}(O(r)).$$

Hence, we conclude that

$$\text{Vect}_r(\mathbb{S}^k) = \pi_{k-1}(O(r))/\pi_0(O(r)) = \pi_{k-1}(O(r))/\mathbb{Z}_2.$$

In order to understand this quotient, one needs to figure out the action of $\pi_0(O(r))$ on $\pi_{k-1}(O(r))$. If $g \in O(r)$, the action by conjugation $i_g : O(r) \to O(r)$, $i_g(h) = ghg^{-1}$, induces an action in homotopy

$$(i_g)_* : \pi_{k-1}(O(r)) \to \pi_{k-1}(O(r)).$$

If g_1 and g_2 belong to the same connected component, then $(i_{g_1})_* = (i_{g_2})_*$. Hence, we obtain an action of $\pi_0(O(r)) = \mathbb{Z}_2$ on $\pi_{k-1}(O(r))$, which is precisely the action above.

For example, if r is odd then $-I$ represents the non-trivial class in $\pi_0(O_r)$. Since the action by conjugation of $-I$ is trivial, we conclude that

$$\text{Vect}_r(\mathbb{S}^k) = \pi_{k-1}(O(r)), \quad \text{if } r \text{ is odd}.$$

For instance, we have

$$\text{Vect}_3(\mathbb{S}^4) = \pi_3(SO(3)) = \pi_3(\mathbb{S}^3) = \mathbb{Z}.$$

On the other hand, when r is even, the action maybe non-trivial. Take for instance $r = 2$, so we have $\pi_1(O(2)) = \mathbb{Z}$. The action of $\pi_0(O_2) = \mathbb{Z}_2$ in $\mathbb{Z}$ is just $\pm 1 \cdot n = \pm n$. Hence, we have

$$\text{Vect}_2(\mathbb{S}^k) = \pi_{k-1}(O(2))/\mathbb{Z}_2 = \pi_{k-1}(\mathbb{S}^1)/\mathbb{Z}_2 = \begin{cases} \mathbb{Z}/\mathbb{Z}_2 & \text{if } k = 2, \\ 0 & \text{if } k \geq 3. \end{cases}$$

Remark 30.1. If a manifold is not of finite type, there still exists a classification of vector bundles over M. In this case, we need to consider the space

$$\mathbb{R}^\infty = \bigoplus_{d=0}^{\infty} \mathbb{R}^d,$$

which is the direct limit of the increasing sequence of vector spaces

$$\cdots \subset \mathbb{R}^d \subset \mathbb{R}^{d+1} \subset \mathbb{R}^{d+2} \subset \cdots$$

This is an example of a so-called *profinite manifold*, a class of infinite-dimensional manifolds sharing many properties with the class of finite-dimensional manifolds.

In $\mathbb{R}^\infty$ we can still consider the Grassmannian

$$\tilde{G}_r(\mathbb{R}^\infty) = G_{\infty-r}(\mathbb{R}^\infty)$$

$$= \{S \subset \mathbb{R}^\infty : \text{linear subspace of codimension } r\}.$$

Over this infinite-dimensional Grassmannian there is a tautological vector bundle $\eta_\infty^r = (\pi, E, \tilde{G}_r(\mathbb{R}^\infty))$, called the **universal bundle of rank** r. It has total space

$$E = \{(S, x) : S \subset \mathbb{R}^\infty \text{ subspace of codimension } r, \; x \in \mathbb{R}^\infty/S\},$$

and projection

$$\pi : E \to \tilde{G}_r(\mathbb{R}^\infty), \; (S, x) \mapsto S.$$

One can show that every vector bundle of rank r over a manifold M is isomorphic to a pullback $\psi^* \eta_\infty^r$ for some classifying map

$$\psi : M \to \tilde{G}_r(\mathbb{R}^\infty).$$

Has before, any two classifying maps are homotopic and one obtains for *any* manifold M a bijection

$$\text{Vect}_r(M) \simeq [M, \tilde{G}_r(\mathbb{R}^\infty)].$$

This approach, via infinite-dimensional Grassmanian, has the advantage of avoiding any reference to "large enough n", as we did before in the case of a manifold of finite type. On the other had, it forces one to deal with vector bundles over infinite-dimensional manifolds.

Exercises

Exercise 30.1
Let $\xi = (\pi, E, M)$ be a vector bundle and $N \subset M$ a closed submanifold. Show that every section $s : N \to E$ over N, admits an extension to a section $\tilde{s} : U \to E$ defined over an open set $U \supset N$.

Exercise 30.2
Let $\psi : M \to G_{n-r}(\mathbb{R}^n)$ be a classifying map for a vector bundle $\xi = (\pi, E, M)$. Show that ψ is obtained from the choice of n global

sections $s'_1, \ldots, s'_n \in \Gamma(\xi)$ generating each fiber E_p, as in the proof of Proposition 30.1.

Exercise 30.3

Let M admit a finite good cover with k open sets and let $n \geq kr$. Show that the map

$$\mathrm{Vect}_r(M) \to [M : G_{n-r}(\mathbb{R}^n)], \quad [\xi] \mapsto f,$$

associating to an isomorphism class of a vector bundle ξ the homotopy class of a classifying map f, and the map

$$[M : G_{n-r}(\mathbb{R}^n)] \to \mathrm{Vect}_r(M), \quad \psi \mapsto \psi^* \eta^r_n,$$

are inverse to each other.

Exercise 30.4

Determine $\mathrm{Vect}_r(\mathbb{S}^1)$, $\mathrm{Vect}_r(\mathbb{S}^2)$ and $\mathrm{Vect}_r(\mathbb{S}^3)$. Give representatives for each equivalence class of vector bundles.

Lecture 31

Connections and Parallel Transport

In general, there is no natural way to differentiate sections of a vector bundle. The reason is that there is no canonical way of comparing fibers of a vector bundle over different points of the base. We need to fix some auxiliary structure on the vector bundle, and this is provided by the notion of connection.

Definition 31.1. A **connection on a vector bundle** $\xi = (\pi, E, M)$ is a map

$$\nabla : \mathfrak{X}(M) \times \Gamma(E) \to \Gamma(E), \ (X, s) \mapsto \nabla_X s,$$

which for any $f \in C^\infty(M)$, $X, X_i \in \mathfrak{X}(M)$ and $s, s_i \in \Gamma(E)$ satisfies:

(i) $\nabla_{X_1+X_2} s = \nabla_{X_1} s + \nabla_{X_2} s$;
(ii) $\nabla_X(s_1 + s_2) = \nabla_X s_1 + \nabla_X s_2$;
(iii) $\nabla_{fX} s = f \nabla_X s$;
(iv) $\nabla_X(fs) = f \nabla_X s + X(f)s$.

Properties (iii) and (iv) show that a connection ∇ can be restricted to any open set $U \subset M$, yielding a connection on $\xi|_U$. On the other hand, the map $X \mapsto \nabla_X$ is $C^\infty(M)$-linear, hence, for any section s defined in a neighborhood U of $p \in M$ and any $\mathbf{v} \in T_p M$, we can define

$$\nabla_{\mathbf{v}} s := \nabla_X s(p) \in E_p,$$

where X is any vector field defined in a neighborhood of p such that $X_p = \mathbf{v}$. Note, however, that $\nabla_{\mathbf{v}} s$ depends on the values of s in a neighborhood of p.

Let $U \subset M$ be a trivializing open set for ξ, so we can choose a basis of sections $\{s_1, \ldots, s_r\}$ for $\xi|_U$. Given any section $s \in \Gamma(\xi)$ we have that

$$s|_U = \sum_{a=1}^{r} f^a s_a,$$

for unique smooth functions $f^a \in C^\infty(U)$. By the properties above, the connection ∇ is then completely determined on the open set U by its effect on the sections s_a, namely, for any vector field $X \in \mathfrak{X}(M)$ we have

$$(\nabla_X s)|_U = \sum_{a=1}^{r} (f^a \nabla_X s_a + X(f^a) s_a).$$

By properties (i) and (iii), there exists a unique $\omega_a^b \in \Omega^1(U)$ expressing the local section $\nabla_X s_a$ in terms of the local basis as

$$\nabla_X s_a = \sum_{b=1}^{r} \omega_a^b(X) s_b.$$

One calls the matrix of 1-forms $\omega = [\omega_b^a]$ the **connection 1-form**. It determines completely the connection on U by the formula

$$(\nabla_X s)|_U = \sum_{a=1}^{r} \left(\sum_{b=1}^{r} f^b \omega_b^a(X) + X(f^a) \right) s_a.$$

The dependence of the connection 1-form on the choice of trivializing sections is discussed in an exercise at the end of this lecture.

Assume, additionally, that U is the domain of a chart $(x^1, \ldots, x^d)$. Then there exists unique functions $\Gamma_{ia}^b \in C^\infty(U)$ such that

$$\nabla_{\frac{\partial}{\partial x^i}} s_a = \sum_{b=1}^{r} \Gamma_{ia}^b s_b.$$

The functions Γ_{ia}^b are called the **Christoffel symbols** of the connection relative to the coordinate systems and basis of local sections. They are related to the connection 1-form by

$$\omega_b^a = \sum_{i=1}^{r} \Gamma_{ib}^a dx^i.$$

If we write $X = \sum_{i=1}^{d} X^i \frac{\partial}{\partial x^i}$, then the local form for the connection becomes

$$(\nabla_X s)|_U = \sum_{a=1}^{r} \sum_{i=1}^{d} \left(\sum_{b=1}^{r} f^b X^i \Gamma_{ib}^a + X^i \frac{\partial f^a}{\partial x^i} \right) s_a.$$

Example 31.1. Recall that the vector bundle $\xi = (\pi, E, M)$ of rank r is trivial if and only if it admits a basis of global sections $\{s_1, \ldots, s_r\}$. For such a trivial vector bundle, each choice of basis determines a unique connection on ξ satisfying

$$\nabla_X s_a := 0, \quad (a = 1, \ldots, r).$$

In other words, the connection 1-form relative to this basis vanishes. Note that this connection depends on the choice of basis.

The collection of all connections on a fixed vector bundle ξ has an affine structure. If $\rho \in C^\infty(M)$ is any smooth function, ∇_1 and ∇_2 are connections, then the affine combination

$$\rho \nabla_1 + (1 - \rho) \nabla_2,$$

also defines a connection on ξ. This fact that allows us to show that connections always exist.

Proposition 31.1. *Every vector bundle $\xi = (\pi, E, M)$ admits a connection.*

Proof. Let $\{U_\alpha\}$ be an open cover of M by trivializing open sets of ξ. The previous example shows that in each U_α we can choose a connection ∇^α. We define a connection ∇ in M by "gluing" these connections: if $\{\rho_\alpha\}$ is a partition of unity subordinated to the cover $\{U_\alpha\}$, then

$$\nabla := \sum_\alpha \rho_\alpha \nabla^\alpha$$

defines a connection on ξ. □

If one equips vector bundles with connections, the usual constructions lead to vector bundles with connections. The proof is left as an exercise.

Proposition 31.2. *Let ξ and ξ' be vector bundles over M, furnished with connections ∇ and ∇'. Then the associated bundles $\xi \oplus \xi'$, ξ^* and $\wedge^k \xi$, have induced connections which are uniquely determined by requiring*

$$\nabla_X(s_1 \oplus s_2) = \nabla_X s_1 \oplus \nabla'_X s_2,$$

$$\nabla_X(s_1 \wedge \cdots \wedge s_k) = \nabla_X s_1 \wedge \cdots \wedge s_k + \cdots + s_1 \wedge \cdots \wedge \nabla_X s_k$$

$$X(\langle s, \eta \rangle) = \langle \nabla_X s, \eta \rangle + \langle s, \nabla_X \eta \rangle.$$

If $\psi : N \to M$ is a smooth map, then $\psi^* \xi$ has a unique connection such that

$$(\nabla_X \psi^* s)(p) = (p, \nabla_{\mathrm{d}_p \psi(X_p)} s), \quad \forall p \in N, s \in \Gamma(\xi).$$

Connections can be used to compare different fibers of a vector bundle. Let $\xi = (\pi, E, M)$ be a vector bundle with a connection ∇. If $c : [0,1] \to M$ is a smooth curve then the pullback bundle $c^* \xi$ has an induced connection which we still denote by ∇. Note that a section s of the bundle $c^* \xi$ is just a section of ξ along c, i.e., a smooth map $s : [0,1] \to E$ such that $\pi(s(t)) = c(t)$, for all $t \in [0,1]$.

Definition 31.2. The **covariant derivative** of a section s along a curve c is the section $D_c s$ along c given by

$$D_c s := \nabla_{\frac{\mathrm{d}}{\mathrm{d}t}} s.$$

A section s along c is called a **parallel section** if it has vanishing covariant derivative: $D_c s = 0$

The operation of covariant derivative enjoys the following properties:

(i) $D_c(s_1 + s_2) = D_c s_1 + D_c s_2$;
(ii) $D_c(fs) = (f \circ c) D_c s + \mathrm{d}f(\dot{c})s$.

One can also express it in a local chart (U, x^i) admitting trivializing sections $\{s_a\}$. Given a curve $c(t)$ in U, if we set $c^i(t) = x^i(c(t))$, we can express any section s along c as $s(t) = \sum_{a=1}^{d} v^a(t) s_a(c(t))$. Then the covariant derivative of s along c is given by

$$D_c s(t) = \sum_{a=1}^{r} \left(\frac{dv^a}{dt}(t) + \sum_{b=1}^{r} \omega_b^a(\dot{c}(t)) v^b(t) \right) s_a(c(t)), \qquad (31.1)$$

where $\omega_b^a \in \Omega^1(U)$ are the components of the connection 1-form.

Remark 31.1. One can define the covariant derivative alternatively as follows. Given a section $s(t)$ along a curve $c(t)$ one can choose a *time-dependent section* $\tilde{s}_t \in \Gamma(E)$ such that

$$\tilde{s}_t(c(t)) = s(t), \quad \forall t \in I.$$

Then the covariant derivative is given by

$$D_c s(t) := \nabla_{c(t)} \tilde{s}_t + \frac{d}{dt} \tilde{s}_t(p) \Big|_{p=c(t)}. \qquad (31.2)$$

One can show that this is independent of the choice of extension $\tilde{s}_t$, either by working in a local chart or by showing that it coincides with our first definition.

Note, in particular, that even for a constant curve $c(t) = p_0$ the covariant derivative along c may not be zero! In fact, in this case, a section along c is just a curve $s : [0,1] \to E_{p_0}$ in the fiber over p_0 and the covariant derivative is the usual derivative of this curve.

Lemma 31.1. *For any curve $c : [0,1] \to M$ and any $v_0 \in E_{c(0)}$, there exists a unique parallel section s along c with initial condition $s(0) = v_0$.*

Proof. Since an interval is contractible, the pullback bundle $c^*\xi$ is trivial. This means that we can find sections $\{s_1, \ldots, s_r\}$ along c such that any section s along c can be uniquely written as $s(t) = \sum_{a=1}^{r} v^a(t) s_a(t)$, for some smooth functions $v^a : [0,1] \to \mathbb{R}$. In particular, if we define $\omega_b^a(t)$ by

$$D_c s_b(t) = \sum_{a=1}^{r} \omega_b^a(t) s_a(t),$$

we find that

$$D_c s = \sum_{a=1}^{r} \left(\frac{dv^a}{dt}(t) + \sum_{b=1}^{r} \omega_b^a(t) v^b(t) \right) s_a(t).$$

Hence, the parallel sections along c are the solutions of the linear system of ODEs:

$$\begin{cases} \frac{dv^a}{dt}(t) = -\sum_{b=1}^{r} \omega_b^a(t) v^b(t), \\ v^a(0) = v_0^a \end{cases} \qquad (a = 1, \ldots, r).$$

Hence, the lemma follows from the well-known results about existence and uniqueness of solutions of linear ODEs with time dependent coefficients. □

Under the conditions of this lemma, we say that the vectors $s(t) \in E_{c(t)}$ are obtained by parallel transport along the curve c. We denote the operation of **parallel transport** along c by

$$\tau_t : E_{c(0)} \to E_{c(t)}, \qquad \tau_t(v_0) := s(t).$$

The next result shows that parallel transport contains all the information about the connection ∇.

Proposition 31.3. *Let $\xi = (\pi, E, M)$ be a vector bundle with a connection ∇ and let $c : [0,1] \to M$ be a smooth curve. Then,*

(i) *Parallel transport $\tau_t : E_{c(0)} \to E_{c(t)}$ along c is a linear isomorphism.*

(ii) *If $\mathbf{v} = c'(0) \in T_{c(0)} M$, then for any section $s \in \Gamma(\xi)$:*

$$\nabla_\mathbf{v} s = \lim_{t \to 0} \frac{1}{t} \left(\tau_t^{-1}(s(c(t))) - s(c(0)) \right).$$

Proof. Since the differential equation defining parallel transport is linear, it depends linearly on the initial conditions, so τ_t is linear. On the other hand, τ_t is invertible, since its inverse is parallel transport along the curve $\bar{c} : [0, t] \to M$, given by $\bar{c}(\varepsilon) = c(t - \varepsilon)$.

For the proof of (ii), first we use Lemma 31.1 to produce a family $\{s_1, \ldots, s_r\}$ consisting of parallel sections along c that for each t

generate the fiber $E_{c(t)}$. Then there are functions $v^a : [0, 1] \to \mathbb{R}$ such that

$$s(c(t)) = \sum_{a=1}^{r} v^a(t) s_a(t),$$

and we find that

$$\lim_{t \to 0} \frac{1}{t} \left(\tau_t^{-1}(s(c(t))) - s(c(0)) \right)$$

$$= \lim_{t \to 0} \sum_{a=1}^{r} \frac{1}{t} \left(v^a(t) \tau_t^{-1}(s_a(t)) - v^a(0) s_a(0) \right)$$

$$= \lim_{t \to 0} \sum_{a=1}^{r} \frac{1}{t} \left(v^a(t) - v^a(0) \right) s_a(0)$$

$$= \sum_{a=1}^{r} \frac{\mathrm{d}v^a}{\mathrm{d}t}(0) s_a(0) = D_c \left(\sum_{a=1}^{r} v^a s_a \right) (0) = \nabla_v s,$$

where in the last line we have used (31.2) and that $D_c s_a = 0$. $\qquad \square$

Consider now the tangent bundle $\xi = TM$ of a manifold M. For a connection ∇ in TM, the notions above have a more geometric meaning. For example, in $M = \mathbb{R}^d$, there is a canonical connection ∇ in $T\mathbb{R}^d = \mathbb{R}^d \times \mathbb{R}^d$, which corresponds to the usual directional derivative. A vector field X (i.e., a section of TM) is parallel for this connection along a curve $c(t)$ if and only if the components of the vectors $X_{c(t)}$ are constant.

For a connection on the tangent bundle TM, there are additional notions that do not make sense for connections on a general vector bundle. This is because a connection on TM differentiates vector fields along vector fields, so we have a more symmetric situation. Here is a first example:

Definition 31.3. Let ∇ be a connection on TM. A **geodesic** is a curve $c(t)$ for which its derivative $\dot{c}(t)$ (a vector field along $c(t)$) is parallel, i.e., that satisfies

$$D_c \dot{c}(t) = 0.$$

If we choose local coordinates $(U, x^1, \ldots, x^d)$, we have trivializing vector fields $\{\frac{\partial}{\partial x^1}, \ldots, \frac{\partial}{\partial x^d}\}$ for $TM|_U$, and we can write:

$$\nabla_{\frac{\partial}{\partial x^i}} \frac{\partial}{\partial x^j} = \sum_k \Gamma_{ij}^k \frac{\partial}{\partial x^k}.$$

A curve $c(t)$ with components $c^i(t) = x^i(c(t))$ is a geodesic if and only these satisfy

$$\frac{\mathrm{d}^2 c^k}{\mathrm{d}t^2}(t) = -\sum_{ij} \Gamma_{ij}^k(c(t)) \frac{\mathrm{d}c^i}{\mathrm{d}t}(t) \frac{\mathrm{d}c^j}{\mathrm{d}t}(t), \quad (k = 1, \ldots, n).$$

Using these equations, it should be clear that given $p_0 \in M$ and $\mathbf{v} \in T_{p_0} M$, there exists a unique geodesic $c(t)$ such that $c(0) = p_0$ and $\dot{c}(0) = \mathbf{v}$. Given $\mathbf{v}$, this geodesic is defined for small time $0 \leq t < \varepsilon$. If we choose $\mathbf{v}$ sufficiently small we can assume that $\varepsilon > 1$ and in this case, we set

$$\exp_{p_0}(\mathbf{v}) := c(1).$$

In this way, we obtain the **exponential map** $\exp_{p_0} : U \to M$, which is defined in a sufficiently small open neighborhood of the origin $U \subset T_{p_0} M$.

Another notion which only makes sense for connections ∇ in TM is the *torsion of a connection*. This is the map $T : \mathfrak{X}(M) \times \mathfrak{X}(M) \to \mathfrak{X}(M)$ defined by

$$T(X, Y) := \nabla_X Y - \nabla_Y X - [X, Y].$$

The properties of ∇ show that T is $C^\infty(M)$-linear in both arguments, so it defines a vector bundle morphism $T : TM \otimes TM \to TM$ called the **torsion tensor** of the connection. A **symmetric connection** is a connection whose torsion vanishes.

In order to give a geometric characterization of the torsion consider a smooth map $\phi : [0, 1] \times [0, 1] \to M$, which one can think as a parameterized surface. Denoting the parameters by (x, y), we have the following maps $[0, 1] \times [0, 1] \to TM$ covering ϕ

$$\frac{\partial \phi}{\partial x} := \phi_* \left(\frac{\partial}{\partial x} \right), \quad \frac{\partial \phi}{\partial y} := \phi_* \left(\frac{\partial}{\partial y} \right).$$

One may think of these maps as vector fields along ϕ. If one fixes y, they give vector fields along the curve $t \mapsto \phi(t, y)$, and similarly if one fixes x. So we may consider the covariant derivatives:

- $D_x \frac{\partial \phi}{\partial y} :=$ covariant derivative along the curve $t \mapsto \phi(t, y)$ at $t = x$;
- $D_y \frac{\partial \phi}{\partial x} :=$ covariant derivative along the curve $t \mapsto \phi(x, t)$ at $t = y$.

Proposition 31.4. *Consider a parameterized surface* $\phi : [0,1] \times [0,1] \to M$. *The torsion of a connection* ∇ *in* TM *satisfies:*

$$D_x \frac{\partial \phi}{\partial y} - D_y \frac{\partial \phi}{\partial x} = T\left(\frac{\partial \phi}{\partial x}, \frac{\partial \phi}{\partial x}\right).$$

Proof. The proof is similar (but simpler!) to the proof of Proposition 32.1, and so is left as an exercise. $\square$

The most classical example of a connection is the *Levi–Civita connection* in the tangent bundle of a Riemannian manifold, which we now describe.

Definition 31.4. Let ξ be a vector bundle over M with a fiber metric $\langle \, , \, \rangle$. A connection on ξ is said to be **compatible with the metric** if

$$\mathcal{L}_X \langle s_1, s_2 \rangle = \langle \nabla_X s_1, s_2 \rangle + \langle s_1, \nabla_X s_2 \rangle,$$

for every vector field $X \in \mathfrak{X}(M)$ and every pair of sections $s_1, s_2 \in \Gamma(\xi)$.

For a Riemannian manifold we have a natural choice of compatible metric.

Proposition 31.5. *Let* $(M, \langle \, , \, \rangle)$ *be a Riemannian manifold. There exists a unique symmetric connection on* TM *compatible with the metric.*

Proof. Let $X, Y, Z \in \mathfrak{X}(M)$ be vector fields in M. The compatibility of ∇ with the metric gives

$$\mathcal{L}_X \langle Y, Z \rangle = \langle \nabla_X Y, Z \rangle + \langle Y, \nabla_X Z \rangle,$$
$$\mathcal{L}_Y \langle Z, X \rangle = \langle \nabla_Y Z, X \rangle + \langle Z, \nabla_Y X \rangle,$$
$$\mathcal{L}_Z \langle X, Y \rangle = \langle \nabla_Z X, Y \rangle + \langle X, \nabla_Z Y \rangle.$$

Adding the first two equations and subtracting the third one, one finds

$$\mathcal{L}_X\langle Y, Z\rangle + \mathcal{L}_Y\langle Z, X\rangle - \mathcal{L}_Z\langle X, Y\rangle$$
$$= 2\langle \nabla_X Y, Z\rangle - \langle X, [Z, Y]\rangle - \langle Y, [Z, X]\rangle - \langle Z, [X, Y]\rangle,$$

where we have used the symmetry of the connection. This relation shows that the two conditions completely determine the connection by the formula

$$\langle \nabla_X Y, Z\rangle = \frac{1}{2}\left(\mathcal{L}_X\langle Y, Z\rangle + \mathcal{L}_Y\langle Z, X\rangle - \mathcal{L}_Z\langle X, Y\rangle\right)$$
$$+ \frac{1}{2}\left(\langle X, [Z, Y]\rangle + \langle Y, [Z, X]\rangle + \langle Z, [X, Y]\rangle\right).$$

On the other, one checks easily that this formula does define a connection on TM which is symmetric and compatible with the metric. $\square$

The connection in the proposition is known as the **Levi–Civita connection** of the Riemannian manifold. This allows to define parallel transport, geodesics, exponential map, etc., for a Riemannian manifold. The fact that this connection comes from a metric leads to additional properties of these concepts. We will not go into any deeper discussion of Riemannian geometry and refer the reader to any textbooks on the subject such as do Carmo (1992) or Gallot *et al.* (2004).

Exercises

Exercise 31.1
Let ξ and ξ' be vector bundles over M, furnished with connections ∇ and ∇'. Show that the associated bundles $\xi \oplus \xi'$, ξ^* and $\wedge^k \xi$ carry unique connections satisfying:

$$\nabla_X(s_1 \oplus s_2) = \nabla_X s_1 \oplus \nabla'_X s_2,$$
$$\nabla_X(s_1 \wedge \cdots \wedge s_k) = \nabla_X s_1 \wedge \cdots \wedge s_k + \cdots + s_1 \wedge \cdots \nabla_X \wedge s_k$$
$$X(\langle s, \eta\rangle) = \langle \nabla_X s, \eta\rangle + \langle s, \nabla_X \eta\rangle.$$

Determine the connection 1-form of these connections in terms of the original connection 1-forms.

Exercise 31.2

Let ξ be a vector bundle over M with a connection ∇. If $\psi : N \to M$ is a smooth map, show that $\psi^*\xi$ has a connection induced from ∇ such that

$$(\nabla_v \psi^* s) = \psi^*(\nabla_{d_p\psi(v)} s), \quad \forall v \in T_pN, s \in \Gamma(\xi).$$

Determine the connection 1-form of the pullback connection in terms of the connection 1-form of the original connection.

Exercise 31.3

Let $\{s_1, \ldots, s_r\}$ and $\{s_1', \ldots, s_r'\}$ be two bases of local sections for a vector bundle $\xi = (\pi, E, M)$ over a common open set $U \subset M$. Denote by $A = (a_i^j) : U \to \mathrm{GL}(r)$ the matrix of change of basis so that $s_i' = \sum_j a_i^j s_j$. Show that the corresponding connection 1-forms ω and ω' are related by

$$\omega' = A^{-1}\omega A + A^{-1}dA.$$

Exercise 31.4

Deduce formula (31.1) for the local expression of the covariant derivative.

Exercise 31.5

Show that the covariant derivative of a section $s(t)$ along a curve $c(t)$, as given in Definition 31.2, can be computed by choosing a time-dependent section extending s and applying formula (31.2). In particular, conclude that this formula does not depend on the choice of extension.

Exercise 31.6

Let ξ be a vector bundle over M with a fiber metric $g := \langle \, , \, \rangle$. Viewing the metric as a section $g \in \Gamma(\otimes^2 E^*)$, verify that the condition that the connection ∇ is compatible with the metric g is equivalent to

$$\nabla_X g = 0, \quad \forall X \in \mathfrak{X}(M).$$

Show that one can always find such a compatible connection ∇.

Exercise 31.7

Let $\xi = (\pi, E, M)$ be a vector bundle with a fiber metric $\langle\,,\,\rangle$. For a connection ∇ in ξ, show that the following are equivalent:

(i) ∇ is compatible with the metric.
(ii) Parallel transport $\tau_t : E_{c(0)} \to E_{c(t)}$ along any curve c is an isometry.
(iii) For any basis of orthonormal trivializing sections the connection 1-form $\omega = [\omega_a^b]$ is a skew-symmetric matrix.

Exercise 31.8

Let $M \subset \mathbb{R}^n$ be an embedded submanifold so that $T_pM \subset \mathbb{R}^n$ has the inner product induced from the standard inner product on $\mathbb{R}^n$. Show that these yield a Riemannian metric g in M, whose associated Levi–Civita connection is given by

$$(\nabla_X Y)(p) = \mathrm{pr}_{T_pM}\left(\mathrm{d}_p Y(X_p)\right),$$

where $\mathrm{pr}_{T_pM} : \mathbb{R}^n \to T_pM$ denotes the orthogonal projection and in the right-side we view $Y \in \mathfrak{X}(M)$ as a map $Y : M \to \mathbb{R}^n$.

Exercise 31.9

Let G be a connected Lie group with Lie algebra $\mathfrak{g}$. Show that there exists a unique torsion-free connection ∇ in TG, which is invariant under left and right translations, and under inversion. Show also that ∇ satisfies the following properties:

(a) For any left-invariant vector fields $X, Y \in \mathfrak{g}$

$$\nabla_X Y = \frac{1}{2}[X, Y];$$

(b) The exponential map of ∇ at the identity $\exp_e$ coincides with the Lie group exponential map $\exp : \mathfrak{g} \to G$;
(c) Parallel transport along the curve $c(t) = \exp(tX)$, $X \in \mathfrak{g}$, is given by

$$\tau_t(\mathbf{v}) = \mathrm{d}L_{\exp(\frac{t}{2}X)} \cdot \mathrm{d}R_{\exp(\frac{t}{2}X)} \cdot \mathbf{v}, \quad \forall \mathbf{v} \in T_eG;$$

(d) The geodesics are translations of the 1-parameter subgroups of G.

Exercise 31.10

A connection ∇ is called **complete** if all its geodesics $\gamma(t)$ are defined for all $t \in \mathbb{R}$. Give an example of a compact manifold M with a non-complete connection.

Note: One can show that the Levi–Civita connection of a compact Riemannian manifold is always complete.

Lecture 32

Curvature and Holonomy

A trivial vector bundle carries natural connections defined in terms of trivializing sections s_i, for which $\nabla s_i = 0$. In general, given some a vector bundle $\xi = (\pi, E, M)$ with a connection ∇, it is not possible to choose a basis of local sections s_i such that $\nabla s_i = 0$. The obstruction is given by the **curvature** of ∇, which is the map

$$R : \mathfrak{X}(M) \times \mathfrak{X}(M) \times \Gamma(\xi) \to \Gamma(\xi)$$

defined by

$$R(X,Y)s = \nabla_X(\nabla_Y s) - \nabla_Y(\nabla_X s) - \nabla_{[X,Y]} s.$$

A simple computation shows that R is $C^\infty(M)$-linear in all the arguments, so we can think of R as a vector bundle map $R : TM \otimes TM \otimes E \to E$. For this reason, one also calls R the **curvature tensor**.

The local expression for the curvature over a chart (U, x^i) admitting a basis of sections $\{s_1, \ldots, s_r\}$ for ξ is

$$R\left(\frac{\partial}{\partial x^i}, \frac{\partial}{\partial x^j}\right) s_a = \sum_{b=1}^{r} R_{ija}^b s_b,$$

where the components R_{ija}^b can be expressed in terms of the Christoffel symbols Γ_{ia}^b by

$$R_{ija}^b = \frac{\partial \Gamma_{ja}^b}{\partial x^i} - \frac{\partial \Gamma_{ia}^b}{\partial x^j} + \sum_{c=1}^{r} \left(\Gamma_{ia}^c \Gamma_{jc}^b - \Gamma_{ja}^c \Gamma_{ic}^b \right).$$

We can also codify the curvature in terms of a matrix of differential forms by setting

$$\Omega_a^b := \sum_{i<j} R_{ija}^b \mathrm{d}x^i \wedge \mathrm{d}x^j.$$

One calls $\Omega = [\Omega_a^b]$ the **curvature 2-form** of the connection. This matrix-valued 2-form is independent of the choice of local coordinates, and it can also be defined from the relation

$$R(X,Y)s_a = \sum_{b=1}^r \Omega_a^b(X,Y)s_b.$$

The dependence of Ω on the choice of trivializing sections is discussed in the Homework.

Theorem 32.1. *For a connection on a vector bundle ξ, the connection 1-form ω and the curvature 2-form Ω associated with some trivializing sections, are related by the **structure equations***

$$\Omega_a^b = \mathrm{d}\omega_a^b + \sum_c \omega_a^c \wedge \omega_c^b \quad \Longleftrightarrow \quad \Omega = \mathrm{d}\omega + \omega \wedge \omega,$$

*and one has the **Bianchi's identity***

$$\mathrm{d}\Omega_a^b = \sum_c \left(\Omega_a^c \wedge \omega_c^b - \omega_a^c \wedge \Omega_c^b \right) \quad \Longleftrightarrow \quad \mathrm{d}\Omega = \Omega \wedge \omega - \omega \wedge \Omega.$$

Proof. Direct computation. □

Let us turn now to the geometric interpretation of curvature in terms of parallel transport. For that we choose a smooth map $\phi : [0,1] \times [0,1] \to M$, as in the discussion leading to Proposition 31.4. We use the same notations as in that proposition. Given a section s of the vector bundle ξ along ϕ, we have the covariant derivatives:

- $D_x s(x,y) :=$ covariant derivative of s along the curve $t \mapsto \phi(t,y)$ at $t = x$.
- $D_y s(x,y) :=$ covariant derivative of s along the curve $t \mapsto \phi(x,t)$ at $t = y$.

We these notations, we have the following interpretation of the curvature tensor.

Proposition 32.1. *Fix a parameterized surface $\phi\colon [0,1]\times[0,1]\to M$. For any section s of ξ along ϕ, the curvature of the connection satisfies*

$$D_x D_y s - D_y D_x s = R\left(\frac{\partial\phi}{\partial x},\frac{\partial\phi}{\partial x}\right)s.$$

Proof. One can choose (x,y)-dependent vector fields $X_{x,y}, Y_{x,y}\in \mathfrak{X}(M)$ extending $\frac{\partial\phi}{\partial x}$ and $\frac{\partial\phi}{\partial y}$, i.e., satisfying

$$X_{x,y}(\phi(x,y)) = \frac{\partial\phi}{\partial x}(x,y), \quad Y_{x,y}(\phi(x,y)) = \frac{\partial\phi}{\partial y}(x,y).$$

We will need the following result whose proof we leave as an exercise:

Lemma 32.1.

$$\left(\frac{\mathrm{d}}{\mathrm{d}y}X_{x,y} - \frac{\mathrm{d}}{\mathrm{d}x}Y_{x,y}\right)\Big|_{\phi(x,y)} = [X_{x,y}, Y_{x,y}]\big|_{\phi(x,y)}.$$

We choose also a (x,y)-dependent section $s_{x,y}\in\Gamma(\xi)$ extending s, so that

$$s_{x,y}(\phi(x,y)) = s(x,y).$$

Using Remark 31.1, we can compute the covariant derivatives of s

$$D_x s(x,y) = \left(\nabla_{X_{x,y}} s_{x,y} + \frac{\mathrm{d}}{\mathrm{d}x}s_{x,y}\right)\Big|_{\phi(x,y)},$$

$$D_y s(x,y) = \left(\nabla_{\beta_{x,y}} s_{x,y} + \frac{\mathrm{d}}{\mathrm{d}y}s_{x,y}\right)\Big|_{\phi(x,y)}.$$

It follows that

$$D_x D_y s(x,y) = \Big(\nabla_{X_{x,y}}\nabla_{Y_{x,y}} s_{x,y} + \frac{\mathrm{d}}{\mathrm{d}x}\nabla_{Y_{x,y}} s_{x,y}$$
$$+ \nabla_{X_{x,y}}\frac{\mathrm{d}s_{x,y}}{\mathrm{d}y} + \frac{\mathrm{d}^2 s_{x,y}}{\mathrm{d}x\mathrm{d}y}\Big)\Big|_{\phi(x,y)},$$

$$D_y D_x s(x,y) = \Big(\nabla_{\beta_{x,y}}\nabla_{X_{x,y}} s_{x,y} + \frac{\mathrm{d}}{\mathrm{d}y}\nabla_{X_{x,y}} s_{x,y}$$
$$+ \nabla_{Y_{x,y}}\frac{\mathrm{d}s_{x,y}}{\mathrm{d}x} + \frac{\mathrm{d}^2 s_{x,y}}{\mathrm{d}y\mathrm{d}x}\Big)\Big|_{\phi(x,y)}.$$

Taking the difference of these two equations, we obtain

$$D_x D_y s(x, y) - D_y D_x s(x, y)$$

$$= \left(\nabla_{X_{x,y}} \nabla_{Y_{x,y}} s_{x,y} - \nabla_{Y_{x,y}} \nabla_{X_{x,y}} s_{x,y} + \nabla_{\frac{d}{dx} Y_{x,y} - \frac{d}{dy} X_{x,y}} s_{x,y} \right) \Big|_{\phi(x,y)}.$$

Finally, applying the lemma above, we obtain

$$D_x D_y s(x, y) - D_y D_x s(x, y) = \left(R(X_{x,y}, Y_{x,y}) s_{x,y} \right) \Big|_{\phi(x,y)}$$

$$= R \left(\frac{\partial \phi}{\partial x}, \frac{\partial \phi}{\partial x} \right) s(x, y).$$

$\square$

A **flat connection** is a connection for which the curvature tensor vanishes. We will often refer to a vector bundle with a flat connection as a **flat vector bundle**. Clearly, if around each point one can choose coordinates and trivializing sections for which the Christoffel symbols vanish, the connection is flat. The converse is also true, as a consequence of the following local normal form for flat bundles.

Corollary 32.1. *Let* $\xi = (\pi, E, M)$ *be a vector bundle of rank* r *with a flat connection* ∇. *For each* $p \in M$, *there exists a base of local sections* $\{s_1, \ldots, s_r\}$ *defined in a neighborhood* U *of* p, *such that*

$$\nabla_X s_i = 0, \quad \forall X \in \mathfrak{X}(M).$$

Hence, $\xi|_U$ *is isomorphic to the trivial vector bundle* ε_U^r *with the canonical flat connection.*

In the case of Riemannian manifolds, Corollary 32.1 takes the following more geometric meaning.

Corollary 32.2. *Let* $(M, \langle\,,\,\rangle)$ *be a Riemannian manifold with vanishing curvature tensor:* $R = 0$. *For each* $p \in M$, *there exists a neighborhood* U *of* p *which is isometric to an open in* $\mathbb{R}^d$ *furnished with the Euclidean metric.*

For the proofs of these corollaries, we refer to the exercises at the end of this lecture. Note that these results describe flat connections locally. To describe what happens with a flat connection globally, we need to introduce the notion of *holonomy* of a connection.

Given a vector bundle $\xi = (\pi, E, M)$ of rank r with a connection ∇ fix a base point $p_0 \in M$. For each closed path $\gamma : [0,1] \to M$ based at p_0, so $\gamma(0) = \gamma(1) = p_0$, parallel transport along the curve $\gamma(t)$ gives a linear isomorphism of E_{p_0}, called the **holonomy** of γ and denoted by

$$H_{p_0}(\gamma) := \tau_1 : E_{p_0} \to E_{p_0}.$$

If we extend this definition in the obvious way to closed paths which are only piecewise smooth, it is clear that

$$H_{p_0}(\gamma_1 \cdot \gamma_2) = H_{p_0}(\gamma_1) \circ H_{p_0}(\gamma_2),$$

where $\gamma_1 \cdot \gamma_2$ denotes the concatenation of the two paths defined by

$$\gamma_1 \cdot \gamma_2(t) := \begin{cases} \gamma_2(2t) & \text{if } 0 \le t \le \frac{1}{2}, \\ \gamma_1(2t-1) & \text{if } \frac{1}{2} \le t \le 1. \end{cases}$$

When the connection is flat holonomy only depends on the homotopy class of the path.

Proposition 32.2. *Given a flat connection, any two path-homotopic closed curves γ_0 and γ_1 have the same holonomy: $H_{p_0}(\gamma_0) = H_{p_0}(\gamma_1)$.*

Proof. One can show that two smooth curves which are C^0 path-homotopic are also smooth path-homotopic (see, e.g., Hirsch, 1994). So let $\gamma : [0,1] \times [0,1] \to M$ be a smooth path-homotopy between γ_0 and γ_1, so that

$$\gamma(t,0) = \gamma_0(t), \quad \gamma(t,1) = \gamma_1(t), \quad \gamma(0,\varepsilon) = \gamma(1,\varepsilon) = p_0.$$

Fix $v_0 \in E_{p_0}$. We define a section $s : [0,1] \times [0,1] \to E$ along $\gamma : [0,1] \times [0,1] \to M$ by

$$s(t,\varepsilon) := \tau_t^{\gamma(\cdot,\varepsilon)}(v_0) = \begin{cases} \text{parallel transport of } v_0 \text{ along} \\ s \mapsto \gamma(s,\varepsilon) \quad \text{with } s \in [0,t]. \end{cases}$$

Note that, by construction, for each fixed ε one has

$$D_t s := D_{\gamma(\cdot,\varepsilon)} s = 0.$$

We claim that for each fixed t one also has

$$D_\varepsilon s := D_{\gamma(t,\cdot)} s = 0.$$

Indeed, since $\gamma(0, \varepsilon) = 0$ and $s(0, \varepsilon) = v_0$, we have

$$D_\varepsilon s(0, \varepsilon) = D_{\gamma(0,\cdot)} s(0, \varepsilon) = \frac{\mathrm{d}}{\mathrm{d}\varepsilon} s(0, \varepsilon) = 0.$$

On the other hand, using Proposition 32.1, we find

$$D_t D_\varepsilon s = R\left(\frac{\partial \gamma}{\partial t}, \frac{\partial \gamma}{\partial \varepsilon}\right) + D_\varepsilon D_t s = 0.$$

Hence, $D_\varepsilon s$ is parallel along the curve $t \mapsto \gamma(t, \varepsilon)$ so we must have $D_\varepsilon s(t, \varepsilon) = 0$, as claimed.

Now, applying our claim, and the fact that $\gamma(1, \varepsilon) = p_0$, we conclude that

$$0 = D_\varepsilon s(1, \varepsilon) = \frac{\mathrm{d}}{\mathrm{d}\varepsilon} s(1, \varepsilon).$$

Hence,

$$\tau_1^{\gamma_0}(v_0) = s(1, 0) = s(1, 1) = \tau_1^{\gamma_1}(v_0).$$

Since $v_0 \in E_{p_0}$ was an arbitrary vector, we conclude that $H_{p_0}(\gamma_0) = H_{p_0}(\gamma_1)$. $\qquad\square$

Every element in $\pi_1(M, p_0)$ has a smooth a representative, so for a flat connection holonomy gives a group homomorphism

$$H_{p_0} : \pi_1(M, p_0) \to \mathrm{GL}(E_{p_0}).$$

This homomorphism is called the **holonomy representation** of ∇ with base point p_0. If $q_0 \in M$ is a different base point in the same connected component of M, we can choose a smooth path $c : [0, 1] \to M$ with $c(0) = p_0$ and $c(1) = q_0$. Then parallel transport along $c(t)$ gives an isomorphism $\tau : E_{p_0} \to E_{q_0}$ and one has

$$H_{q_0} = \tau \circ H_{p_0} \circ \tau^{-1}.$$

Hence, the holonomy representations for different base points in the same connected component are related by conjugacy.

Theorem 32.2. *Let M be a connected manifold with base point $p_0 \in M$, there is a $1 : 1$ correspondence*

{isomorphism classes of flat vector bundles of rank r over M}

$$\xleftarrow{\sim} \mathrm{Hom}(\pi_1(M, p_0), \mathrm{GL}(r))/\mathrm{GL}(r).$$

where $\mathrm{GL}(r)$ acts on $\mathrm{Hom}(\pi_1(M, p_0), \mathrm{GL}(r))$ by conjugation.

Proof. We already know that a flat vector bundle (ξ, ∇) induces a representation of the fundamental group, namely the holonomy representation

$$H_{p_0} : \pi_1(M, p_0) \to \mathrm{GL}(E_{p_0}).$$

Fixing a basis for the fiber E_{p_0}, we obtain a group homomorphism

$$H_{p_0} : \pi_1(M, p_0) \to \mathrm{GL}(r).$$

Two different basis for E_{p_0} are related by conjugation of an element of $\mathrm{GL}(r)$. It follows that two isomorphic vector bundles (ξ_1, ∇_1) and (ξ_1, ∇_2) induce homomorphisms which are related by conjugation too. Hence, one can associate to an isomorphism class of vector bundles an element in the quotient

$$\mathrm{Hom}(\pi_1(M, p_0), \mathrm{GL}(r))/\mathrm{GL}(r).$$

Conversely, given a representation $H : \pi_1(M, p_0) \to \mathrm{GL}(r)$ representing some element in this quotient, we construct a flat vector bundle as follows: on the one hand, the representation gives an action of $\pi(M, p_0)$ in $\mathbb{R}^r$. On the other hand, the fundamental group $\pi_1(M, p_0)$ acts in the universal cover $\widetilde{M}$ by deck transformations: identifying $\widetilde{M}$ with the set of homotopy classes of paths $[c]$ with initial point $c(0) = p_0$, the action of $\pi_1(M, p_0)$ in $\widetilde{M}$ is given by concatenation:

$$\pi_1(M, p_0) \times \widetilde{M} \to \widetilde{M}, \ ([\gamma], [c]) \mapsto [\gamma \cdot c].$$

Since this action is proper and free, the resulting diagonal action of $\pi_1(M, p_0)$ in $\widetilde{M} \times \mathbb{R}^r$ is also proper and free. Hence, the quotient space $E = (\widetilde{M} \times \mathbb{R}^r)/\pi_1(M, p_0)$ is a manifold, and we have the projection

$$\pi : E \to M, [[c], \mathbf{v}] \mapsto c(1).$$

The triple $\xi = (\pi, E, M)$ is a vector bundle. Moreover, the canonical flat connection on $\widetilde{M} \times \mathbb{R}^r$ induces a connection on ξ for which the holonomy with base point p_0 is precisely $H : \pi_1(M, p_0) \to \mathrm{GL}(r)$. Finally, one checks that given two homomorphisms $H_0, H_1 : \pi_1(M, p_0) \to \mathrm{GL}(r)$ in the same conjugacy class this construction produces isomorphic flat vector bundles. $\square$

Remark 32.1. The space appearing in the previous result is an example of a **character variety**. More general, given a Lie group G

and a finitely generated group Γ, the G-character variety of Γ is the space of equivalence classes of group homomorphisms:

$$\text{Hom}(\Gamma, G)/G.$$

Exercises

Exercise 32.1
Show that the connection 1-form and the curvature 2-form of a connection satisfy the structure equations and Bianchi's identity of Theorem 32.1.

Exercise 32.2
Prove Lemma 32.1.

Exercise 32.3
Let $\{s_1, \ldots, s_r\}$ and $\{s'_1, \ldots, s'_r\}$ be two bases of local sections for a vector bundle $\xi = (\pi, E, M)$ over a common open set $U \subset M$. Denote by $A = (a_i^j) : U \to \text{GL}(r)$ the matrix of change of basis so that $s'_i = \sum_j a_i^j s_j$. Show that the corresponding curvature 2-forms Ω and Ω' are related by

$$\Omega' = A^{-1}\Omega A.$$

Exercise 32.4
Show that if ∇ is a flat connection on a vector bundle $\xi = (\pi, E, M)$, then around every point $p \in M$ one can find a local basis of flat sections for ξ.

Hint: Using Exercise 3 in the previous section and the previous exercise, show that the condition $\omega' = 0$ defines an integrable distribution in $U \times \text{GL}(r)$, so one can apply Frobenius.

Exercise 32.5
Let $(M, \langle\ ,\ \rangle)$ be a Riemannian manifold whose curvature tensor vanishes: $R = 0$. Show that for each $p \in M$, there exists a neighborhood U isometric to an open in $\mathbb{R}^d$ with the Euclidean metric.

Exercise 32.6
Let G be a connected Lie group with Lie algebra $\mathfrak{g}$. As in Exercise 31.9, consider the unique connection ∇ in TG which for any two left

invariant vector fields $X, Y \in \mathfrak{g}$ satisfies

$$\nabla_X Y = \frac{1}{2}[X, Y].$$

Show that this connection has a curvature tensor

$$R(X, Y) \cdot Z = \frac{1}{4}[[X, Y], Z], \quad \forall X, Y, Z \in \mathfrak{g}).$$

Exercise 32.7
On the trivial line bundle over $\mathbb{S}^1$, fix a base point and a non-vanishing section e. Find the holonomy representation for the connection defined by $\nabla_{\frac{\partial}{\partial\theta}} e = e$.

Exercise 32.8
On the tangent bundle of the 2-torus $\mathbb{T}^2$ with coordinates (θ^1, θ^2) consider the flat connection defined by

$$\nabla_{\frac{\partial}{\partial\theta^1}} \frac{\partial}{\partial\theta^1} = \frac{\partial}{\partial\theta^2}, \quad \nabla_{\frac{\partial}{\partial\theta^1}} \frac{\partial}{\partial\theta^2} = \nabla_{\frac{\partial}{\partial\theta^2}} \frac{\partial}{\partial\theta^1} = \nabla_{\frac{\partial}{\partial\theta^2}} \frac{\partial}{\partial\theta^2} = 0.$$

Find the holonomy representation for a base point of your choice.

Exercise 32.9
Let $\mathcal{F}$ be a foliation of a manifold M and consider the normal bundle $\nu(\mathcal{F}) = TM/T\mathcal{F}$. Given a vector field tangent to the foliation $X \in \mathfrak{X}(\mathcal{F})$ and a section $s \in \Gamma(\nu(\mathcal{F}))$ define:

$$\nabla_X s := \mathrm{pr}_{\nu(L)}[X, Y],$$

where $Y \in \mathfrak{X}(M)$ is any vector field such that $\mathrm{pr}_{\nu(L)}(Y) = s$. Show the following:

(a) ∇ is well-defined, i.e., it does not depend on the choice of lift Y of s.
(b) ∇ satisfies all the properties in the definition of a connection, except that it only differentiates along the directions of $\mathcal{F}$.
(c) L, ∇ induces a flat connection ∇^L on the normal bundle of each leaf L.

The connection ∇^L is called the **Bott connection** and its holonomy is called the **linear holonomy** of the leaf L.

Exercise 32.10

Let L be a manifold, $\tilde{L}$ its universal covering space, and $\rho : \pi_1(L) \to$ $GL(V)$ a representation. The diagonal action of $\pi_1(L)$ on $\tilde{L} \times V$ is proper and free so

$$M := (\tilde{L} \times V)/\pi_1(L)$$

is a smooth manifold.

(a) Show that the foliation $\{\tilde{L} \times v : v \in V\}$ projects onto a foliation $\mathcal{F}$ of M having $L \simeq (\tilde{L} \times \{0\})/\pi_1(L)$ as a leaf.
(b) Find the linear holonomy of the leaf L (see previous exercise).

Lecture 33

The Chern–Weil Homomorphism

We saw in the previous section that a flat vector bundle is globally characterized by its holonomy representation. We will now study vector bundles with connections not necessarily flat, a situation that is more complicated but more interesting. Eventually, we will see that one can use a connection on a vector bundle to construct cohomology classes which are invariants of the vector bundle, and which characterize certain properties of the vector bundle up to isomorphism.

Let $\pi : E \to M$ be a vector bundle. We consider differential forms in M with values in E, which we denote by

$$\Omega^\bullet(M; E) := \Gamma(\wedge^k T^*M \otimes E).$$

So a differential form of degree k with values in E is a $C^\infty(M)$-multilinear alternating map

$$\omega : \underbrace{\mathfrak{X}(M) \times \cdots \times \mathfrak{X}(M)}_{k\text{-times}} \to \Gamma(E).$$

In particular, $\Omega^0(M; E)$ is the space $\Gamma(E)$ of global sections of the vector bundle $\pi : E \to M$. Note that we also have

$$\Omega^\bullet(M; E) = \Omega^\bullet(M) \otimes \Gamma(E),$$

where $\otimes$ denotes here the tensor product of $C^\infty(M)$-modules. This last interpretation shows that we have a well-defined wedge product $\omega \wedge \eta \in \Omega^{k+l}(M; E)$, for any $\omega \in \Omega^k(M)$ and $\eta \in \Omega^l(M; E)$.

A choice of connection ∇ in $\pi : E \to M$ allows us to take the differential of E-valued differential forms as follows. The connection determines an operator $d_\nabla : \Omega^0(M; E) \to \Omega^1(M; E)$ through the formula

$$(d_\nabla s)(X) = \nabla_X s.$$

The map d_∇ is $\mathbb{R}$-linear and satisfies the Leibniz identity

$$d_\nabla(fs) = df \otimes s + f d_\nabla s.$$

One can extend d_∇ to arbitrary forms by requiring that for any form $\omega \in \Omega^\bullet(M)$ and section $s \in \Gamma(E)$ the following general Leibniz identity holds

$$d_\nabla(\omega \otimes s) = d_\nabla(\omega) \otimes s + (-1)^{\deg \omega} \omega \wedge d_\nabla(s). \qquad (33.1)$$

The following proposition gives an explicit expression for the linear operator d_∇.

Proposition 33.1. *Given a connection ∇ on a vector bundle E, for $\omega \in \Omega^k(M; E)$ define $d_\nabla \omega \in \Omega^{k+1}(M; E)$ by*

$$d_\nabla \omega(X_0, \ldots, X_k) := \sum_{i=0}^{k+1} (-1)^i \nabla_{X_i}(\omega(X_0, \ldots, \widehat{X}_i, \ldots, X_k))$$

$$+ \sum_{i<j} (-1)^{i+j} \omega([X_i, X_j], X_0, \ldots, \widehat{X}_i, \ldots, \widehat{X}_j, \ldots, X_k). \qquad (33.2)$$

This defines a linear operator $d_\nabla : \Omega^\bullet(M; E) \to \Omega^{\bullet+1}(M; E)$ which is uniquely determined by the following two properties:

(i) *For any 0-form $s \in \Gamma(E)$, one has $(d_\nabla s)(X) = \nabla_X s$.*
(ii) d_∇ *is $\mathbb{R}$-linear and satisfies the Leibniz identity (33.1).*

Proof. One checks easily that the operator d_∇ defined by (33.2) satisfies (i) and (ii). Since any E-valued k-form η can be written as a linear combination:

$$\eta = \sum_{i=1}^{l} \omega_i \otimes s_i \quad (\omega_i \in \Omega^k(M), \ s_i \in \Gamma(E)),$$

it is clear that (i) and (ii) determined completely d_∇. $\qquad\square$

Example 33.1. Let ∇^0 and ∇ be two linear connections on E. Note that $\mathrm{End}(E)$ is a vector bundle and we can define a $\mathrm{End}(E)$-valued form $\omega \in \Omega^1(M, \mathrm{End}(E))$ by

$$\omega(X)(s) := \nabla_X s - \nabla_X^0 s.$$

Conversely, given a linear connection ∇^0 and a $\mathrm{End}(E)$-valued form $\omega \in \Omega^1(M, \mathrm{End}(E))$, one obtains a new connection by setting

$$\nabla_X s := \nabla_X^0 s + \omega(X)(s).$$

In this way we see that, after fixing a reference connection ∇^0, there is a 1:1 correspondence between linear connections on E and elements of $\Omega^1(M, \mathrm{End}(E))$. In other words, the linear connections on E form an affine space modeled on the vector space $\Omega^1(M, \mathrm{End}(E))$.

If $\{s_1, \ldots, s_r\}$ is a basis of local sections for E over an open set $U \subset M$, one has a flat connection ∇^0 on $E|_U$ determined by

$$\nabla_X^0 s_a = 0, \quad (a = 1, \ldots, r).$$

Hence, given any connection ∇ one obtains $\mathrm{End}(E)$-valued form $\omega \in \Omega^1(U, \mathrm{End}(E))$, which on the local basis $\{s_1, \ldots, s_r\}$ has the form

$$\omega(X)(s_a) = \sum_{b=1}^{r} \omega_a^b(X) s_b,$$

for certain 1-forms $\omega_a^b \in \Omega^1(U)$. The matrix $[\omega_a^b]$ is, of course, the connection 1-form we saw before. Moreover, if $\alpha \in \Omega^k(M, \mathrm{End}(E))$, then $\alpha|_U = \sum_{b=1}^{r} \alpha^a s_a$ with $\alpha^b \in \Omega^k(U)$ and one finds

$$(\mathrm{d}_\nabla \alpha)|_U = \sum_{b=1}^{r} \left(\mathrm{d}\alpha^b + \sum_{a=1}^{r} \alpha^a \wedge \omega_a^b \right) s_b.$$

Note that, in general, $\mathrm{d}_\nabla^2 \neq 0$, so d_∇ is not a differential. In fact, the curvature of ∇ can be seen as the failure in d_∇ being a differential.

Proposition 33.2. *Let ∇ be a connection on a vector bundle $\xi = (\pi, E, M)$ with curvature tensor R. Then*

(i) *For any 0-form $s \in \Gamma(E)$*

$$d_\nabla^2 s(X,Y) = R(X,Y)s, \quad (X,Y \in \mathfrak{X}(M)).$$

(ii) *Viewing the curvature as a 2-form $R \in \Omega^2(M, \operatorname{End} E)$, for the connection on $\operatorname{End}(E)$ induced by ∇:*

$$d_\nabla R = 0. \tag{33.3}$$

Proof. Using the definition of d_∇ one finds that

$$d_\nabla^2 s(X,Y) = \nabla_X(d_\nabla s(Y)) - \nabla_Y(d_\nabla s(X)) - d_\nabla s([X,Y]))$$
$$= \nabla_X(\nabla_Y s) - \nabla_Y(\nabla_X s) - \nabla_{[X,Y]}s = R(X,Y)s.$$

The proof of (ii) is left as an exercise..   □

Remark 33.1. The previous result shows that d_∇ is a differential if and only if the connection is flat. In this case, one calls the cohomology of the complex $(\Omega^\bullet(M;E), d_\nabla)$ the **de Rham cohomology of M with coefficients in E** and denotes it by $H^\bullet(M;E)$. Notice that the usual de Rham cohomology corresponds to the case where $E = M \times \mathbb{R}$ is the trivial flat line bundle.

The Bianchi identity (33.3) can be used to define certain cohomology classes. For that we need first to recall that for a finite dimensional vector space V one has the following canonical identification between the homogeneous polynomials and the multilinear symmetric functions:

(i) Every k-multilinear symmetric map $P : V \times \cdots \times V \to \mathbb{R}$ determines a degree k homogeneous polynomial $\widetilde{P} : V \to \mathbb{R}$ by the formula

$$\widetilde{P} : v \mapsto P(v, \ldots, v).$$

(ii) Conversely, every homogeneous polynomial $\widetilde{P} : V \to \mathbb{R}$ of degree k determines a k-multilinear symmetric map $P : V \times \cdots \times V \to \mathbb{R}$ by polarization

$$P(v_1, \ldots, v_k) = \frac{1}{k!} \frac{\partial}{\partial t_1} \cdots \frac{\partial}{\partial t_k} \widetilde{P}(t_1 v_1 + \cdots + t_k v_k) \Big|_{t_1 = \cdots = t_k = 0}.$$

These correspondences are inverse to each other, and the usual product of polynomials corresponds to the product of k-multilinear, symmetric maps, defined by

$$P_1 \circ P_2(v_1, \ldots, v_{k+l})$$

$$= \frac{1}{(k+l)!} \sum_{\sigma \in S_{k+l}} P_1(v_{\sigma(1)}, \ldots, v_{\sigma(k)}) P_2(v_{\sigma(k+1)}, \ldots, v_{\sigma(k+l)}).$$

Example 33.2. If one fixes a base $\xi^1, \ldots, \xi^r$ for V^*, then one can think of the polarization of the polynomial $\widetilde{P} : V \to \mathbb{R}$ as follows. One can write

$$\widetilde{P}(v) = \sum_{i_1, \ldots, i_k = 1}^{r} a_{i_1 \cdots i_k} \xi^{i_1}(v) \cdots \xi^{i_k}(v),$$

where the coefficients $a_{i_1 \cdots i_k}$ are *symmetric* in the indices. Then the corresponding k-multilinear, symmetric map $P : V \times \cdots \times V \to \mathbb{R}$ is given by

$$P(v_1, \ldots, v_k) = \sum_{i_1, \ldots, i_k = 1}^{r} a_{i_1 \cdots i_k} \xi^{i_1}(v_1) \cdots \xi^{i_k}(v_k).$$

For example, let $V = \mathbb{R}^3$ with linear coordinates (x, y, z). The homogeneous polynomial of degree 2

$$\widetilde{P}(x, y, z) := x^2 + xy + z^2 = x^2 + \frac{1}{2}(xy + yx) + z^2,$$

corresponds to the bilinear symmetric map

$$P(v, w) = v_1 w_1 + \frac{1}{2}(v_1 w_2 + v_2 w_1) + v_3 w_3.$$

We are interested in the case where $V = \mathfrak{g}$ is the Lie algebra of a Lie group G. We will denote by $I^k(G)$ the space of k-multilinear, symmetric maps $P : \mathfrak{g} \times \cdots \times \mathfrak{g} \to \mathbb{R}$ which are invariant under the adjoint action

$$P(\operatorname{Ad} g \cdot v_1, \ldots, \operatorname{Ad} g \cdot v_k) = P(v_1, \ldots, v_k), \quad \forall g \in G,\ v_1, \ldots, v_k \in \mathfrak{g}.$$

If we let

$$I(G) = \bigoplus_{k=0}^{\infty} I^k(G),$$

we obtain a ring for the usual addition and the symmetric product of multilinear symmetric maps. Under the correspondence above, we can identify $I(G)$ with the algebra of polynomials in $\mathfrak{g}$ which are Ad-invariant.

For now, we are only interested in the case where $G = GL(r)$, so that $\mathfrak{g} = \mathfrak{gl}(r)$ is the space of all $r \times r$-matrices. In this case, the adjoint action is given by matrix conjugation

$$\operatorname{Ad} A \cdot X = AXA^{-1}, \quad A \in GL(r),\ X \in \mathfrak{gl}(r).$$

Then the invariance condition is just invariance under conjugation, i.e,

$$P(AX_1 A^{-1}, \ldots, AX_k A^{-1}) = P(X_1, \ldots, X_k),$$

which must hold for any invertible matrix $A \in GL(r)$ and any $X_1, \ldots, X_k \in \mathfrak{gl}(r)$.

Example 33.3. A collection of $\operatorname{Ad}_{GL(r)}$-invariant polynomials on $\mathfrak{gl}(r)$ can be obtained by taking traces of powers

$$X \mapsto \operatorname{tr}(X^k).$$

Actually, these polynomials generate the ring of $\operatorname{Ad}_{GL(r)}$-invariant polynomials. We will come back to this issue in the next lecture.

Returning to the discussion of vector bundles with connection, the key remark is now the following result.

Proposition 33.3. *Let ∇ be a connection on a rank r vector bundle E over M. Every element $P \in I^k(GL(r))$ determines a map*

$$P : \Omega^{\bullet}(M; \otimes^k \operatorname{End}(E)) \to \Omega^{\bullet}(M), \quad \omega \mapsto P \circ \omega, \tag{33.4}$$

which satisfies

$$\mathrm{d}P = P\mathrm{d}_\nabla.$$

Proof. Note that if $s_1, \ldots, s_r$ is a base of local of sections of E then for any section $A \in \Gamma(\mathrm{End}(E))$, we have

$$As_i = \sum_{j=1}^{r} A_i^j s_j,$$

for some functions A_i^j. Given $P \in I^k(GL(r))$, we define $P : \Gamma(\otimes^k \mathrm{End}(E)) \to C^\infty(M)$ by

$$P(A_1 \otimes \cdots \otimes A_k) := P([(A_1)_i^j], \cdots, [(A_k)_i^j]).$$

The invariance condition shows that this expression is independent of the choice of base of local of sections, so this map is well-defined. A degree l form $\omega \in \Omega^l(M; \otimes^k \mathrm{End}(E))$ can be seen as an l-multilinear alternating map

$$\omega : \mathfrak{X}(M) \times \cdots \times \mathfrak{X}(M) \to \Gamma(\otimes^k \mathrm{End}(E)),$$

so composing with P determines an l-multilinear alternating map

$$P \circ \omega : \mathfrak{X}(M) \times \cdots \times \mathfrak{X}(M) \to C^\infty(M).$$

Hence, $P(\omega) \in \Omega^l(M)$, and an elementary computation using the definitions of d, d_∇ and the fact that P is multilinear, shows that $\mathrm{d}P = P\mathrm{d}_\nabla$. $\qquad\square$

Now let R denote the curvature tensor of the connection ∇. The k-symmetric power of R is the element $R^k \in \Omega^{2k}(M; \otimes^k \mathrm{End}(E))$ defined by

$$R^k(X_1, \ldots, X_{2k}) := \frac{1}{(2k)!} \sum_{\sigma \in S_{2k}} (-1)^\sigma R(X_{\sigma(1)}, X_{\sigma(2)})$$

$$\otimes \cdots \otimes R(X_{\sigma(2k-1)}, X_{\sigma(2k)}).$$

Therefore, given $P \in I^k(GL(r))$ we obtain a differential form $P(R^k) \in \Omega^{2k}(M)$. If one fixes a local basis of sections $\{s_1, \ldots, s_r\}$

and lets $\Omega = [\Omega_a^b]$ denote the curvature 2-form of the connection relative to this basis, this form is given explicitly by

$$P(R^k)(X_1, \ldots, X_{2k})$$
$$= \frac{1}{(2k)!} \sum_{\sigma \in S_{2k}} (-1)^\sigma P(\Omega(X_{\sigma(1)}, X_{\sigma(2)}), \ldots, \Omega(X_{\sigma(2k-1)}, X_{\sigma(2k)})).$$

Using this expression, one checks that if $P_1 \in I^k(GL(r))$ and $P_2 \in I^l(GL(r))$, then

$$P_1 \circ P_2(R^{k+l}) = P_1(R^k) \wedge P_2(R^l) \in \Omega^{2(k+l)}(M).$$

On the other hand, applying Bianchi's identity (33.3), one finds

$$\mathrm{d}P(R^k) = P(\mathrm{d}_\nabla R^k) = kP(R^{k-1}\mathrm{d}_\nabla R) = 0.$$

We conclude that $P(R^k) \in \Omega^{2k}(M)$ is a closed form.

Theorem 33.1 (Chern–Weil). *Let ∇ be a connection on a rank r vector bundle E over M, with curvature tensor R. The map $I(GL(r)) \to H(M)$ defined by*

$$I^k(GL(r)) \to H^{2k}(M), \quad P \longmapsto [P(R^k)],$$

is a ring homomorphism. This homomorphism is independent of the choice of connection.

Proof. All that it remains to be checked is that the homomorphism is independent of the choice of connection. For that we claim that, if ∇^0 and ∇^1 are two connections on E and $P \in I^k(GL(r))$, then the differential forms $P(R_{\nabla_0}^k)$ and $P(R_{\nabla_1}^k)$ differ by an exact form.

To prove the claim, consider the projection $p : M \times [0,1] \to M$. The pullback bundle p^*E carries a connection ∇ defined by requiring that on pullback sections

$$\nabla_{\frac{\partial}{\partial t}} p^*s = 0, \quad \nabla_X p^*s := tp^*(\nabla_X^1 s) + (1-t)p^*(\nabla_X^0 s), \quad (X \in \mathfrak{X}(M)).$$

On the other hand, we have an *integration along the fibers* of p

$$\int_0^1 : \Omega^\bullet(M \times [0,1]) \to \Omega^{\bullet-1}(M),$$

which is explicitly given by

$$\left(\int_0^1 \omega\right)(X_1,\dots,X_{l-1}) = \int_0^1 \omega\left(\frac{\partial}{\partial t},X_1,\dots,X_{l-1}\right)dt.$$

One defines the **Chern–Simons transgression form** by setting

$$P(\nabla_0,\nabla_1) := k\int_0^1 P(R_\nabla^k) \in \Omega^{2k-1}(M). \tag{33.5}$$

We leave as an exercise to check that

$$dP(\nabla_0,\nabla_1) = P(R_{\nabla_1}^k) - P(R_{\nabla_0}^k),$$

so the claim follows. $\qquad\qquad\qquad\qquad\qquad\qquad\qquad\square$

The ring homomorphism given by the previous result

$$CW[\xi] : I(GL(r)) \to H^\bullet(M),$$

is called the **Chern–Weil homomorphism** of the vector bundle $\xi = (\pi, E, M)$. This homomorphism depends only on the isomorphism class of ξ. This follows from the following more general result, which expresses the functoriality of the Chern–Weil homomorphism relative to pullbacks. This will be very useful later in the study of characteristic classes of vector bundles. The proof is left for the exercises.

Proposition 33.4. *Let $\psi : N \to M$ be a smooth map and let $\xi = (\pi, E, M)$ be a vector bundle of rank r. For any $P \in I^\bullet(GL(r))$ and any connection ∇ one has*

$$\psi^* P(R_\nabla^k) = P(R_{\psi^*\nabla}^k).$$

Hence, the following diagram is commutative

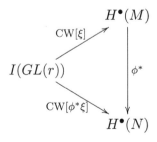

Exercises

Exercise 33.1
Prove Bianchi's identity (33.3).

Exercise 33.2
Let $E = M \times \mathbb{R}$ be the trivial vector bundle over M with global section e. Given a connection ∇ on E let $\omega \in \Omega^1(M)$ be the (global) connection 1-form defined by

$$\nabla_X s = \omega(X)s.$$

Show that ∇ is flat if and only if ω is closed. For a flat ∇, prove that

(a) The complex $(\Omega^\bullet(M; E), d_\nabla) = (\Omega^\bullet(M), d_\nabla)$ is given by

$$d_\nabla \alpha = d\alpha + \omega \wedge \alpha.$$

(b) $H^\bullet(M; E)$ is isomorphic to $H^\bullet(M)$ whenever the cohomology class $[\omega]$ trivial.

Hint: If $\omega = df$, consider the map $\Phi : \Omega^\bullet(M; E) \to \Omega^\bullet(M)$, $\alpha \mapsto e^f \alpha$.

Exercise 33.3
Let E be a vector bundle over M of rank r. Given $P \in I^k(GL(r))$, show that the map $P : \Omega^\bullet(M; \otimes^k \operatorname{End}(E)) \to \Omega^\bullet(M)$ given by (33.4) satisfies $dP = Pd_\nabla$.

Exercise 33.4
Show, by direct computation, that the Chern–Simons transgression form (33.5) satisfies

$$dP(\nabla_0, \nabla_1) = P(R_{\nabla_1}^k) - P(R_{\nabla_0}^k).$$

Exercise 33.5
Let $\psi : N \to M$ be a smooth map and $\xi = (\pi, E, M)$ a vector bundle of rank r with a connection ∇. Show that, for all $P \in I^\bullet(GL(r))$, one has

$$\psi^* P(R_\nabla^k) = P(R_{\psi^* \nabla}^k).$$

Exercise 33.6

Let M be a manifold of dimension ≤ 3. Show that the Chern–Weil homomorphism of any vector bundle over M is trivial.

Hint: Show that the curvature form Ω of a connection that preserves a metric is a skew-symmetric matrix relative to a orthonormal frame. Then use that a $\mathrm{Ad}_{GL(r)}$-invariant polynomial of degree one is of the form $P(X) = a \operatorname{tr} X$, for some $a \in \mathbb{R}$.

Exercise 33.7

Show that the Chern–Weil homomorphism of the tangent bundle of $\mathbb{S}^4$ is trivial.

Hint: Consider the invariant polynomials $\sigma_1(X) = \operatorname{tr} X$, $\sigma_2(X) = \operatorname{tr} X^2$, and the Levi–Civita connection of the metric induced by the canonical embedding $\mathbb{S}^4 \subset \mathbb{R}^4$.

Lecture 34

Characteristic Classes

A cohomology class in the image of the Chern–Weil homomorphism is called a **characteristics class** of ξ. There are certain canonical characteristic classes that arise from natural choices of elements in the ring of invariant polynomials $I(GL(r))$.

The Pontrjagin Classes of a Real Vector Bundle

We have already observed that traces of powers yield invariant polynomials. One can show that any homogeneous polynomial $P \in I^k(GL(r))$ can be written as a $\mathbb{R}$-linear combination of invariant polynomials of the form:

$$X \mapsto \operatorname{tr}(X^{k_1}) \cdots \operatorname{tr}(X^{k_s}), \quad k_1 + \cdots + k_s = k.$$

However, these are not algebraically independent.

Theorem 34.1. *The coefficients of the characteristic polynomial*

$$\det(\lambda I - X) = \lambda^r + \sigma_1(X)\lambda^{r-1} + \cdots + \sigma_r(X) \quad (X \in \mathfrak{gl}(r)),$$

are algebraically independent and generate the ring $I(GL(r))$.

Remark 34.1. The coefficients $\sigma_k : \mathfrak{gl}(r) \to \mathbb{R}$ can be expressed using the elementary symmetric functions. Recall that if $x_1, \ldots, x_r$

denote r indeterminates then the polynomial $p(x) = \prod_{i=1}^{r}(x - x_i)$ with coefficients in the field of fractions $\mathbb{R}(x_1, \ldots, x_n)$ is given by

$$p(x) = \prod_{i=1}^{r}(x - x_i) = x^r - s_1 x^{r-1} + \cdots + (-1)^r s_r,$$

where the coefficients are the elementary symmetric functions

$$s_1 = \sum_i x_i, \quad s_2 = \sum_{i<j} x_i x_j, \quad \ldots \quad s_r = x_1 \cdots x_r.$$

Applying this to the characteristic polynomial, one obtains

$$\sigma_1(X) = -\operatorname{tr} X,$$

$$\sigma_2(X) = +\frac{1}{2}\left((\operatorname{tr} X)^2 - \operatorname{tr} X^2\right), \ldots, \quad \sigma_r(X) = (-1)^r \det X.$$

One can show that the field $\mathbb{R}(x_1, \ldots, x_r)$ is a Galois extension of the field $\mathbb{R}(s_1, \ldots, s_r)$ with Galois group the symmetric group S_n. In other words, any symmetric expression in the indeterminates $x_1, \ldots, x_r$ is a polynomial in the elementary symmetric functions $s_1, \ldots, s_r$. Applying this to the invariant polynomials, one obtains the theorem above. Note that in this discussion one can replace $\mathbb{R}$ by $\mathbb{C}$, or any other field of characteristic zero.

The previous discussion suggest to apply the Chern–Weil homomorphism to the invariant polynomials $\sigma_1, \ldots, \sigma_r$. Before we do that, let us recall that one can equip any vector bundle ξ with a fiber metric, and then one can choose a connection ∇ compatible with the metric. We leave as an exercise to check that for such a connection the curvature 2-form relative to an orthonormal frame is always skew-symmetric

$$\Omega = -\Omega^T.$$

Since for a skew-symmetric X one has $\sigma_{2k+1}(X) = 0$, it follows that for such a metric connection one has

$$\sigma_{2k+1}(R^{2k+1}) = 0.$$

This explains why in the following definition we only consider even-dimensional classes.

Definition 34.1. The **Pontrjagin classes** of a vector bundle $\xi = (\pi, E, M)$ of rank r are

$$p_k(\xi) := \left[\sigma_{2k} \left(\left(\frac{1}{2\pi} R \right)^{2k} \right) \right] \in H^{4k}(M), \quad (k = 1, 2, \ldots),$$

where R is the curvature of any connection ∇ in ξ. The **total Pontrjagin class** of the vector bundle ξ is

$$p(\xi) = 1 + p_1(\xi) + \cdots + p_{[r/2]}(\xi),$$

where $[r/2]$ denotes the largest integer less or equal to $r/2$.

Remark 34.2. The normalization factor $\frac{1}{2\pi}$ is included so that the Pontrjagin classes belong to the image of the natural homomorphism

$$H^\bullet(M, \mathbb{Z}) \to H^\bullet(M).$$

If one fixes a local basis of sections $\{s_1, \ldots, s_r\}$ and lets $\Omega = [\Omega_a^b]$ denote the curvature 2-form of the connection relative to this basis, one obtains that the kth Pontrjagin class is (locally) represented by the closed $4k$-form

$$p_k(\xi) = \frac{1}{(2\pi)^k (2k)!} \sum_{i_1 < i_2 < \cdots < i_{2k}} \sum_{\sigma \in S_{2k}} (-1)^\sigma \Omega_{\sigma(i_1)}^{i_1} \wedge \cdots \wedge \Omega_{\sigma(i_{2k})}^{i_{2k}}.$$

This type of formula, although being very explicit, it is not the most effective way to determine these characteristic classes in specific examples. It is usually more efficient to apply the basic properties of the Pontrjagin classes that will be discussed in the next paragraphs.

Proposition 34.1. *Let M be a smooth manifold, ξ and η vector bundles over M. The Pontrjagin classes satisfy:*

(i) $p(\xi \oplus \eta) = p(\xi) \cup p(\eta)$;
(ii) $p(\psi^* \xi) = \psi^* p(\xi)$, *for any smooth map $\psi : N \to M$;*
(iii) $p(\xi) = 1$, *if ξ admits a flat connection.*

The proof follows from the construction of the Pontrjagin classes and is left as an exercise.

The Pontrjagin classes $p_i = p_i(TM)$ of the tangent bundle give an important invariant of a smooth manifold M. Although from its definition it seems that these classes are only invariants of diffeomorphism type, Novikov proved that these classes are in fact topological invariants: two smooth manifolds which are homeomorphic have the same Pontrjagin classes p_i. Here it is important that we are dealing with classes in de Rham cohomology. Using classifying bundles, one can also define integral cohomology versions of the Pontrjagin classes. The integral Pontrjagin classes of TM are not topological invariants.

Over a compact oriented manifold of dimension $\dim M = 4m$ one can also define **Pontrjagin numbers** of a vector bundle ξ. One chooses non-negative integers $a_1, \ldots, a_{[r/2]}$ such that

$$4(a_1 + 2a_2 + \cdots + [r/2]a_{[r/2]}) = 4m,$$

and defines a Pontrjagin number of ξ by

$$\int_M p_1^{a_1}(\xi) \wedge p_2^{a_2}(\xi) \wedge \cdots \wedge p_{[r/2]}^{a_{[r/2]}}(\xi).$$

The **Pontrjagin numbers of** M, where M is compact, oriented, of dimension $4m$ are, by definition, the Pontrjagin numbers of its tangent bundle. For example, a compact, oriented manifold of dimension 4 has only one Pontrjagin number $\int_M p_1$, while in dimension 8 there are two Pontrjagin numbers, namely,

$$\int_M p_1^2, \quad \int_M p_2.$$

Example 34.1. Let $M = \mathbb{S}^d \hookrightarrow \mathbb{R}^{d+1}$ and denote by $\nu(\mathbb{S}^d) = T_{\mathbb{S}^d}\mathbb{R}^{d+1}/T\mathbb{S}^d$ the normal bundle of $\mathbb{S}^d$. Note that the Whitney sum

$$T\mathbb{S}^d \oplus \nu(\mathbb{S}^d) = T_{\mathbb{S}^d}\mathbb{R}^{d+1},$$

is the trivial vector bundle over $\mathbb{S}^d$. On the other hand, the normal bundle $\nu(\mathbb{S}^d)$ is also trivial, for it is a line bundle which admits a nowhere vanishing section. By properties (i) and (iii) in the proposition, we conclude that $p(T\mathbb{S}^d) = 1$. Note that $\mathbb{S}^d$ has trivial tangent bundle only for $d = 1, 3, 7$.

Example 34.2. Let $M = \mathbb{CP}^d$. Recall that we have $\mathbb{CP}^d = \mathbb{S}^{2d+1}/\mathbb{S}^1$, where $\mathbb{S}^{2d+1} \subset \mathbb{C}^{d+1}$ and $\mathbb{S}^1$ acts by complex multiplication: $\theta \cdot z = e^{i\theta}$. The Euclidean metric in $\mathbb{C}^{d+1} = \mathbb{R}^{2d+2}$ induces a Riemannian metric in $\mathbb{S}^{2d+1}$ which is invariant under the $\mathbb{S}^1$-action. Hence, this induces a Riemannian metric in the quotient $\mathbb{CP}^d = \mathbb{S}^{2d+1}/\mathbb{S}^1$, called the **Fubini–Study metric**. One can use the connection associated with this metric to compute the Pontrjagin classes $p(T\mathbb{CP}^d)$.

For example, in the exercises we sketch how in the case of $\mathbb{CP}^2$ one finds that $\mathbb{CP}^2$ with its canonical orientation (the one induced from the standard orientation of $\mathbb{S}^5$) has Pontrjagin number

$$\int_M p_1 = 3.$$

The Chern Classes of a Complex Vector Bundle

So far, all our vector bundles were *real* vector bundles. It is also useful to consider **complex vector bundles** $\xi = (\pi, E, M)$, where the fibers E_x are now complex vector spaces of complex dimension r and the transition functions are maps

$$g_{\alpha\beta} : U_\alpha \cap U_\beta \to GL(r, \mathbb{C}).$$

Every complex vector bundle of rank r can be viewed as a real vector bundle of rank $2r$ equipped with a complex structure J, i.e., an endomorphism of (real) vector bundles $J : \xi \to \xi$ such that $J^2 = -\mathrm{id}$. The complex structure J and the complex structure in the fibers determine each other by

$$(a + ib)\mathbf{v} = a\mathbf{v} + bJ(\mathbf{v}), \quad \forall \mathbf{v} \in E.$$

On a complex vector bundle ξ one can consider $\mathbb{C}$-**connections**, i.e., connections ∇ such that for each vector field $X \in \mathfrak{X}(M)$ the map $s \mapsto \nabla_X s$ is $\mathbb{C}$-linear:

$$\nabla_X(\lambda s) = \lambda \nabla_X s, \quad \forall \lambda \in \mathbb{C}, s \in \Gamma(\xi).$$

Using the endomorphism J, this condition can be expressed as

$$\nabla_X(Js) = J\nabla_X s, \quad \forall s \in \Gamma(\xi), X \in \mathfrak{X}(M).$$

Hence, a $\mathbb{C}$-connection is just an ordinary connection which is compatible with the complex structure J, i.e., that satisfies

$$\nabla J = 0.$$

We leave as an exercise to check that any complex vector bundle admits a $\mathbb{C}$-connection.

The connection 1-form ω and the curvature 2-form Ω of a $\mathbb{C}$-connection relative to any local $\mathbb{C}$-basis of sections, defined over an open $U \subset M$, are matrices of complex-valued forms

$$\omega = [\omega_a^b] \in \Omega^1(U, \mathfrak{gl}(r, \mathbb{C})), \quad \Omega = [\Omega_a^b] \in \Omega^2(U, \mathfrak{gl}(r, \mathbb{C})).$$

Hence, using a $\mathbb{C}$-connection, one defines the **Chern–Weil homomorphism** much the same way as in the real case, obtaining now a ring homomorphism into the *complex-valued* de Rham cohomology

$$I(GL(r, \mathbb{C})) \to H^\bullet(M, \mathbb{C}).$$

Again, the ring of invariant polynomials $I(GL(r, \mathbb{C}))$ is generated by the elementary invariant polynomials now viewed as polynomials $\sigma_1, \ldots, \sigma_r$ in $\mathfrak{gl}(r, \mathbb{C})$

$$\det(\lambda I - X) = \lambda^r + \sigma_1(X)\lambda^{r-1} + \cdots + \sigma_r(X)\lambda + \sigma_r(X), \quad X \in \mathfrak{gl}(r, \mathbb{C}).$$

Therefore, one can define a new set of characteristic classes.

Definition 34.2. Let $\xi = (\pi, E, M)$ be a complex vector bundle of rank r. For $k = 1, \ldots, r$, the kth **Chern class** of ξ is

$$c_k(\xi) = \left[\sigma_k \left(\left(\frac{1}{2\pi i} R \right)^k \right) \right] \in H^{2k}(M),$$

where R is the curvature of any $\mathbb{C}$-connection ∇ in ξ. The **total Chern class** of ξ is the sum

$$c(\xi) = 1 + c_1(\xi) + \cdots + c_r(\xi) \in H(M).$$

Note that, *a priori*, the Chern classes are cohomology classes lying in *complex* de Rham cohomology $H^\bullet(M, \mathbb{C})$. However, the presence of i in the normalization factor makes them real cohomology classes. To see this, we use the following lemma which is the complex analogue of the fact that real vector bundles admit fiber metrics and compatible connections. The proof is left as an exercise.

Lemma 34.1. *Every complex vector bundle $\xi = (\pi, E, M)$ admits a fiber Hermitian metric $h = \langle \cdot, \cdot \rangle$ and a compatible $\mathbb{C}$-connection ∇, i.e., one satisfying $\nabla h = 0$.*

Choosing a connection as in the lemma, for any orthonormal $\mathbb{C}$-basis of local sections $\{s_1, \ldots, s_r\}$ of E, the connection 1-form ω and the curvature 2-form Ω take values in the Lie algebra

$$\mathfrak{u}(r) = \{X \in \mathfrak{gl}(r, \mathbb{C}) : X + \overline{X}^T = 0\}.$$

In particular, the eigenvalues of Ω are purely imaginary. Hence, $i\Omega$ has real eigenvalues and it follows that $\sigma_k((R/2\pi i)^k)$ is a real form, showing that the Chern classes are real cohomology classes, as claimed.

Similar to the real case, the Chern classes enjoy the following properties.

Proposition 34.2. *Let M be a smooth manifold, ξ and η complex vector bundles over M. The Chern classes satisfy:*

(i) $c(\xi \oplus \eta) = c(\xi) \cup c(\eta)$;
(ii) $c(\psi^* \xi) = \psi^* c(\xi)$, *for any smooth map $\psi : N \to M$;*
(iii) $c(\xi) = 1$, *if ξ admits a flat $\mathbb{C}$-connection;*
(iv) $c(\gamma_1^1) = 1 - \mu$ *where μ denotes the canonical orientation of $\mathbb{CP}^1$.*

Remark 34.3. One can show that properties (i)–(iv) above determine completely the Chern class.

Proof. We leave the proof of properties (i)–(iii) to the exercises. To prove (iv), we define a $\mathbb{C}$-connection ∇ on the canonical (complex) line bundle γ_1^1 over $\mathbb{CP}^1 = \mathbb{S}^2$ as follows. First, since γ_1^1 is a subbundle of the trivial bundle,

$$\gamma_1^1 \subset \mathbb{CP}^1 \times \mathbb{C}^2,$$

a section of γ_1^1 can be viewed as map $s : \mathbb{CP}^1 \to \mathbb{C}^2$. We define the connection ∇ by

$$(\nabla_X s)(p) := \mathrm{pr}_{E_p}(\mathrm{d}_p s(X)),$$

where $E_p \subset \mathbb{C}^2$ is the fiber over p, and $\mathrm{pr}_{E_p} : \mathbb{C}^2 \to E_p$ denotes the projection relative to the standard Hermitian inner product on $\mathbb{C}^2$.

The bundle γ_1^1 trivializes over the open set

$$U_0 := \{[z_0 : z_1] \in \mathbb{CP}^1 : z_0 \neq 0\}.$$

Namely, we have a non-vanishing section $s : U_0 \to \gamma_1^1$ defined by

$$s([1 : z]) := ([1, z], (1, z)).$$

The $\mathbb{C}$-valued connection 1-form of ∇ is defined by $\nabla_X s = \omega(X)s$. Denoting by (x, y) the coordinates on U_0 given by $z = x + iy$, a straightforward computation yields

$$\omega = \frac{1}{1 + x^2 + y^2} \left((x dx + y dy) + i(-y dx + x dy) \right).$$

It follows from the structure equations that the curvature 2-form is

$$\Omega = d\omega = \frac{2i}{(1 + x^2 + y^2)^2} \, dx \wedge dy.$$

We conclude that the 1st Chern class $c_1(\gamma_1^1)$ is represented by the closed 2-form

$$\sigma_1 \left(\frac{1}{2\pi i} R \right) = -\frac{1}{2\pi i} \Omega = -\frac{dx \wedge dy}{\pi(1 + x^2 + y^2)^2}.$$

To prove (iv), observe that, since U_0 is an open dense set and $(U_0, (x, y))$ is a positive chart, we have

$$\int_M c_1(\gamma_1^1) = -\int_{\mathbb{R}^2} \frac{1}{\pi(1 + x^2 + y^2)^2} dx \, dy$$

$$= -\int_0^{2\pi} \int_0^{+\infty} \frac{r \, dr}{\pi(1 + r^2)^2} \, d\theta$$

$$= -\int_0^{+\infty} \frac{2r \, dr}{(1 + r^2)^2} = \frac{1}{1 + r^2} \Big|_0^{+\infty} = -1.$$

$\square$

One natural way of obtaining complex vector bundles is to start with a **complex manifold** M. Such a manifold is specified by an atlas $\{(U_\alpha, \phi_\alpha)\}$, where the charts are homeomorphisms

$$\phi_\alpha : U_\alpha \to \mathbb{C}^d, \quad x \mapsto (z_\alpha^1(x), \ldots, z_\alpha^d(x))$$

and the transition functions are holomorphic maps

$$\phi_\beta \circ \phi_\alpha^{-1} : \phi_\alpha(U_\alpha \cap U_\beta) \to \phi_\beta(U_\alpha \cap U_\beta),$$

defined on open subsets of $\mathbb{C}^d$. Such charts are called *holomorphic charts*. If we write the coordinates in a holomorphic chart as $z_\alpha^k = x_\alpha^k + iy_\alpha^k$, then we obtain real charts $(x_\alpha^k, y_\alpha^k) : U_\alpha \to \mathbb{R}^{2d}$. Hence, every complex manifold of dimension d has an underlying real smooth structure of dimension $2d$. A basic example of a complex, compact, manifold is the complex projective space $\mathbb{CP}^d$.

For a complex manifold M the tangent bundle TM is a complex vector bundle over M, viewed as a real manifold. This can be seen either by constructing local $\mathbb{C}$-trivializations, using holomorphic charts, or by observing that there is a well defined endomorphism $J : TM \to TM$ with $J^2 = -\mathrm{Id}$, which in local holomorphic coordinates $z_\alpha^k = x_\alpha^k + iy_\alpha^k$ is given by

$$J\left(\frac{\partial}{\partial x^k}\right) = \frac{\partial}{\partial y^k}, \quad J\left(\frac{\partial}{\partial y^k}\right) = -\frac{\partial}{\partial x^k}.$$

Similarly, the cotangent bundle and all the associated bundles are also complex vector bundles over M. Hence, one can define the Chern classes of these bundles. For example, you are asked to show in one exercise at the end of this lecture that the total Chern class of the complex projective space is

$$c(T\mathbb{CP}^d) = (1+a)^{d+1},$$

where $a \in H^2(\mathbb{CP}^d)$ is an appropriate generator.

A holomorphic map preserves the canonical orientation of $\mathbb{C}^d$, so every complex manifold has a canonical orientation. Hence, for a compact complex manifold M of (complex) dimension d, one can define **Chern numbers** by

$$\int_M c_1^{a_1} \wedge c_2^{a_2} \wedge \cdots \wedge c_d^{a_d},$$

where $c_i = c_i(TM)$ and $a_1, \ldots, a_d$ are any non-negative integers such that

$$2(a_1 + 2a_2 + \cdots + da_d) = 2d.$$

Another class of examples of complex vector bundles arises by complexification of a real vector bundle. If $\xi = (\pi, E, M)$ is a real vector bundle of rank r we can form its tensor product with the trivial *real* rank 2 vector bundle $M \times \mathbb{C} \to M$. The resulting bundle, denoted $\xi \otimes \mathbb{C}$, is a real vector bundle of rank 2r admitting the endomorphism $J : \xi \otimes \mathbb{C} \to \xi \otimes \mathbb{C}$ given by

$$J(v \otimes \lambda) := v \otimes i\lambda.$$

Since $J^2 = -\mathrm{Id}$, this defines a complex structure in $\xi \otimes \mathbb{C}$. One calls the resulting complex vector bundle $\xi \otimes \mathbb{C}$ the **complexification** of ξ.

Proposition 34.3. *Let ξ be a real vector bundle. Then the Pontrjagin classes of ξ and the Chern classes of $\xi \otimes \mathbb{C}$ are related by*

$$p_k(\xi) = (-1)^k c_{2k}(\xi \otimes \mathbb{C}).$$

Proof. Immediate from the formulas defining them! $\qquad\qquad\square$

Our discussion of the Pontrjagin classes suggests that the odd classes $c_{2k+1}(\xi \otimes \mathbb{C})$ vanish. To see this, one defines the **complex conjugate** of a complex vector bundle $\xi = (\pi, E, M)$ to be the complex vector bundle $\bar{\xi}$ which, as a real vector bundle, coincides with ξ, but with a complex structure $J_{\bar{\xi}} := -J_\xi$. Note, e.g., that the identity map id: $\xi \to \bar{\xi}$ satisfies

$$\mathrm{id}(\lambda \mathbf{v}) = \bar{\lambda}\, \mathrm{id}(\mathbf{v}), \quad \forall \mathbf{v} \in E, \lambda \in \mathbb{C}.$$

Proposition 34.4. *Let $\xi = (\pi, E, M)$ be a complex vector bundle. The Chern classes of ξ and $\bar{\xi}$ are related by $c_k(\bar{\xi}) = (-1)^k c_k(\xi)$ so that*

$$c(\bar{\xi}) = 1 - c_1(\xi) + c_2(\xi) - \cdots + (-1)^r c_r(\xi).$$

Proof. Let ∇ be a $\mathbb{C}$-connection on ξ. It defines also a $\mathbb{C}$-connection on $\bar{\xi}$ which we denote by $\bar{\nabla}$. If one fixes local trivializing sections $\{s_1, \ldots, s_r\}$ for ξ, then we have

$$\nabla_X s_a = \sum_b \omega_a^b(X) s_b, \quad \bar{\nabla}_X s_a = \sum_b \bar{\omega}_a^b(X) s_b.$$

Hence, the curvature 2-forms of these two connections relative to this basis are related by

$$\Omega_{\bar{\nabla}}(X, Y) = \overline{\Omega_{\nabla}(X, Y)},$$

and it follows that

$$\sigma_k\left(\left(\frac{1}{2\pi i}R_{\overline{\nabla}}\right)^k\right) = \sigma_k\left(\overline{\left(-\frac{1}{2\pi i}R_\nabla\right)^k}\right) = (-1)^k\sigma_k\left(\overline{\left(\frac{1}{2\pi i}R_\nabla\right)^k}\right)$$

$$= (-1)^k\sigma_k\left(\left(\frac{1}{2\pi i}R_\nabla\right)^k\right)$$

so that $c_k(\bar{\xi}) = (-1)^k c_k(\xi)$. $\qquad\qquad\qquad\qquad\qquad\square$

The complexification $\xi \otimes \mathbb{C}$ and its conjugate complex vector bundle $\overline{\xi \otimes \mathbb{C}}$ are isomorphic complex vector bundles. An explicit isomorphism is given by the complex conjugation map

$$\xi \otimes \mathbb{C} \to \overline{\xi \otimes \mathbb{C}}, \quad v \otimes \lambda \mapsto v \otimes \bar{\lambda}.$$

Hence, from Proposition 34.4, we deduce the following.

Corollary 34.1. *Let $\xi = (\pi, E, M)$ be a real vector bundle. Then*

$$c_k(\xi \otimes \mathbb{C}) = 0, \quad \text{if } k \text{ is odd.}$$

Different choices of invariant function lead to other interesting characteristic classes. For example, the invariant function χ : $\mathfrak{gl}(r, \mathbb{C}) \to \mathbb{C}$ given by

$$\chi(X) := \text{tr}(\exp(X)),$$

gives rise to the **Chern character** of the vector bundle

$$\text{ch}(\xi) = \left[\chi\left(\frac{1}{2\pi i}R\right)\right] \in H^\bullet(M).$$

The Chern character satisfies

$$\text{ch}(\xi_1 \oplus \xi_1) = \text{ch}(\xi_1) + \text{ch}(\xi_2), \quad \text{ch}(\xi_1 \otimes \xi_1) = \text{ch}(\xi_1) \cup \text{ch}(\xi_2),$$

i.e., it is a semi-ring homomorphism. For this reason, it is important in K-theory. Other examples of characteristic classes include the Todd class of a complex vector bundle that appears in the Hirzebruch–Riemann–Roch formula in algebraic geometry, or the L-class of a real vector bundle that appears in Hirzebruch's signature formula in differential topology — see, e.g., Hirzebruch (1995) or Husemoller (1994). For a leisurely historical note on Chern classes see also Hirzebruch (2011).

The Euler Class of an Oriented Real Vector Bundle

The presence of extra data on a vector bundle can also lead to special characteristic classes. For example, the **Euler class** of an *oriented* vector bundle $\xi = (\pi, E, M)$ can be viewed as a characteristic class. For that, fix a fiberwise metric g and a connection ∇ compatible with the metric g. Then for any local positive orthonormal basis of sections $\{s_1, \ldots, s_r\}$ the corresponding connection 1-form ω takes values in the Lie algebra $\mathfrak{so}(r)$ consisting of all skew-symmetric matrices. If we change to a new basis of sections $\{s'_1, \ldots, s'_r\}$ the two bases are related by

$$s'_a = \sum_{b=1}^{r} A^b_a s_b, \quad A = [A^b_a] : U \to SO(r).$$

Hence, we now look for invariant functions in $I(SO(r))$ to produce characteristic classes.

The restriction of the elementary invariant polynomials σ_k to $\mathfrak{so}(r)$ give obvious elements in $I(SO(r))$. When r is odd, one can show that these generate all invariant polynomials, but when r is even, this is not true anymore and one needs to add an extra polynomial to obtain a set of generators. This can already be seen for $r = 2$.

Example 34.3. The Lie algebra

$$\mathfrak{so}(2) = \left\{ \begin{pmatrix} 0 & x \\ -x & 0 \end{pmatrix} : x \in \mathbb{R} \right\} \subset \mathfrak{gl}(2, \mathbb{R})$$

is abelian, so the invariance condition is empty. The elementary polynomial $\sigma_1(X) = -\operatorname{tr} X$ restricts to zero, while $\sigma_2(X) = \det X$ restricts to a perfect square

$$\det(X) = x^2.$$

We also have the degree 1 invariant polynomial $\operatorname{Pf} : \mathfrak{so}(2) \to \mathbb{R}$ defined by

$$\operatorname{Pf}(X) = x,$$

which is not generated by $\{\sigma_1, \sigma_2\}$.

One can define for any $r = 2m \geq 2$ an analogous invariant polynomial Pf, called the **Pfaffian**. For that, observe that any skew-symmetric matrix $X \in \mathfrak{so}(2m)$ is conjugate to a block diagonal matrix

$$X = ADA^T, \quad D = \begin{pmatrix} S_1 & & & \\ & S_2 & & \\ & & \ddots & \\ & & & S_m \end{pmatrix}, \quad \text{where } S_k = \begin{pmatrix} 0 & x_k \\ -x_k & 0 \end{pmatrix}.$$

It follows that the determinant of X is a perfect square

$$\det(X) = \left(\det(A) \prod_{i=1}^{m} x_k \right)^2.$$

and one defines the Pfaffian of X to be the function given by

$$\mathrm{Pf}(X) := \det(A) \prod_{i=1}^{m} x_k.$$

The fact that this is a well-defined degree m polynomial follows from the following explicit formula, whose proof is left as an exercise

$$\mathrm{Pf}(X) = \frac{1}{2^m m!} \sum_{\sigma \in S_{2m}} (-1)^\sigma \prod_{k=1}^{m} X_{\sigma(2k-1)\sigma(2k)}.$$

On the other hand, given $B \in SO(2m)$, so that $B^{-1} = B^T$ and $\det B = 1$, we find

$$\mathrm{Pf}(BXB^{-1}) = \det(BA) \prod_{i=1}^{m} x_k = \det(A) \prod_{i=1}^{m} x_k = \mathrm{Pf}(X).$$

Hence, $\mathrm{Pf} \in I^m(SO(2m))$. One can show that the invariant polynomials $\{\sigma_2, \sigma_4, \ldots, \sigma_{2m}, \mathrm{Pf}\}$ are algebraically independent and generate $I(SO(2m))$. They define characteristic classes of an oriented vector bundle.

Theorem 34.2. *Let $\xi = (\pi, E, M)$ be an oriented vector bundle of rank $r = 2m$. Then its Euler class $e(\xi)$ is represented by the form*

$$\mathrm{Pf}\left(\left(\frac{1}{2\pi} R \right)^m \right) \in \Omega^{2m}(M),$$

where R the curvature tensor of any connection ∇ compatible with a fiberwise metric.

We will not go into any more details in the theory of characteristic classes and we will omit the proof of this result. We refer the reader to Husemoller (1994), Kobayashi and Nomizu (1996), or Milnor and Stasheff (1974).

Exercises

Exercise 34.1

Show that every complex vector bundle $\xi = (\pi, E, M)$ admits a $\mathbb{C}$-connection ∇ compatible with a fiber Hermitian metric $h = \langle \cdot, \cdot \rangle$.

Exercise 34.2

Prove the properties of the Pontrjagin classes and the Chern classes stated in Propositions 34.1 and 34.2.

Exercise 34.3

Let $\xi = (\pi, E, M)$ be a complex vector bundle. Show that its $\mathbb{C}$-dual $\xi^* = \operatorname{Hom}(\xi, \mathbb{C})$ is a complex vector bundle and that their Chern classes are related by

$$c_k(\xi^*) = (-1)^k c_k(\xi).$$

Hint: Use a fiber Hermitian metric.

Exercise 34.4

Let γ_d^1 be the canonical complex line bundle over $\mathbb{CP}^d$. Show that

$$c(\gamma_d^1) = 1 - a,$$

where $a \in H^2(\mathbb{CP}^d)$ is an appropriate generator.

Exercise 34.5

Denote by $\varepsilon_{\mathbb{CP}^d}^{d+1} = \mathbb{CP}^d \times \mathbb{C}^{d+1} \to \mathbb{CP}^d$, the trivial complex vector bundle equipped with the standard Hermitian inner product h on the fibers. Let $(\gamma_d^1)^\perp \subset \varepsilon_{\mathbb{CP}^d}^{d+1}$ denote the h-orthogonal bundle to the canonical complex line bundle γ_d^1, so that

$$\varepsilon_{\mathbb{CP}^d}^{d+1} = \gamma_d^1 \oplus (\gamma_d^1)^\perp.$$

(a) Show that there is an isomorphism of complex vector bundles:

$$T\mathbb{CP}^d \simeq \mathrm{Hom}_{\mathbb{C}}(\gamma_d^1, (\gamma_d^1)^{\perp}).$$

(b) Show that there are isomorphisms of complex vector bundles:

$$T\mathbb{CP}^d \oplus \varepsilon_{\mathbb{CP}^d}^1 \simeq \mathrm{Hom}_{\mathbb{C}}(\gamma_d^1, \varepsilon_{\mathbb{CP}^d}^{d+1}) = \mathrm{Hom}_{\mathbb{C}}(\gamma_d^1, \varepsilon_{\mathbb{CP}^d}^1 \oplus \cdots \oplus \varepsilon_{\mathbb{CP}^d}^1).$$

(c) Conclude that the total Chern class of the tangent bundle to $\mathbb{CP}^d$ is

$$c(T\mathbb{CP}^d) = (1 + a)^{d+1},$$

where $a \in H^2(\mathbb{CP}^d)$ is an appropriate generator.

Exercise 34.6
Let $\xi = (\pi, E, M)$ be an oriented vector bundle of rank r. Show that

$$e(\xi)^2 = p_{[r/2]}(\xi).$$

Exercise 34.7
Prove that if a compact, oriented, manifold M of dimension $4m$ can be embedded in $\mathbb{R}^{4m+1}$ then all its Pontrjagin classes must vanish: $p(TM) = 1$.

Hint: The normal bundle $\nu(M)$ is trivial.

Exercise 34.8
Two oriented manifolds M_1 and M_2 are said to be cobordant if $\dim M_1 = \dim M_2$ and there exists an oriented manifold with boundary N such that, as oriented manifolds,

$$\partial N = M_1 - M_2,$$

where $-M_2$ denotes M_2 with the opposite orientation. Show that if M_1 and M_2 are compact oriented cobordant manifolds of dimension $4m$ then they must have the same Pontrjagin numbers.

Hint: Show first that if $M = \partial N$, where N is compact, oriented, then the Pontrjagin numbers of M must vanish. For this, choose a connection ∇ on N with the property that $\nabla_X Y$ is tangent to ∂N whenever X and Y are tangent to ∂N.

Lecture 35

Fiber Bundles

Bundles with fiber which are not vector spaces also occur frequently in Differential Geometry. We will study them briefly in these last two lectures.

Let $\pi : E \to M$ be a surjective submersion. A **trivializing chart** for π with fiber type F is a pair (U, ϕ), where $U \subset M$ is an open set and $\phi : \pi^{-1}(U) \to U \times F$ is a diffeomorphism such that the following diagram commutes

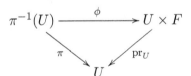

Each fiber $E_p = \pi^{-1}(p)$ is diffeomorphic to F via the diffeomorphism

$$\phi^p : E_p \xrightarrow{\phi} \{p\} \times F \longrightarrow F .$$

Using this map, we can write $\phi(\mathbf{v}) = (p, \phi^p(\mathbf{v}))$ if $\mathbf{v} \in E_p$. Given two trivializing charts (U_α, ϕ_α) and (U_β, ϕ_β), we have a transition map

$$\phi_\alpha \circ \phi_\beta^{-1} : (U_\alpha \cap U_\beta) \times F \to (U_\alpha \cap U_\beta) \times F,$$

$$(p, f) \mapsto (p, \phi_\alpha^p \circ (\phi_\beta^p)^{-1}(f)).$$

The second component of this map yields the **transition functions**

$$g_{\alpha\beta} : U_\alpha \cap U_\beta \to \mathrm{Diff}(F), \quad g_{\alpha\beta}(p) := \phi_\alpha^p \circ (\phi_\beta^p)^{-1}.$$

If one is given a covering of M by trivializing charts $\{(U_\alpha, \phi_\alpha) : \alpha \in A\}$, this leads to a cocycle $\{g_{\alpha\beta}\}$ with values in the group $\mathrm{Diff}(F)$, generalizing the cocycles for vector bundles discussed in Lecture 27. We would like this to determine the fiber bundle and recover the bundle from the cocycle. However, $\mathrm{Diff}(F)$ is infinite dimensional which poses some difficulties. For this reason, we will restrict our attention to fiber bundles for which the transition functions take values in a finite-dimensional Lie group $G \subset \mathrm{Diff}(F)$, so we consider cocycles

$$g_{\alpha\beta} : U_\alpha \cap U_\beta \to G \subset \mathrm{Diff}(F).$$

Equivalently, we assume that we have an *effective* action of a Lie group G on F and that the transition functions take the form:

$$\phi_\alpha^p \circ (\phi_\beta^p)^{-1}(f) = g_{\alpha\beta}(p) \cdot f,$$

for a map $g_{\alpha,\beta} : U_\alpha \cap U_\beta \to G$. Our formal definition of a G-fiber bundle is then the following.

Definition 35.1. Let G be a Lie group and $G \times F \to F$ a smooth, effective, action. A **G-fiber bundle** over M with fiber type F is a triple $\xi = (\pi, E, M)$, where $\pi : E \to M$ is a smooth map admitting a collection of trivializing charts $\mathcal{C} = \{(U_\alpha, \phi_\alpha) : \alpha \in A\}$ with fiber type F, satisfying the following properties:

(i) $\{U_\alpha : \alpha \in A\}$ is an open cover of M: $\bigcup_{\alpha \in A} U_\alpha = M$;

(ii) For any $\alpha, \beta \in A$ there are smooth maps $g_{\alpha\beta} : U_\alpha \cap U_\beta \to G$ such that the transition functions take the form

$$(p, f) \mapsto (p, g_{\alpha\beta}(p) \cdot f);$$

(iii) The collection $\mathcal{C}$ is maximal: if (U, ϕ) is a trivializing chart of fiber type F with the property that for every $\alpha \in A$, there exist $g_\alpha : U \cap U_\alpha \to G$ such that

$$\phi^p \circ (\phi_\alpha^p)^{-1}(f) = g_\alpha(p) \cdot f, \quad \forall f \in F,$$

then $(U, \phi) \in \mathcal{C}$.

We shall use the same notation as in the case of vector bundles, so we have the **total space**, the **base space**, and the **projection** of the G-fiber bundle. Also, one calls G the **structure group** of the fiber bundle. Given a G-fiber bundle ξ a subcollection of charts of $\mathcal{C}$ which still covers M is called an **atlas** or a **trivialization** of ξ. We define a **section** over an open set U in the obvious way and we denote the set of all sections over U by $\Gamma_U(E)$. Although a fiber bundle always has local sections, it may fail to have global sections.

A morphism of G-fiber bundles can be defined in a fashion similar to the definition of a morphism of vector bundles, where we replace $GL(r)$ by the structure group G.

Definition 35.2. Let $\xi = (\pi, E, M)$ and $\xi' = (\pi', E', M')$ be two G-fiber bundles with the same fiber F and structure group G. A **morphism of G-fiber bundles** consists of a pair of maps

$$
\begin{array}{ccc}
E & \xrightarrow{\ \Psi\ } & E' \\
\pi \downarrow & & \downarrow \pi' \\
M & \xrightarrow[\ \psi\ ]{} & M'
\end{array}
$$

such that for each $p \in M$, the map between the fibers

$$\Psi^p := \Psi|_{E_p} : E_p \to E'_q, \quad (q = \psi(p)),$$

satisfies

$$\phi'^q_\beta \circ \Psi^p \circ (\phi^p_\alpha)^{-1} \in G,$$

for any trivializations $\{\phi_\alpha\}$ of ξ and $\{\phi'_\beta\}$ of ξ'.

In this way, we have the category of fiber bundles with fiber type F and structure group G. Just like in the case of vector bundles, we shall also distinguish between *equivalence* and *isomorphism* of G-fiber bundles, according to whether the base map is the identity map or not.

Among the most important classes of G-fiber bundles we have:

• **Vector bundles:** In this case, the fiber F is a vector space and the structure group is the group of linear invertible transformations $G = GL(V)$. These are precisely the bundles that we have studied in the previous lectures.

- **Principal G-bundles:** In this case, the fiber F is itself a Lie group G and the structure group is the same Lie group G acting on itself by translations $G \times G \to G$, $(g, h) \mapsto gh$. Principal bundles play a central role among all G-fiber bundles for reasons that will be clear later.

The set of transition functions associated with an atlas of a G-fiber bundle completely determined the bundle. The discussion is entirely analogous to the case of vector bundles. First, if $\xi = (\pi, E, M)$ is a G-fiber bundle, the transition functions $g_{\alpha\beta} : U_\alpha \cap U_\beta \to G$ relative to some trivialization $\{(U_\alpha, \phi_\alpha)\}$ satisfy the **cocycle condition**

$$g_{\alpha\beta}(p) g_{\beta\gamma}(p) = g_{\alpha\gamma}(p), \quad (p \in U_\alpha \cap U_\beta \cap U_\gamma).$$

We say that two cocycles $\{g_{\alpha\beta}\}$ and $\{g'_{\alpha\beta}\}$ are **equivalent** if there exist smooth maps $\lambda_\alpha : U_\alpha \to G$ such that

$$g'_{\alpha\beta}(p) = \lambda_\alpha(p) \cdot g_{\alpha\beta}(p) \cdot \lambda_\beta^{-1}(p), \quad (p \in U_\alpha \cap U_\beta).$$

One checks easily the following analogue of Proposition 27.1.

Proposition 35.1. *Let M be a manifold and G a Lie group acting on another smooth manifold F. Given a cocycle $\{g_{\alpha\beta}\}$ with values in G, subordinated to a covering $\{U_\alpha\}$ of M, there exists a G-fiber bundle $\xi = (\pi, E, M)$ with fiber type F which admits an atlas $\{\phi_\alpha\}$, for which the set of transition functions is $\{g_{\alpha\beta}\}$. Two equivalent cocycles determine isomorphic G-fiber bundles.*

Let $\xi = (\pi, E, M)$ be a G-fiber bundle with fiber type F and let $\{g_{\alpha\beta}\}$ be a cocycle associated with a trivialization $\{\phi_\alpha\}$ of ξ. If $H \subset G$ is a Lie subgroup, we say that the **structure group of ξ can be reduced to H** if the cocycle is equivalent to a cocycle $\{g'_{\alpha\beta}\}$ which take values in H

$$g'_{\alpha\beta} : U_\alpha \cap U_\beta \to H \subset G.$$

We will see later how to describe this notion independently of choice of trivializations. The next examples illustrate how the structure group and its possible reductions are intimately related to geometric properties of the bundle.

Example 35.1. G-fiber bundle $\xi = (\pi, E, M)$ with fiber type F and is called **trivial** if it is isomorphic to the trivial bundle pr $: M \times F \to M$. This is the case if and only if its structure group can be reduced to the trivial group $\{e\}$.

Example 35.2. We saw before that a vector bundle of rank r is orientable if and only if its structure group can be reduced to $GL^+(r)$. Similarly, a vector bundle admits a fiber metric if and only if its structure group can be reduced to $O(r)$ (and by the polar decomposition, this can always be achieved). A further reduction of its structure group to $SO(r)$ amounts to an additional choice of orientation for the bundle (which may or may not be possible).

Remark 35.1. The specification of the structure group is crucial. For example, a G-cocycle may take values in a subgroup $H \subset G$ and be trivial as a G-cocycle, but not as an H-cocycle. An example is discussed in the exercises.

The cocycles associated with a G-fiber bundle, as well as the notion of equivalence of cocycles, do not make any use of a G-action on F. For this reason principal G-bundles play a fundamental role among all G-fiber bundles. In fact, they can be used to build any other G-fiber bundle. At the level of cocycles one has:

- Given a principal G-bundle $\xi = (\pi, P, M)$, a trivialization $\{\phi_\alpha\}$ of ξ determines a cocycle $\{g_{\alpha\beta}\}$ with values in G. If G acts in F we obtain a G-fiber bundle $\xi_F = (\pi, E, M)$ with fiber type F.
- Conversely, given a G-fiber bundle $\xi_F = (\pi, E, M)$ and fixing a trivialization $\{\phi_\alpha\}$ of ξ_F, the associated cocycle $\{g_{\alpha\beta}\}$ takes values in G. Since G acts on itself by translations this cocycle defines a principal G-bundle $\xi = (\pi, P, M)$.

To make this more explicit, we observe that principal G-bundles can also be described more succinctly as follows.

Proposition 35.2. *A fiber bundle $\xi = (\pi, P, M)$ is a principal G-bundle if and only if there exists a right action $P \times G \to P$ satisfying the following properties*:

(i) *The action is free and proper.*
(ii) *M is diffeomorphic to P/G and under this identification $\pi\colon P \to M \simeq P/G$ is the quotient map.*

(iii) *The local trivializations (U, ϕ) are G-equivariant:* $\phi^p(\mathbf{v} \cdot g) = \phi^p(\mathbf{v})g$.

Proof. Given a principal G-bundle $\xi = (\pi, P, M)$ one constructs a right action $P \times G \to P$ working on trivializing charts (U, ϕ): the action of G on $\pi^{-1}(U)$ is defined by

$$u \cdot g := \phi^{-1}(p, \phi^p(u)g), \quad (p = \pi(u)).$$

One checks easily that this definition is independent of the choice of trivialization. The rest of the statements are left as an exercise. $\square$

Conversely, by an exercise in Lecture 17, if one is given a free and proper right action $P \times G \to P$, one obtains a principal G-bundle $\xi = (\pi, P, M)$ by setting $M = P/G$ and letting $\pi : P \to M$ be the quotient map. We will call a *free* and *proper* right action a **principal action**. So principal bundles amount simply to principal actions.

If one is given a principal action $P \times G \to P$ and an action $G \times F \to F$ one can form the **associated fiber bundle** $\xi_F = (\pi_F, E, M)$. This is the G-fiber bundle with total space

$$E := P \times_G F,$$

the quotient space for the right action of G on $P \times F$ defined by

$$(u, f) \cdot g := (u \cdot g, g^{-1} \cdot f)$$

(recall that G acts on the right in P and on the left in F). The projection map $\pi_F : E \to M$ is given by: $\pi_F([u, f]) = \pi(u)$, where $\pi : P \to M$ is the quotient map of the action.

These descriptions of principal G-bundles and the associated bundles allows one to give many examples of G-fiber bundles.

Example 35.3. For any Lie group G and manifold M, the trivial principal G-bundle over M is $\mathrm{pr}_M : M \times G \to M$. Sections of this bundle are just smooth maps $M \to G$. Moreover, if G acts on some space F, then the associated bundle is the trivial bundle $\mathrm{pr}_M : M \times F \to M$.

Example 35.4. For any Lie group G and any closed subgroup $H \subset G$ the right action of H on G is principal, so the quotient $G \to G/H$ is a principal H-bundle. For example, if we let $\mathbb{S}^3$ be the group of unit quaternions and let $\mathbb{S}^1 \subset \mathbb{S}^3$ be the subgroup of unit complex numbers, then we obtain a principal $\mathbb{S}^1$-bundle, which is easily seen to be isomorphic to the Hopf fibration.

Example 35.5. If $\pi : \tilde{M} \to M$ is the universal covering space of a manifold M, the triple $(\pi, \tilde{M}, M)$ is a principal bundle with structure group the fundamental group $\pi_1(M)$, equipped with the discrete topology. More generally, if $H \subset \pi_1(M)$ is a normal subgroup then the covering space $P := \tilde{M}/H \to M$ with group of deck transformations $G := \pi_1(M)/H$ is also a principal G-bundle.

Example 35.6. Let M be a manifold of dimension d. The **frame bundle** is the principal bundle $\pi : F(M) \to M$ with structure group $GL(d)$ whose fiber over $p \in M$ consists of the set of all ordered basis (i.e., frames) of $T_p M$

$$F(M)_p = \{(\mathbf{v}_1, \ldots, \mathbf{v}_d) : \mathbf{v}_1, \ldots, \mathbf{v}_d \text{ is a basis of } T_p M\}.$$

The group $GL(d)$ acts principally on the right on $F(M)$: if $u = (\mathbf{v}_1, \ldots, \mathbf{v}_d)$ is a frame and $A = (a_i^j)$ is an invertible matrix, then $u \cdot A = (\mathbf{w}_1, \ldots, \mathbf{w}_d)$ is the frame

$$\mathbf{w}_i = \sum_{j=1}^{d} a_i^j \mathbf{v}_j, \quad (i = 1, \ldots, d).$$

The group $GL(d)$ acts (on the left) in $\mathbb{R}^d$ by matrix multiplication. Hence, we can form an associated fiber bundle with fiber $\mathbb{R}^d$, i.e., a vector bundle. We leave it as an exercise to check that this bundle is canonically isomorphic to the tangent bundle $T(M)$. Similarly, one obtains the cotangent bundle, exterior bundles, tensor bundle, etc., if one considers instead the natural actions of $GL(d)$ on $(\mathbb{R}^d)^*$, $\wedge^k \mathbb{R}^d$, $\otimes^k \mathbb{R}^d$, etc.

Example 35.7. More generally, for any (real) vector bundle $\pi : E \to M$ of rank r, the frames of the fibers form a principal bundle with structure group $GL(r)$, denoted $F(E)$. For the usual action of $GL(r)$ on $\mathbb{R}^r$ one obtains an associated bundle to $F(E)$ with fiber $\mathbb{R}^r$, which is canonically isomorphic to the original vector bundle $\pi : E \to M$. Similarly, one can obtain as associated bundles E^*, $\wedge^k E$, $\otimes^k E$, etc. One can also consider complex vector bundles and the bundle of complex frames where $GL(d)$ is replaced by $GL(d, \mathbb{C})$.

The description of a G-fiber bundle as the associated bundle to a principal G-bundle, allows one to express properties of the G-fiber bundle in terms of the principal G-bundle. The next proposition illustrates this.

Proposition 35.3. *Let* $\xi = (\pi, P, M)$ *be a principal G-bundle and* $G \times F \to F$ *a smooth action. The sections of the associated bundle* $\xi_F = (\pi, E, M)$ *are in one-to-one correspondence with the G-equivariant maps* $h : P \to F$.

Proof. The total space of the associated bundle is

$$E = P \times_G F = (P \times F)/G.$$

An element $v \in E_p$ is an equivalence class in $P_p \times_G F$, which can be written as

$$v = [(u, h_p(u))], \quad \forall u \in P_p,$$

for a unique, G-equivariant, map $h_p : P_p \to F$. The G-equivariance means that $h_p(u \cdot g) = g^{-1} \cdot h_p(u)$, for all $g \in G$. Hence, a section $s : M \to E$ can be written in the form

$$s(p) = [(u, h(u))] \quad \text{with } u \in P \text{ such that } \pi(u) = p,$$

for a unique G-equivariant map $h : P \to F$. Conversely, any G-equivariant map $h : P \to F$ determines a section of ξ_F through this formula. □

A general G-fiber bundle $\xi_F = (\pi, E, M)$ may not have any sections. Moreover, if it admits a section, the bundle need not be trivial (consider, e.g., vector bundles). However, a principal G-bundle is trivial if and only if it admits a section, a fact which we leave as an exercise. Another important general fact, which we will not prove, is the following.

Theorem 35.1. *Let* $\xi_F = (\pi, E, M)$ *be a G-fiber bundle with contractible fiber F. Then* ξ_F *admits a section and any two sections of* ξ_F *are homotopic.*

In order to understand the issue of reduction of the structure group without referring to cocycles, it is convenient to enlarge the concept of morphism of principal bundles to allow for different structure groups.

Definition 35.3. Let $\xi' = (\pi', P', M')$ be a principal G'-bundle, $\xi = (\pi, P, M)$ a principal G-bundle and $\phi : G' \to G$ a Lie group homomorphism. A ϕ-**morphism** $\Psi : \xi' \to \xi$ is a map $\Psi : P' \to P$ such that

$$\Psi(u \cdot g) = \Psi(u)\phi(g), \quad \forall u \in P', \quad g \in G'.$$

A ϕ-morphism of principal bundles $\Psi : \xi' \to \xi$ takes fibers to fibers so it covers a smooth map $\psi : M' \to M$ making the following diagram commute

$$
\begin{array}{ccc}
P' & \xrightarrow{\ \Psi\ } & P \\
{\scriptstyle \pi'}\downarrow & & \downarrow{\scriptstyle \pi} \\
M' & \xrightarrow[\ \psi\ ]{} & M
\end{array}
$$

If $\Psi : P' \to P$ and $\phi : G' \to G$ are both embeddings, one can identify P' and G' with its images $\Psi(P') \subset P$ and $H := \Phi(G') \subset G$. We then say that ξ' is a **subbundle** of the principal bundle ξ. When $M' = M$ and $\psi =$ id we say that ξ' is a **reduced subbundle** of ξ. It is not hard to check that this matches the notion of reduction of the structure group from G to H that we have introduced before using cocycles.

Example 35.8. If M carries a Riemannian structure, then we can consider the **orthogonal frame bundle** whose fiber is

$$OF(M)_p = \{(\mathbf{v}_1, \ldots, \mathbf{v}_d) \text{ an orthonormal basis of } T_pM\}.$$

This is a principal $O(d)$-bundle, which is a reduced subbundle of $F(M)$, obtained by reduction of the structure group from $GL(d)$ to $O(d)$. In general, a reduction of $F(M)$ to a closed subgroup $G \subset GL(d)$ is called a G-**structure** on M. We leave the details as an exercise.

Exercises

Exercise 35.1
Give a proof of Proposition 35.1

Exercise 35.2
Consider the covering of $M = \mathbb{S}^1$ by the open sets:

$$U_\pm = \{(x,y) \in \mathbb{R}^2 : x^2 + y^2 = 1\} - \{(\pm 1, 0)\}.$$

Define a cocycle $\{g_\pm\}$ (with only one element!) relative to this covering by

$$g_\pm(x,y) = \begin{cases} I & \text{if } (x,y) \in y > 0, \\ -I & \text{if } (x,y) \in y < 0. \end{cases}$$

where I is the 2×2 identity matrix Show that this cocycle defines:

(a) a G-fiber bundle with fiber type $\mathbb{S}^1$ and structure group $SO(2)$ which is isomorphic, as an $SO(2)$-bundle, to the trivial bundle.
(b) a G-fiber bundle with fiber type $\mathbb{S}^1$ and structure group $\mathbb{Z}_2 = \{I, -I\}$ which is not isomorphic, as a $\mathbb{Z}_2$-bundle, to the trivial bundle.

Exercise 35.3
Complete the proof of Proposition 35.2.

Exercise 35.4
Given a principal action $P \times G \to P$ and an action $G \times F \to F$ show that the associated fiber bundle $\xi_F = (\pi_F, E, M)$ satisfies the axioms of a G-fiber bundle of Definition 35.1.

Exercise 35.5
Show that a principal bundle is trivial if and only if it has a global section.

Note: This exercise is a very special case of the next exercise.

Exercise 35.6

Let $\xi = (\pi, P, M)$ be a principal G-bundle and $H \subset G$ a closed subgroup. Since G acts in the quotient G/H, there is an associate bundle $\xi_{G/H} = (\pi', P \times_G (G/H), M)$. Show that this bundle can be identified with the quotient $(\pi', P/H, M)$, where $\pi' : P/H \to M$ is the map induced by $\pi : P \to M$, and that the following statements are equivalent:

(a) The structure group of ξ can be reduced to H;
(b) The associated bundle $\xi_{G/H}$ has a section;
(c) There exists a G-equivariant map $h : P \to G/H$.

Exercise 35.7

Let M be a Riemannian manifold and let $\pi : OF(M) \to M$ be the principal $O(d)$-bundle formed by the orthogonal frames:

$$OF(M)_p = \{(\mathbf{v}_1, \dots, \mathbf{v}_d) \text{ an orthonormal base of } T_p M\}.$$

Show that $OF(M)$ is a reduced bundle of $F(M)$ relative to the natural inclusion $O(d) \subset GL(d)$. Conversely, prove that any reduction of the frame bundle $F(M)$ to a $O(d)$-bundle $P \to M$ is canonically isomorphic to the orthogonal frame bundle $OF(M)$ for a unique Riemannian metric on M.

Exercise 35.8

Let M be a manifold of dimension $2d$. An **almost complex structure** on M is an endomorphism $J : TM \to TM$ such that $J^2 = -\mathrm{Id}$, so such a structure makes TM into a complex vector bundle. Consider the natural inclusion $GL(d, \mathbb{C}) \subset GL(2d)$, arising from the canonical identification $\mathbb{C}^d \simeq \mathbb{R}^{2d}$. Show that a reduction of $F(M)$ to a $GL(d, \mathbb{C})$-bundle is canonically isomorphic to the bundle of complex frames for a unique almost complex structure J on M.

Principal Fiber Bundles

If $\xi = (\pi, P, M)$ is a principal G-bundle and $G \times F \to F$ is a smooth action, it is natural to expect that *any functorial construction in the associated bundle $\xi_F = (\pi, E, M)$ can be expressed in terms of ξ and F*. We have seen examples of this principle in the last lecture — see, e.g., Proposition 35.3. As another instance of this principle, we will now discuss a notion of connection on a principal G-bundle, which, as we will see later, induces a connection on any associated vector bundle.

Definition 36.1. Let $\xi = (\pi, P, M)$ be a principal G-bundle. A **principal bundle connection** in ξ is a distribution $H \subset TP$ such that

(i) H is horizontal, i.e., for all $u \in P$

$$T_u P = H_u \oplus \ker d_u \pi;$$

(ii) H is G-invariant, i.e., for all $g \in G$, and $u \in P$

$$H_{ug} = (R_g)_* H_u,$$

where $R_g : P \to P$, $u \mapsto ug$, denotes translation by g.

Given a connection H on a principal G-bundle $\xi = (\pi, P, M)$, we call H_u the **horizontal space** and $V_u := \ker d_u \pi$ the **vertical space** at $u \in P$. An arbitrary tangent vector $\mathbf{v} \in T_u P$ has a unique decomposition

$$\mathbf{v} = h(\mathbf{v}) + v(\mathbf{v}) \text{ where } h(\mathbf{v}) \in H_u, v(\mathbf{v}) \in V_u.$$

Hence, any vector field $X \in \mathfrak{X}(P)$ on the total space of the bundle also splits into a horizontal vector field $h(X)$ and a vertical vector field $v(X)$.

Example 36.1. Let $\xi_F = (\pi, E, M)$ be a vector bundle furnished with a connection ∇. Denote by $\xi = (\pi, F(E), M)$ the bundle of frames of ξ_F (see Example 35.6). If $u = (\mathbf{v}_1, \ldots, \mathbf{v}_r) \in F(E)$ is a frame and $c : I \to M$ is a curve with $c(0) = \pi(u)$, then the vector fields $X_1, \ldots, X_r$ along $c(t)$ obtained by parallel transport of $\mathbf{v}_1, \ldots, \mathbf{v}_r$ determine a curve $u(t) = (X_1(t), \ldots, X_r(t))$ in $F(E)$. We consider all the curves $u(t)$ obtained in this way and we define the subspace

$$H_u := \left\{ u'(0) \in T_u F(E) : \text{ for all curves } u(t) \right\}.$$

This yields a C^∞ distribution $u \mapsto H_u$ which satisfies conditions (i) and (ii) of Definition 36.1. Hence, every connection ∇ in a vector bundle determines a principal bundle connection H in the corresponding bundle of frames.

Let $\xi = (\pi, P, M)$ be a principal G-bundle. The G-action on P induces an infinitesimal Lie algebra action $\psi : \mathfrak{g} \to \mathfrak{X}(P)$. For each $X \in \mathfrak{g}$ the vector field $\psi(X)$ is vertical, and the map $X \mapsto \psi(X)|_u$ gives a linear isomorphism $\mathfrak{g} \simeq V_u$. Therefore, given a principal bundle connection H, for any $\mathbf{v} \in T_u P$ we can take its vertical component $v(\mathbf{v})$ and find a unique $X_{\mathbf{v}} \in \mathfrak{g}$ is such that $\psi(X_{\mathbf{v}})|_u = v(\mathbf{v})$.

Definition 36.2. The **connection 1-form** of a principal bundle connection H on $\xi = (\pi, P, M)$ is the $\mathfrak{g}$-valued 1-form $\omega \in \Omega^1(P; \mathfrak{g})$ given by

$$\omega(\mathbf{v}) := X_{\mathbf{v}}.$$

Note that $\omega(\mathbf{v}) = 0$ iff $\mathbf{v}$ is a horizontal vector, so ω uniquely determines the distribution H. Indeed, the following proposition states that the connection 1-form completely characterizes the connection H. The proof is left as an exercise.

Proposition 36.1. *Let* $\xi = (\pi, P, M)$ *be a principal G-bundle. Given a principal bundle connection H on P its connection 1-form* ω *satisfies*:

(i) $\omega(\psi(X)) = X$, *for all* $X \in \mathfrak{g}$;
(ii) $(R_g)_*\omega = \mathrm{Ad}_{g^{-1}}\,\omega$, *for all* $g \in G$.

Conversely, if $\omega \in \Omega^1(P; \mathfrak{g})$ *satisfies* (i) *and* (ii), *there exists a unique principal bundle connection H in P whose connection 1-form is* ω.

We leave as an exercise to show that the previous description of principal bundle connections in terms of connections 1-forms implies the existence of connections on any principal bundle using a partition of unity.

Let us introduce now the curvature of a connection H on a principal G-bundle $\xi = (\pi, P, M)$. For that, we define the *exterior covariant derivative* associated with H to be the differential operator $D : \Omega^k(P; \mathfrak{g}) \to \Omega^{k+1}(P; \mathfrak{g})$ given by

$$(D\theta)(X_0, \dots, X_k) = (\mathrm{d}\theta)(h(X_0), \dots, h(X_k)), \quad (X_0, \dots, X_k \in \mathfrak{X}(P)).$$

Definition 36.3. The **curvature 2-form** of a connection H with connection 1-form $\omega \in \Omega^1(P, \mathfrak{g})$ is the $\mathfrak{g}$-valued 2-form

$$\Omega := D\omega \in \Omega^2(P, \mathfrak{g}).$$

Fix a trivialization $\{(U_\alpha, \phi_\alpha)\}$ of the principal G-bundle $\xi = (\pi, P, M)$. Then we have local sections

$$s_\alpha : U_\alpha \to P, \ p \mapsto \phi_\alpha^{-1}(p, e),$$

where $e \in G$ denotes the identity element. The connection 1-form ω determines a family of **local connection 1-forms**

$$\omega_\alpha := (s_\alpha)^*\omega \in \Omega^1(U_\alpha; \mathfrak{g}).$$

On the other hand, the curvature 2-form Ω determines a family of **local curvature 2-forms**

$$\Omega_\alpha := (s_\alpha)^*\Omega \in \Omega^2(U_\alpha; \mathfrak{g}).$$

An exercise at the end of this lecture discusses how these local forms are related to overlaps.

Example 36.2. We saw in Example 36.1 that a connection ∇ on a vector bundle $\pi : E \to M$ determines a connection H on the principal $GL(r)$-bundle of frames $\pi : F(E) \to M$. The associated connection 1-form and curvature 2-form takes values in the Lie algebra $\mathfrak{gl}(r)$.

A trivialization $\{(U_\alpha, \phi_\alpha)\}$ for the vector bundle $\pi : E \to M$ yields also a trivialization for the bundle of frames $\pi : F(E) \to M$. The matrices of connection 1-forms $\omega_\alpha = [\omega_a^b]$ and curvature 2-forms $\Omega_\alpha = [\Omega_a^b]$ associated with ∇ agree with the local connection 1-form and curvature 2-form of the principal connection H.

In order to discuss the structure equations of a principal connection, we define the bracket of two $\mathfrak{g}$-valued 1-forms $\eta_1, \eta_2 \in \Omega^1(P, \mathfrak{g})$ to be the $\mathfrak{g}$-valued 2-form $[\eta_1, \eta_2]$ given by

$$[\eta_1, \eta_2](X, Y) := [\eta_1(X), \eta_2(Y)] - [\eta_1(Y), \eta_2(X)].$$

Theorem 36.1. *Let H be a connection on a principal G-bundle $\xi = (\pi, P, M)$, with connection 1-form ω and curvature 2-form Ω. Then the following hold:*

(i) *Structure equation:* $\Omega = \mathrm{d}\omega + \frac{1}{2}[\omega, \omega]$.
(ii) *Bianchi's identity:* $D\Omega = 0$.

There is also a notion of *parallel transport* for principal bundle connections. Given a connection H on a principal G-bundle $\xi = (\pi, P, M)$, if $X \in \mathfrak{X}(M)$ is a vector field on the base M, there exists a unique vector field $\widetilde{X}$ in the total space P which is horizontal and is π-related to X:

$$\pi_* \widetilde{X} = X.$$

One calls $\widetilde{X} \in \mathfrak{X}(P)$ the **horizontal lift** of $X \in \mathfrak{X}(M)$. The following result, stating the most important properties of the horizontal lift, follows immediately from the definitions.

Proposition 36.2. *Let $X, Y \in \mathfrak{X}(M)$ and $f \in C^\infty(M)$. Then,*

(i) $\widetilde{X} + \widetilde{Y}$ *is the horizontal lift of $X + Y$;*
(ii) $(\pi^* f)\widetilde{X}$ *is the horizontal lift of fX;*
(iii) $h([\widetilde{X}, \widetilde{Y}])$ *is the horizontal lift of $[X, Y]$.*

Note that, by property (iii), the vector field

$$[\widetilde{X}, \widetilde{Y}] - \widetilde{[X,Y]} \in \mathfrak{X}(P),$$

is vertical. This leads to a geometric interpretation of curvature, whose proof we also leave as an exercise.

Theorem 36.2. *Let H be a connection on a principal G-bundle $\xi = (\pi, P, M)$, with curvature 2-form $\Omega \in \Omega^2(P; \mathfrak{g})$. For any local section $s : U \to P$ and vector fields $X, Y \in \mathfrak{X}(U)$ we have*

$$(s^*\Omega)(X,Y)_p^* = \left([\widetilde{X}, \widetilde{Y}] - \widetilde{[X,Y]}\right)_{s(p)}.$$

A **flat connection** is a connection whose curvature form vanishes identically. Since the horizontal lifts $\widetilde{X}$ of vector fields X in $\mathfrak{X}(M)$ generate the horizontal distribution of the connection, we deduce the following.

Corollary 36.1. *A connection is flat if and only if its horizontal distribution is integrable.*

In order to define parallel transport we define the **horizontal lift of of a curve** $c : I \to M$ to be a curve $u : I \to P$ such that $\pi(u(t)) = c(t)$ and $\dot{u}(t)$ is horizontal for all $t \in I$.

Proposition 36.3. *Let H be a connection on a principal G-bundle $\xi = (\pi, P, M)$. If $c : I \to M$ is a curve and $u_0 \in \pi^{-1}(c(0))$, there exists a unique horizontal lift $u : I \to P$ of $c(t)$ with $u(0) = u_0$.*

Proof. Local triviality of the bundle shows that we can always lift $c(t)$ to a curve $v : I \to P$, such that $v(0) = u_0$ and $\pi(v(t)) = c(t)$. The horizontal lift $u : I \to P$ through u_0, if it exists, takes the form

$$u(t) = v(t)g(t),$$

for some curve $g : I \to G$ with $g(0) = e$. If ω denotes the connection 1-form of H, differentiating the last expression gives

$$\omega(\dot{u}(t)) = \mathrm{Ad}_{g(t)^{-1}} \omega(\dot{v}(t)) + g(t)^{-1}\dot{g}(t),$$

where $t \mapsto g(t)^{-1}\dot{g}(t) := \mathrm{d}_{g(t)}L_{g(t)^{-1}}\dot{g}(t)$ is a curve in the Lie algebra $\mathfrak{g}$. The curve $u(t)$ will be horizontal iff $g(t)$ satisfies the equation

$$g(t)^{-1}\dot{g}(t) = -\mathrm{Ad}_{g(t)^{-1}} \omega(\dot{v}(t)).$$

Hence, the proof is completed by applying the following lemma.

Lemma 36.1. *Let G be a Lie group with Lie algebra $\mathfrak{g}$. If $t \mapsto X(t)$ is a curve in $\mathfrak{g}$, then there exists a unique curve $g : I \to G$, with $g(0) = e$, satisfying:*

$$g(t)^{-1}\dot{g}(t) = X(t), \quad (t \in I).$$

The equation appearing in the lemma is a linear o.d.e. with time dependent coefficients. By general results about such o.d.e. it has solutions for all times t for which the coefficients are defined. We leave the details to the reader. $\square$

Using the previous proposition, given a curve $c : I \to M$, we can proceed to define **parallel transport** along c to be the map $\tau_t : P_{c(0)} \to P_{c(t)}$ given by

$$\tau_t(u_0) := u(t),$$

where $u(t)$ is the unique horizontal lift $u : I \to P$ of $c(t)$ such that $u(0) = u_0$. Note that we can also define parallel transport along curves which are only piecewise smooth, by making parallel transport successively along its smooth components.

Proposition 36.4. *Parallel transport along a piecewise smooth curve $c : [0, 1] \to M$ commutes with the G-action, i.e.,*

$$\tau_t \circ R_g = R_g \circ \tau_t, \quad \forall g \in G.$$

Moreover,

(i) *τ_1 is an isomorphism with inverse parallel transport along $\bar{c}(t) := c(1 - t)$.*

(ii) *If c_1 and c_2 are piecewise smooth curves and $c_1(1) = c_2(0)$, then parallel transport along the concatenation $c_1 \cdot c_2$ coincides with the composition of the parallel transports along c_1 and c_2.*

Proof. The first statement follows by observing that R_g takes horizontal curves to horizontal curves. The rest is obvious. $\square$

Let $\xi = (\pi, P, M)$ be a principal G-bundle and fix a base point $p_0 \in M$. The **holonomy group** $\Phi(p_0)$ of a connection on ξ consists of all the isomorphisms $\tau_1 : P_{p_0} \to P_{p_0}$ obtained by performing parallel transport along piecewise smooth curves $c : I \to M$ with $c(0) = c(1) = p_0$. Choosing $u_0 \in \pi^{-1}(p_0)$, one can identify the group $\Phi(p_0)$

with a subgroup $\Phi(u_0) \subset G$ by associating to $\tau \in \Phi(p_0)$ the unique element $g \in G$ such that $\tau(u) = u_0 g$. Given two points $u_0, u_0' \in \pi^{-1}(p_0)$ there exists a unique element $g_0 \in G$ such that $u_0' = u_0 g_0$, and we have

$$\Phi(u_0') = g_0 \Phi(u_0) g_0^{-1}.$$

Hence, the subgroups $\Phi(u)$, as u varies in $\pi^{-1}(p_0)$, are all conjugate.

The holonomy of a connection is, in a sense, a global version of the curvature of the connection. This can be made precise as follows. We refer to Kobayashi and Nomizu (1996) for a proof.

Theorem 36.3 (Ambrose–Singer). *Let H be a connection on a principal G-bundle $\xi = (\pi, P, M)$ with curvature 2-form Ω. Given $u \in P$ denote by $P(u) \subset P$ the set of all $u' \in P$ which can be connected to u through a horizontal curve. The holonomy group $\Phi(u)$ is a Lie subgroup of G with Lie algebra*

$$\{\Omega_{u'}(\mathbf{v}, \mathbf{w}) : u' \in P(u), \mathbf{v}, \mathbf{w} \in H_{u'}\} \subset \mathfrak{g}.$$

Let $\xi = (\pi, P, M)$ be a principal G-bundle and assume one is given a linear action $G \times \mathbb{R}^r \to \mathbb{R}^r$. The resulting associated bundle $E := P \times_G \mathbb{R}^r \to M$ is then a vector bundle, which we denote by $\xi_{\mathbb{R}^R}$. Given a connection on the principal bundle ξ, parallel transport in $P \to M$ induces a parallel transport operation in the associated bundle $\xi_{\mathbb{R}^r}$ as we explain next.

If $c : I \to M$ is a piecewise smooth curve, a horizontal lift of $c(t)$ in the associated bundle is, by definition, a curve $\mathbf{v}(t) \in E$ of the form

$$\mathbf{v}(t) = [(u(t), \mathbf{v})] \in P \times_G \mathbb{R}^r := E,$$

where $u(t)$ is a horizontal lift of $c(t)$ in P. It is easy to see that for any $\mathbf{v}_0 \in E_{c(0)}$ there exists a unique horizontal lift $\mathbf{v}(t)$ of $c(t)$ such that $\mathbf{v}(0) = \mathbf{v}_0$. As before, one can then define the parallel transport $\tau_t : E_{c(0)} \to E_{c(1)}$ along $c(t)$.

Now let s be a section of $\xi_{\mathbb{R}^r}$. Given $\mathbf{v} \in T_p M$ let $c : I \to M$ be a curve such that $c(0) = p$ and $\dot{c}(0) = \mathbf{v}$. The covariant derivative of s in the direction $\mathbf{v}$ is

$$\nabla_{\mathbf{v}} s := \lim_{t \to 0} \frac{1}{t} \left[\tau_t^{-1}(s(c(t))) - s(p) \right] \in E_p. \tag{36.1}$$

It is easy to check that this definition is independent of the choice of curve c. Also, we define the covariant derivative of a section s along a vector field $X \in \mathfrak{X}(M)$ to be the section

$$(\nabla_X s)(p) := \nabla_{X_p} s.$$

Proposition 36.5. *The covariant derivative* $\nabla : \mathfrak{X}(M) \times \Gamma(E) \to \Gamma(E)$ *associated with a connection* H *in* ξ *is a vector bundle connection on* $\xi_{\mathbb{R}^r}$.

Proof. One needs to check that the operator defined by (36.1) satisfies the defining properties of a vector bundle connection. For that, one uses the following alternative definition of (36.1). Given a section s of $\xi_{\mathbb{R}^r}$ let $h : P \to \mathbb{R}^r$ be the corresponding G-equivariant map — see Proposition 35.3. Then for any vector field X, the map

$$\mathcal{L}_{\widetilde{X}} h := (\mathcal{L}_{\widetilde{X}} h^1, \ldots, \mathcal{L}_{\widetilde{X}} h^r) : P \to \mathbb{R}^r,$$

is G-equivariant, where $\widetilde{X} \in \mathfrak{X}(P)$ denotes the horizontal lift of X.

Lemma 36.2. *Let* $X \in \mathfrak{X}(M)$ *be a vector field and* s *a section of* $\xi_{\mathbb{R}^r}$ *corresponding to a* G-*equivariant map* $h : P \to \mathbb{R}^r$. *Then the* G-*equivariant map* $\mathcal{L}_{\widetilde{X}} h : P \to \mathbb{R}^r$ *corresponds to the section* $\nabla_X s$ *defined by* (36.1).

Using this lemma, one checks easily that ∇ satisfies the properties of a connection. $\qquad\square$

As a corollary, we obtain the correspondence between the principal bundle connections and vector bundle connections mentioned at the beginning of this lecture.

Corollary 36.2. *Let* $E \to M$ *be a vector bundle. Every principal bundle connection* H *on the bundle of frames* $F(E) \to M$ *determines a vector bundle connection* ∇ *on* E. *Moreover, every vector bundle connection on* E *arises in this way from a unique principal bundle connection on* $F(E)$.

Proof. The first part of the theorem follows from the proposition. The second part follows from Example 36.1. $\qquad\square$

This correspondence suggests that the theory of characteristics classes for vector bundles, that we studied before, can be generalized to principal bundles. Indeed, if H is a connection in the principal G-bundle $\xi = (\pi, P, M)$ with curvature form Ω, one defines the **Chern–Weil homomorphism** similarly to the way it was defined for vector bundles

$$\mathrm{CW}[\xi] : I^k(G) \to H^{2k}(M), \ P \mapsto [P(\Omega^k)],$$

This homomorphism is independent of the choice of connection.

One can use the Chern–Weil homomorphism to associate characteristics classes to the principal bundle. For example, if ξ is a principal bundle with structure group $GL(r, \mathbb{R})$ the **Pontrjagin classes** of ξ are obtained by considering the elementary symmetric polynomials

$$p_k(\xi) := \left[\sigma_{2k} \left(\frac{1}{2\pi} \Omega \right)^{2k} \right] \in H^{4k}(M).$$

These classes coincide with the Pontrjagin classes of the associated vector bundle $\xi_{\mathbb{R}^r}$, arising from the canonical action $GL(r, \mathbb{R})$ on $\mathbb{R}^r$. Similarly, if ξ is a principal bundle with structure group $GL(r, \mathbb{C})$, the **Chern classes** of ξ are given by

$$c_k(\xi) := \left[\sigma_k \left(\frac{1}{2\pi i} \Omega \right)^k \right] \in H^{2k}(M).$$

These classes coincide with the Chern classes of the associated complex vector bundle $\xi_{\mathbb{C}^r}$, arising from the canonical action $GL(r, \mathbb{C})$ on $\mathbb{C}^r$.

Exercises

Exercise 36.1

Show that a principal G-bundle is trivial if and only if it admits a section. Moreover, given any principal G-bundle $\xi = (\pi, P, M)$:

(a) Prove that there is an open cover $\{U_\alpha : \alpha \in A\}$ of M over which ξ admits local sections $s_\alpha : U_\alpha \to P$;

(b) Given an open cover as in (a), show that any two sections s_α and s_β with overlapping domains are related by

$$s_\beta(p) = s_\alpha(p)g_{\alpha\beta}(p) \quad (p \in U_\alpha \cap U_\beta),$$

for unique functions $g_{\alpha\beta} : U_\alpha \cap U_\beta \to G$ forming a cocycle.

Exercise 36.2
Show that a principal G-bundle always admits a connection.

Exercise 36.3
If H is a connection on a principal G-bundle $\xi = (\pi, P, M)$, with curvature 2-form $\Omega \in \Omega^2(P; \mathfrak{g})$. For any local section $s : U \to P$, show that

$$(s^*\Omega)(X, Y)^* = [\tilde{X}, \tilde{Y}] - \widetilde{[X, Y]}.$$

Exercise 36.4
Let $\xi = (\pi, P, M)$ be a principal G-bundle with connection H, and denote by ω its connection 1-form and by Ω its curvature 2-form. Let $\{U_\alpha : \alpha \in A\}$ be an open cover of M over which ξ admits local sections $s_\alpha : U_\alpha \to P$ and denote by $g_{\alpha\beta} : U_\alpha \cap U_\beta \to G$ the associated cocycle (see the previous exercise). Show that the local connection 1-forms $\omega_\alpha = s_\alpha^*\omega$ and local curvature 2-forms $\Omega_\alpha = s_\alpha^*\Omega$ satisfy

$$\omega_\beta = \mathrm{Ad}_{g_{\alpha\beta}^{-1}} \omega_\alpha + g_{\alpha\beta}^* \omega_{MC}, \quad \Omega_\beta = \mathrm{Ad}_{g_{\alpha\beta}^{-1}} \Omega_\alpha,$$

where $\omega_{MC} \in \Omega^1(G, \mathfrak{g})$ is the Maurer–Cartan form of G, defined by

$$\omega_{MC}(v) = d_g L_{g^{-1}}(v) \quad (v \in T_g G).$$

Exercise 36.5
Let G be a Lie group with Lie algebra $\mathfrak{g}$. If $t \mapsto X(t)$ is a curve in $\mathfrak{g}$, show that there exists a unique curve $g : I \to G$, with $g(0) = e$, satisfying

$$g(t)^{-1}\dot{g}(t) = X(t), \quad (t \in [0, 1]).$$

Exercise 36.6
Complete the details of the proof of Proposition 36.5 and, in particular, prove Lemma 36.2.

Exercise 36.7

Let $\pi : P \to M$ be a principal $\mathbb{S}^1$-bundle and identify the Lie algebra of $\mathbb{S}^1$ with $i\mathbb{R}$.

(a) Given a connection on P with curvature 2-form Ω show that there exists a real closed 2-form $\omega \in \Omega^2(M)$ such that $\frac{1}{2\pi i}\Omega = \pi^*\omega$;

(b) Check that the cohomology class $[\omega] \in H^2(M)$ coincides with the first Chern class $c_1(P)$;

(c) Conclude that the Chern class $c_1(P)$ is trivial if and only if P admits a flat connection.

Bibliography

Bott, R. and Tu, L. W. (1982). *Differential Forms in Algebraic Topology*. Graduate Texts in Mathematics, Vol. 82 (Springer-Verlag, New York, Berlin). https://mathscinet.ams.org/mathscinet-getitem?mr=658304.

Candel, A. and Conlon, L. (2000). *Foliations. I.* Graduate Studies in Mathematics, Vol. 23 (American Mathematical Society, Providence, RI). https://doi.org/10.1090/gsm/023.

Cohen, R. L. (1985). The immersion conjecture for differentiable manifolds. *Ann. Math. (2)* **122**(2), 237–328. https://doi.org/10.2307/1971304.

do Carmo, M. P. A. (1992). *Riemannian Geometry*. Mathematics: Theory & Applications (Birkhäuser Boston, Inc., Boston, MA). Translated from the second Portuguese edition by Francis Flaherty. https://mathscinet.ams.org/mathscinet/article?mr=1138207.

Dubrovin, B. A., Fomenko, A. T., and Novikov, S. P. (1992). *Modern Geometry—Methods and Applications. Parts I, II and III*, 2nd edn. Graduate Texts in Mathematics, Vol. 93 (Springer-Verlag, New York). Part I. The geometry of surfaces, transformation groups, and fields. Translated from the Russian by Robert G. Burns. https://mathscinet.ams.org/mathscinet/article?mr=1138462.

Duistermaat, J. J. and Kolk, J. A. C. (2000). *Lie Groups*. Universitext (Springer-Verlag, Berlin). https://doi.org/10.1007/978-3-642-56936-4.

Freedman, M. H. and Luo, F. (1989). *Selected Applications of Geometry to Low-Dimensional Topology*. University Lecture Series, Vol. 1 (American Mathematical Society, Providence, RI). Marker Lectures in the Mathematical Sciences held at the Pennsylvania State University, University Park, Pennsylvania, February 2–5, 1987. https://doi.org/10.1090/ulect/001.

Gallot, S., Hulin, D., and Lafontaine, J. (2004). *Riemannian Geometry*. 3rd edn. Universitext (Springer-Verlag, Berlin). https://doi.org/10.1007/978-3-642-18855-8.

Hatcher, A. (2002). *Algebraic Topology* (Cambridge University Press, Cambridge). https://mathscinet.ams.org/mathscinet-getitem?mr=1867354.

Helgason, S. (2001). *Differential Geometry, Lie Groups, and Symmetric Spaces.* Graduate Studies in Mathematics, Vol. 34 (American Mathematical Society, Providence, RI). Corrected reprint of the 1978 original. https://doi.org/10.1090/gsm/034.

Hirsch, M. W. (1994). *Differential Topology.* Graduate Texts in Mathematics, Vol. 33 (Springer-Verlag, New York). Corrected reprint of the 1976 original. https://mathscinet.ams.org/mathscinet-getitem?mr=1336822.

Hirzebruch, F. (1995). *Topological Methods in Algebraic Geometry,* English edn. Classics in Mathematics (Springer-Verlag, Berlin). Translated from the German and Appendix One by R. L. E. Schwarzenberger, Appendix Two by A. Borel. Reprint of the 1978 edition.

Hirzebruch, F. (2011). Why do I like Chern? And why do I like Chern classes? *Frontiers of Mathematical Sciences* (International Press, Somerville, MA), pp. 61–67.

Humphreys, J. E. (1978). *Introduction to Lie Algebras and Representation Theory.* Graduate Texts in Mathematics, Vol. 9 (Springer-Verlag, New York-Berlin). Second printing, revised. https://mathscinet.ams.org/mathscinet-getitem?mr=499562.

Husemoller, D. (1994). *Fibre Bundles,* 3rd edn. Graduate Texts in Mathematics, Vol. 20 (Springer-Verlag, New York). https://doi.org/10.1007/978-1-4757-2261-1.

Kelley, J. L. (1975). *General Topology.* Graduate Texts in Mathematics, Vol. 27 (Springer-Verlag, New York-Berlin). Reprint of the 1955 edition (Van Nostrand, Toronto, Ont.). https://mathscinet.ams.org/mathscinet-getitem?mr=0370454.

Kobayashi, S. and Nomizu, K. (1996). *Foundations of Differential Geometry. Vol. I, II.* Wiley Classics Library (John Wiley & Sons, Inc., New York). Reprint of the 1963 original (A Wiley-Interscience Publication). https://mathscinet.ams.org/mathscinet-getitem?mr=1393940.

Lee, J. M. (2013). *Introduction to Smooth Manifolds,* 2nd edn. *Graduate Texts in Mathematics,* Vol. 218 (Springer, New York). https://mathscinet.ams.org/mathscinet-getitem?mr=2954043.

Levy, A. (2002). *Basic Set Theory* (Dover Publications, Inc., Mineola, NY). Reprint of the 1979 original [Springer, Berlin; MR0533962 (80k:04001)]. https://mathscinet.ams.org/mathscinet-getitem?mr=1924429.

Manolescu, C. (2014). Triangulations of manifolds. *ICCM Not.* **2**(2), 21–23. https://doi.org/10.4310/ICCM.2014.v2.n2.a2.

Milnor, J. (2015). Introduction to exotic spheres. *Bull. Am. Math. Soc. (N.S.)* **52**(4), 593–602.

Milnor, J. W. and Stasheff, J. D. (1974). *Characteristic Classes.* Annals of Mathematics Studies, Vol. 76 (Princeton University Press, Princeton, NJ; University of Tokyo Press, Tokyo). https://mathscinet.ams.org/mathscinet-getitem?mr=440554.

Moerdijk, I. and Mrcun, J. (2003). *Introduction to Foliations and Lie Groupoids.* Cambridge Studies in Advanced Mathematics, Vol. 91 (Cambridge University Press, Cambridge). https://doi.org/10.1017/CBO9780511615450.

Samelson, H. (1990). *Notes on Lie Algebras*, 2nd edn. Universitext (Springer-Verlag, New York). https://doi.org/10.1007/978-1-4613-9014-5.

Serre, J.-P. (2006). *Lie Algebras and Lie Groups. Lecture Notes in Mathematics*, Vol. 1500 (Springer-Verlag, Berlin). 1964 lectures given at Harvard University, Corrected fifth printing of the second edition (1992). https://mathscinet.ams.org/mathscinet-getitem?mr=2179691.

Sharpe, R. W. (1997). *Differential Geometry*. Graduate Texts in Mathematics, Vol. 166 (Springer-Verlag, New York). Cartan's generalization of Klein's Erlangen program. With a foreword by S. S. Chern. https://mathscinet.ams.org/mathscinet-getitem?mr=1453120.

Spivak, M. (1979). *A Comprehensive Introduction to Differential Geometry*, 2nd edn., Vols. I–IV (Publish or Perish, Inc., Wilmington, Del.). https://mathscinet.ams.org/mathscinet-getitem?mr=532830.

Taubes, C. H. (2011). *Differential Geometry: Bundles, Connections, Metrics and Curvature*. Oxford Graduate Texts in Mathematics, Vol. 23 (Oxford University Press, Oxford). https://doi.org/10.1093/acprof:oso/9780199605880.001.0001.

Varadarajan, V. S. (1984). *Lie Groups, Lie Algebras, and Their Representations*. Graduate Texts in Mathematics, Vol. 102 (Springer-Verlag, New York). Reprint of the 1974 edition. https://doi.org/10.1007/978-1-4612-1126-6.

Warner, F. W. (1983). *Foundations of Differentiable Manifolds and Lie Groups*. Graduate Texts in Mathematics, Vol. 94 (Springer-Verlag, New York-Berlin). Corrected reprint of the 1971 edition. https://mathscinet.ams.org/mathscinet-getitem?mr=722297.

Weibel, C. A. (1994). *An Introduction to Homological Algebra. Cambridge Studies in Advanced Mathematics*, Vol. 38 (Cambridge University Press, Cambridge). https://doi.org/10.1017/CBO9781139644136.

Whitney, H. (1944). The self-intersections of a smooth n-manifold in $2n$-space. *Ann. Math. (2)* **45**, 220–246. https://doi.org/10.2307/1969265.

Index